THE INTERPRETATION OF IGNEOUS ROCKS

THE INTERPRETATION
OF IGNEOUS ROCKS

K. G. Cox, J. D. Bell and R. J. Pankhurst

London
GEORGE ALLEN & UNWIN
Boston   Sydney

First published in 1979

GEORGE ALLEN & UNWIN LTD
40 Museum Street, London WC1A 1LU

©K. G. Cox, J. D. Bell and R. J. Pankhurst, 1979

British Library Cataloguing in Publication Data

Cox, Keith Gordon
   The interpretation of igneous rocks.
   1. Rocks, Igneous
   I. Title   II. Bell, J D   III. Pankhurst, R J
   552'.1   QE461   78-41057

   ISBN 0-04-552015-1
   ISBN 0-04-552016-X Pbk

Typeset in 10 on 12 point Times by Preface Limited, Salisbury
and printed and bound in Great Britain at
William Clowes & Sons Limited, Beccles and London

Now how Elephants should come to be buried in Churches, is a Question not easily answered . . .
Robert Plot (1677)

Preface

Our aim in writing this book is to try to show how igneous rocks can be persuaded to reveal some of the secrets of their origins. The data of igneous rocks consist of field relations, texture, mineralogy, and geochemistry. Additionally, experimental petrology tells us how igneous systems might be expected to behave. Working on this material we attempt to show how hypotheses concerning the origins and evolution of magmas are proposed and tested, and thus illuminate the interesting and fundamental problems of petrogenesis. The book assumes a modest knowledge of basic petrography, mineralogy, classification, and regional igneous geology. It has a role complementary to various established texts, several of which are descriptively good and give wide coverage and evaluation of petrogenetic ideas in various degrees of detail. Existing texts do not on the whole, however, deal with methodology, though this is one of the more important aspects of the subject. At first sight it may appear that the current work is a guidebook for the prospective research worker and thus has little relevance for the non-specialist student of geology. We hope this will prove to be far from the case. The methodological approach has an inherent interest because it can provide the reader with problems he can solve for himself, and as an almost incidental consequence he will acquire a satisfying understanding. Moreover, if the study of igneous rocks has any value at all for the non-specialist, it is at least as likely to lie in the scientific discipline of its methods as in any other aspect of the subject.

This book is not built round any specific course. There is much in it that would be of benefit to and accessible to any undergraduate student of geology; but at the same time postgraduate students and even teachers of petrology may find it useful. In short, it is offered to petrologists of all ages.

The authors are particularly grateful to R. W. Cleverly, P. J. Betton and D. Kiernan-Walker for their attempts to minimise the number of mistakes in the manuscript, and to Mrs P. Jackson for her excellent typing.

Oxford
1978

K. G. Cox
J. D. Bell
R. J. Pankhurst

Contents

List of Tables

1
Fractionation in igneous processes

Introduction

The study of igneous rocks, that is to say rocks which have solidified from molten material, is one of the most fundamental branches of geology. In many areas of the world igneous rocks are not particularly abundant, nor are they remarkable for the richness of their associated mineral deposits, but all the rocks we see at the surface of the earth have igneous processes somewhere in their past history. Every sedimentary and metamorphic rock must ultimately trace its ancestry back to an igneous source.

The systematic investigation of igneous rocks began in the mid-nineteenth century with the invention of the thin section by H. C. Sorby. This made possible the rise of microscopical petrography culminating towards the end of the century in the great descriptive studies by Zirkel and Rosenbusch. In the early twentieth century, with a firm descriptive basis to build on, petrologists such as N. L. Bowen and Alfred Harker began to turn their attentions to the genetic relationships of igneous rocks. Much effort has subsequently been devoted to trying to understand the way in which different rock types have been formed, and how an initially formed melt, or magma, may become compositionally differentiated to give rise to a variety of different rock types as ultimate products. More recently, igneous rocks have been studied in order to extract the information they give about processes at depth and the nature of the upper mantle, what may be thought of as an applied stage in the history of the subject.

The purpose of this book is to provide an introduction to the methodology of genetic igneous petrology, that is, the way in which data derived from igneous rocks can be interpreted to give information bearing on the evolution of magmas, on igneous processes in general, and on the nature of the source materials from which magmas are formed. The importance of such studies is emphasised by the fact that igneous activity provides us with the only samples available with which we may directly study the composition of the interior of the earth.

The concept of fractionation

One of the most obvious facts about igneous rocks is that they are extremely variable both in mineralogy and chemical composition. This leads petrologists to think automatically in evolutionary terms, like zoologists and botanists. Yet it is not immediately obvious that inanimate materials have a capacity to evolve, until one contemplates the variety of igneous rocks and asks how the individual types may have come to be created. One is obliged to postulate either that they were all originally created different, or that some processes exist which have the capacity to generate variety from an original uniform starting product. The latter view is, of course, evolutionary and is the one that appeals to the scientist who, after all, would have little of interest to do if he were to adopt any other standpoint.

With regard to rocks there is some debate as to what constitutes a suitable uniform starting product. The earth might have been compositionally uniform when first formed or, if it were not, possibly there was a stage in the evolution of the solar system when compositional uniformity existed throughout it. For present purposes this debate is immaterial (its resolution is perhaps one of the ultimate aims of the earth and planetary sciences) as long as the idea that an evolutionary model requires a parental material is established. **Fractionation** may thus be defined as the formation of a variety of substances from an initial, single, parental material. Any process which causes this to happen will be referred to as a fractionation process, and it is extremely important to note that *no specific mechanism is implied* by this term. The term **differentiation** is also commonly used in petrology and is synonymous with fractionation. Both terms have often been used to imply that the process concerned is fractional crystallisation, but such lack of precision is undesirable.

In igneous petrology fractionation consists first and foremost of the production of variation in chemical composition. Some elements, e.g. Mg, show great variation in igneous rocks and may thus be said to be strongly fractionated. Others, e.g. Si, are less variable (the range of Si variation in common igneous rocks lies within a factor of about two) and are thus less strongly fractionated. With regard to mechanisms of fractionation, every process has central to it the migration or transport of atoms of a particular element relative to others. Only by such means may differing bulk chemical compositions be formed from a single starting product. This idea forms the basis of the classification of fractionation processes which follows.

Classification of fractionation processes

In most natural circumstances likely to be of interest to the igneous petrologist, materials can exist as solid, liquid, or vapour. In some marginal circumstances, where processes involve super-critical fluids which may contain large amounts of dissolved material, there may be a lack of distinction between liquid and vapour. However, in normal igneous processes a clear distinction usually exists. Thus a division of systems on the basis of the states of matter they contain forms a convenient and logical basis for the consideration of the way elements may migrate during fractionation processes. Of the possible combinations of solid, liquid, and vapour, we shall consider only those which contain liquid since these are the only systems which concern igneous petrology directly. Table 1.1 lists the possibilities and assigns them to their specialist fields.

Systems involving liquid only. No natural magmatic body is ever completely independent of its containing wall rocks but it is reasonable to suppose that wall-rock influence is negligible in the central parts of some magma bodies, which may therefore be considered as completely liquid systems. If no solid material is present this may be either because solid material once present has been mechanically removed in some way, or because the temperature of the liquid is too high for any solid to be in equilibrium with it. Any solid present at an earlier stage must in this case have dissolved. In either case the only mechanism available for fractionation, if the body consists of a single homogeneous liquid, is that of **diffusion**. Elements may migrate through the body at different rates

Table 1.1   States of matter in geological systems

States of matter in the system	*Types of geological process concerned*
solid only	metamorphic
liquid only	igneous
vapour only	vulcanological
solid + liquid	igneous
solid + vapour	metasomatic
liquid + vapour	igneous
solid + liquid + vapour	igneous

in response to thermal or pressure gradients and thus produce a variable, that is a fractionated, liquid body. Elements may also migrate differentially in response to a compositional gradient once this has been established by external means, for example by reaction of the margins of the body with wall rocks. Diffusion differentiation is only known to be geologically important on a small scale and it is not certain whether it can operate to produce substantial amounts of differentiated rocks. The problem is made more difficult by the fact that natural systems consisting only of liquid appear to be very rare. Thus if diffusion differentiation takes place it probably operates largely in the presence of other more powerful fractionation processes such as fractional crystallisation (see below) and the effects are largely masked.

In contrast to the case noted above, some liquid-only systems may contain two immiscible liquids in which immiscibility has been induced by, say, a fall of temperature. Immiscible liquids initially exist as droplets of the minor phase suspended in the major, but if a density contrast exists between the phases the possibility of migration upwards and coagulation of the less dense phase exists. Since the immiscible phases must have different compositions, a powerful mechanism of differentiation is thus available. Several lines of evidence suggest that immiscible-liquid fractionation does in fact operate in the natural environment, but it appears to be restricted to a relatively small compositional range of magmas. Immiscibility between sulphide melts and silicate melts is well established and may have considerable importance in the formation of some magmatic sulphide ore bodies. Evidence comes from both experimental petrology and from the interpretation of petrographic features as immiscible droplets. However, in most silicate systems investigated experimentally there is no evidence of liquid immiscibility between one type of silicate melt and another, the main exceptions being certain siliceous melts which do not on the whole coincide compositionally with common rock types. Furthermore, as Bowen (1928) pointed out, petrographic evidence of liquid immiscibility is rare. Thus, although important in certain cases, differentiation by liquid immiscibility does not appear to be a widespread phenomenon.

Field evidence appears from time to time which demonstrates *apparent* immiscibility between two silicate melts, the mixed acid–basic intrusions from various localities in the British Tertiary province and Iceland being good examples (Blake *et al.* 1965). In these intrusions 'pillows' of basaltic rock are enclosed in a matrix of granophyre, the former having distinct chilled edges against the latter. The evidence that such bodies were emplaced as magmas consisting of large globules of basic magma dispersed in a matrix of acid magma is difficult to dispute.

Nevertheless since experimental work suggests that such compositions should be miscible, it is probably safest to interpret the occurrences as representing the accidental mixing of two magmas immediately prior to emplacement, and the solidification of the body before mixing had had time to take place.

To distinguish genuine immiscibility from this type of temporary disequilibrium immiscibility, the term thermodynamic immiscibility is sometimes used for the former.

Systems involving solid + liquid. Fractionation processes involving solid + liquid (=crystal–liquid fractionation) systems are of extreme importance because they are capable of effecting gross compositional changes in all types of magma. Furthermore, the conditions necessary for their operation are encountered by every magma, that is to say during partial melting when magma is first formed, during the migration of magma when it is in contact with a variety of wall rocks, and during the final crystallisation of magma. In all these circumstances liquid is in contact with solid material and an ideal mechanism of fractionation is available.

Partial melting. Geophysical evidence suggests that with the exception of the outer core the earth normally consists of solid material. Thus any magma must originate by the melting of pre-existing solid rock. Melting may be induced by a local increase of temperature, by decompression, or by the influx of a mobile constituent such as water which depresses the melting point of the solid. Melting of the source rock is, however, probably rarely complete, and it is probable that most melts coalesce to form discrete magma bodies and migrate away from their source regions leaving some form of refractory residue behind. Since partial melts rarely if ever have the same chemical composition as the source material it follows that the separation of a melt from its source constitutes a fractionation process. Partial melting at depth to give rise to large bodies of magma is not, however, a process which can be observed directly, though minor examples taking place at near-surface levels are commonplace. Nevertheless fractionation by partial melting is believed to be one of the dominating igneous processes and much of this book is concerned with it, especially the ways in which its properties can be predicted and its operation in natural cases inferred.

Fractional crystallisation. When a magma solidifies it does so over a temperature range rather than at a specific temperature. The temperature at which crystallisation begins is called the **liquidus** temperature, and there is some lower temperature, at which crystallisation is complete, termed the **solidus** temperature. Any magma which exists with a temperature between the solidus and liquidus in general consists of a mixture of liquid and solid, that is crystals suspended in liquid. The solid material in equilibrium with the liquid may consist of one or more minerals and these do not normally have, either individually or collectively, the same chemical composition as the liquid. For example, many basaltic magmas lying at temperatures just below the liquidus consist of olivine crystals plus liquid. The olivine usually has a much higher Mg-content than the liquid, lower Si, and is virtually free of Ca and Al, both of which are major constituents of the liquid. Thus a mechanism which removes the crystals from the liquid changes the bulk composition of the magma. This is an example of fractional crystallisation which is a process of major importance, firstly because all magmas as they cool must at some stage pass through the appropriate temperature interval, and secondly because crystal–liquid separation mechanisms appear to operate frequently and rather effectively in nature. The term fractional crystallisation does not itself, however, imply the precise mechanism of crystal–liquid separation. Certainly **gravitative separation** of crystals is a well-known phenomenon and many intrusions (known as **layered intrusions**), representing high-level magma chambers now exposed by erosion, provide ample evidence that crystals have settled to the floor of the chamber. In ideal cases such intrusions solidify from the base upwards as a 'sedimentary' accumulation of precipitated crystals.

Such rocks are termed **cumulates** and although they are strictly igneous in origin their bulk compositions in no sense represent the compositions of original liquids. However, while cumulates are forming, the liquid remaining in the chamber must by inference change its composition progressively, and may be tapped off and erupted from time to time as a varying series of lavas. The distinction here between cumulates which do *not* represent liquid compositions and lavas which frequently do, is of paramount importance in igneous petrology. They are distinctly opposite sides of the same coin. Suppose, for example, that the basalt magma with olivine crystals mentioned above were subjected to gravitative separation of olivine. A cumulate of almost pure dunite (largely olivine but with a little basaltic material trapped between the crystals of olivine) would form at the base of the chamber. If such a rock were collected and analysed it would be quite wrong to infer that a dunitic liquid had once existed – correct interpretation of the evidence would depend on recognition of the rock as a cumulate, an important topic which is discussed in Chapter 12.

Gravitative separation of crystals from liquid normally involves the settling out of the former, but evidence is occasionally obtained of separation by upward flotation of crystals. This appears to affect only low-density minerals such as leucite, and occasionally perhaps feldspars.

Several other mechanisms for crystal–liquid separation are theoretically possible but their importance is more difficult to evaluate than the gravitative process. **Flowage differentiation** (Bhattacharji & Smith 1964) is a process by which solid particles are concentrated towards the centre of liquid flowing in a narrow conduit. If the flow is arrested and the magma solidifies *in situ*, the resulting igneous body is a dyke with phenocrysts concentrated towards the centre. However, such bodies can form in several other ways, for example by the initial intrusion of crystal-free liquid followed by crystal-rich liquid, and it is difficult to demonstrate the operation of flowage differentiation with logical certainty in natural occurrences. Nucleation of crystals on conduit walls is another theoretically effective fractionation mechanism. If a magma is cooling and crystallising olivine for example, and the wall rocks through which it is passing also contain olivine, some of the material coming out of solution may nucleate on the olivine of the walls. This might constitute a very effective mechanism for removing crystalline material from a liquid but is at the moment unproven. Any non-gravitative crystal removal mechanism is, however, likely to be difficult to demonstrate because gravitational separation probably affects most crystal-bearing liquids and hence obscures the evidence of other processes. Fortunately, for many petrological purposes lack of knowledge of the precise mechanism by which crystals and liquids have become separated is not of major importance because the geochemical effects of the separation, often the point of interest in a petrological study, do not depend upon it.

Systems involving liquid + vapour. Magmas contain variable amounts of dissolved volatile constituents of which the most abundant are H_2O and CO_2. Under pressure these will remain in solution but pressure reduction, or their increase of concentration in the liquid as volatile-free phases crystallise out, will cause them to be evolved as bubbles. Bubbles trapped by the solidification of the liquid are preserved as **vesicles**, a characteristic feature of most lavas, except those erupted on the deep sea floor under a considerable pressure from the overlying water. Plutonic rocks solidifying at high levels in the crust also show cavities interpreted as vapour bubbles (**druses** or **miarolitic cavities**) but they are comparatively rare. Druses do not normally show the smooth, rounded outlines of volcanic vesicles and they are usually lined by well-formed crystals of the host rock projecting into them.

The separation of bubbles from their parent liquid is a relatively efficient fractionation mechanism in so far as the original liquid is heavily depleted in its volatile constituents. Minor amounts of metallic ions, particularly the alkali metals, can also be carried in the vapour. Although the fractionation of relatively non-volatile constituents in this fashion can be demonstrated by the study of active volcanoes and their emitted gases, it is by no means certain that such processes of fractionation (often termed **volatile transfer**) are particularly important in their effects on the bulk composition of most igneous rocks, though they have often been thought to be. The difficulty, a familiar one in petrology, lies in detecting the effects of one process in the presence of others, a topic discussed in somewhat more detail in the next section. This is not to minimise the importance of the volatile constituents in petrogenesis in general, the topic under discussion here being simply concerned with the ability of escaping volatiles to remove other materials from the magma or to transfer them from one part of the magma to another. In this sense, as a fractionation mechanism, volatile transfer has not yet been demonstrated to be of great effectiveness.

Systems involving liquid + solid + vapour. Many magmatic systems under low confining pressures are capable of existing in this condition. Most of the fractionation processes so far discussed are thus potentially able to operate simultaneously (i.e. diffusion through the liquid, liquid immiscibility, crystal–liquid fractionation, volatile transfer). Fractionation by partial melting is not likely to be involved because partial melting is presumed normally to take place under too high a pressure for vapour to be present. Of the possible processes, however, *one*, namely crystal–liquid fractionation, has been shown by countless studies to be highly effective. Thus the most reasonable position for an investigating scientist to adopt is to try to establish how much of the chemical variation can be attributed to this process, before attempting to assign any remaining unexplained chemical features to other processes. There is, of course, no guarantee that such a procedure will lead to the truth but it has the advantage of being a disciplined approach with clearly stated assumptions.

Open systems

In the thermodynamic sense a system is defined as any part of the universe we wish to isolate for the purposes of consideration. Systems can be classified as open or closed depending whether material is allowed to pass into or out of the system. Many of the systems considered so far have been closed in that no material has been introduced after the formation of the magma concerned, and, although material has in many examples been removed from the liquid as cumulates etc., it may still be considered to be within the system. In nature, however, it is important to consider a number of open system possibilities, as follows.

Wall-rock reaction. Once a magma is formed and begins to migrate towards the surface under the influence of gravity it may come into contact with wall rocks which are substantially different from its source material and with which it is not in chemical equilibrium. If sufficient time is available, reactions between the liquid and the enclosing solid will take place and thus modify the compositions of both. It is convenient to think of such a process as involving multiple source materials in the production of the magma by partial melting, though often the original magma is said to be **contaminated**. In the original melting event, the source contributes certain materials to the liquid and retains others in the refractory residue. Each time the magma re-equilibrates with a new wall rock this process is repeated. Some material may be lost to the wall rock and new material may be gained by the liquid, depending on the complexities of the equilibria concerned. The petrologist relies, in his attempts to unravel the histories of magmas, largely on the fact that chemical relationships within a magma and within related suites of magmas retain some sort of 'memory' of the events which have influenced them. Re-equilibration with rocks other that the source tends, however, to be a memory-erasing process so that the petrologist can often do no more than attempt to work back to the conditions of last equilibration rather than to the original source. Trace element studies can, however, to some extent be used to attempt to overcome this problem (see Chs 14 and 15).

Hybrid rocks. The term hybrid is applied to rocks which have a mixed parentage either because the original magma has incorporated fragments of some other rock (the process of **assimilation**) or because two magmas have become mixed together. Hybrid rocks are sometimes easy to recognise, particularly when, as in the first case, they contain recognisable, undigested blocks of foreign material (xenoliths). With time, however, fragments will tend to react with the enclosing magma, dissolve, and become dispersed. However, unless the magma becomes completely homogenised, streaks and patches of varying composition remain, and xenoliths, although considerably transformed by reaction with the magma, may still be recognisable. Hybrids formed by mixing of two magmas may sometimes be detected by the presence of two suites of phenocrysts of types not normally expected to be in equilibrium with each other. Hybrid rocks are fairly common, particularly in the marginal facies of plutonic intrusions. Hybridisation followed by homogenisation and re-equilibration of the crystalline phases present would, of course, be extraordinarily difficult to detect, and the role in petrogenesis of such a process is very much unknown. Much difficulty arises from the fact that common rock types, for example basalt and granite, may in any case be related to each other by closed system fractionation processes. For example granitic liquids may be formed by the advanced fractional crystallisation of some types of basaltic magma. In such a process magmas are formed which in many respects are compositionally intermediate between basalt and granite. Hybrid rocks formed by the incorporation of granite into basaltic magma have similar compositional characters and the two types of intermediate magma are thus difficult to distinguish. Nevertheless considerations of phase relationships and energy requirements do place reasonable constraints on the way hybridisation may be expected to happen, at least with regard to rocks formed by the assimilation of solid country rocks by magma.

Post-solidification alteration. Compositional alteration of igneous rocks once they have solidified is a widespread phenomenon and is a result of interaction with groundwater, seawater, or hydrothermal fluids. Common effects include the leaching and removal of some constituents, the oxidation of Fe^{2+} to Fe^{3+}, and the introduction of water and CO_2 to give hydrated minerals and carbonates. Studies of stable isotopes (see Ch. 15) have been widely used to investigate such alteration effects.

Summary

The processes discussed above include most of the ways in which important compositional changes can be induced in magmas. The composition observed in an igneous rock is the end product of the operation on the original source material of some or all of these processes.

Six important variables may be distinguished which summarise the possible fractionation events which the history of an igneous rock may include:

(a) *The composition of the source.* This is clearly one of the most important variables, and a knowledge of the nature of the source is often the object of petrological studies.

(b) *Partial melting*. The pressure and temperature conditions under which partial melting takes place control, acting on a given source material, the composition of the original magma. A magma produced by partial melting and not subjected to any other fractionation process is frequently termed a **primary magma**. Much petrological discussion has concerned the question of whether primary magmas ever reach the surface of the earth.

(c) *Re-equilibration with host rocks*. This raises the possibility of multiple sources, and can disguise or remove chemical features inherited from the original partial melting event. Re-equilibration may in some cases be a continuous process, not separated distinctly from the original melting event.

(d) *Fractionation en route to the surface*. This involves principally the possibilities of fractional crystallisation, liquid immiscibility, volatile transfer, and diffusion in the liquid. Any of these processes, if allowed to operate significantly, will remove the primary character of the magma.

(e) *Open system processes operating on the magma*. This includes assimilation of wall rocks, and mixing with other magmas.

(f) *Post-solidification alteration*. Included here are weathering and various metamorphic changes of open system type.

In the following pages we first describe the chemical variation which igneous rocks show, and then go on to discuss ways in which the data, both petrographic and geochemical, that the rocks provide, can be used to attempt to unravel some of the complex history of magmatic evolution.

2
Compositional variation in magmas

In this chapter we survey the range of chemical compositional variation shown by magmas and discuss some of the ways in which it can be represented. The raw material of such a study consists of analyses of volcanic rocks, since these may be taken to be near approximations to liquid compositions, from which they differ mainly in the content of volatile constituents such as H_2O and CO_2 which may be lost during eruption. Plutonic rocks do not always give good indications of liquid compositions because of the role of crystal accumulation processes (see Ch. 13).

Chemical analyses of rocks are usually expressed as weight per cents of *oxides* for major elements (Si, Ti, Al, Fe, Mn, Mg, Ca, Na, K, P) and as *parts per million* (ppm) for trace elements. Major element analysis is usually only carried out for cations (as listed above) and it is assumed, with the exception of Fe, that they are accompanied by an equivalent amount of oxygen. Making allowance for the oxygen, an analysis for the major cations will sum to approximately 100% (usually between about 99% and 101%). The general state of oxidation of rocks is, however, such that Fe exists both as Fe^{2+} and Fe^{3+}. Normal practice therefore consists of carrying out analyses for total Fe and Fe^{2+} separately; thus Fe^{3+} is derived by calculation. Only extremely rarely is the combination of Fe (metal) with Fe^{2+} found in natural rocks, though it is, of course, characteristic of the iron meteorites. Occasionally rocks are analysed for anions other than O, e.g. S, Cl and F. When these are determined the analysis total must be reduced to make allowance for the fact that some of the O assumed to be present is in fact replaced by an equivalent weight of these other elements. The remaining major constituent of most analyses is H_2O+, that is to say water present in a combined state within the rock (e.g. as a constituent of amphibole or other hydrous minerals). Other water, present simply as dampness within the powder analysed, is also often quoted, under the symbol

H₂O−. In some analyses the total volatile content is simply but only approximately expressed as 'loss on ignition'.

Inter-element correlations

Volcanic rocks show a great range of variation in most of the major constituents, but such variation is far from random and shows a series of rational patterns which we shall now explore. The analyses used for this study are taken from the tables of Carmichael, Turner and Verhoogen (1974), and for the purposes of Figures 2.1–2.3, all the analyses of volcanic rocks have been used, with the exception of the ocean floor basalts and the ultrapotassic lavas (Carmichael *et al.* 1974, Tables 8.1, 8.2 and 10.4). The basalts are excluded only because the remaining tables contain quite enough basalts for our purposes, and the ultrapotassic rocks are excluded for reasons explained later in this chapter. The remaining analyses cover the range of compositional variation of volcanic rocks adequately, though a small number from other sources have been added to make up deficiencies in the fields of trachyandesites and picrite basalts (see Figs. 2.1 and 2.2).

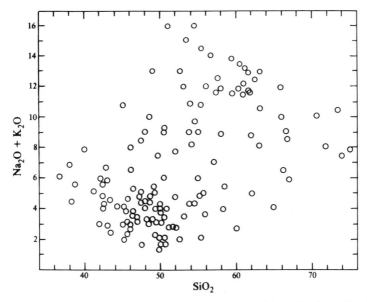

Figure 2.1 Plot of total alkalis versus silica for a wide selection of volcanic rocks (data from Carmichael, Turner & Verhoogen 1974).

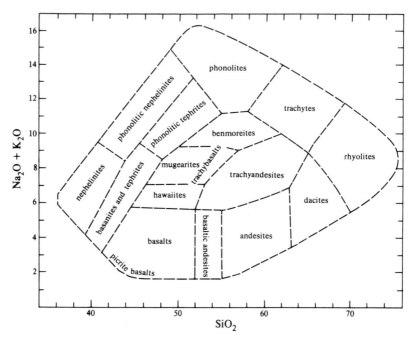

Figure 2.2 Nomenclature of normal (i.e. non-potassic) volcanic rocks. The boundaries are not intended to be sharp, the fields labelled being intended only to show the approximate areas in which different types plot. For the nomenclature of equivalent potassic types, see Table 2.1.

Alkalis, silica, and nomenclature. Two of the most important and useful chemical parameters to be considered are SiO_2 and $Na_2O + K_2O$ ('total alkalis'). For the compiled analyses these are shown plotted against each other in Figure 2.1 which indicates the fairly precisely defined field occupied by the great majority of volcanic rocks. The broad scatter of points furthermore suggests that the two parameters may be regarded as essentially *independent* of each other. In Figure 2.2, the names of rock types are added to the diagram and are based almost exclusively on those given to each rock by the original investigators. If it is found, as indeed it mainly is, that all the rocks with a particular name plot in a relatively restricted field, it follows that the content of SiO_2 and total alkalis is capable of giving much more information about the rock than might at first be assumed. A term such as rhyolite, for example, may have been chosen by the original investigator for a number of reasons, but in many cases (if glassy rocks are excluded) will almost certainly have taken mineralogical features such as the presence of abundant quartz and alkali feldspar, and the comparative rareness or absence of minerals such as olivine, pyroxene, amphibole and calcic plagioclase into consideration. These are the accepted features of rhyolite, and any gross deviation from them would have resulted in a different name having been chosen. *From this it follows that if it is possible to use the SiO₂ and total alkali contents to name a rock, then it is also possible to predict the main features of the rest of its chemistry.* While on the subject of nomenclature, however, several features of Figure 2.2 deserve comment. Firstly this figure is *not* intended to form a classificatory system but it has been necessary to add field boundaries to it to indicate the approximate extents of the fields into which various named rock types may normally be expected to fall. (The paper by Le Maitre (1976) includes an alkali/silica diagram for each of the rock types given in Appendix 2. These diagrams give an excellent indication of the amount of scatter which rocks with a given name may be expected to show. The names used are throughout those used by the authors originally describing the rocks.) Nevertheless it is impossible not to regard such a diagram as having some degree of classificatory function and therefore a brief digression about some of the fields is necessary.

The main point of disagreement between Figure 2.2 and the named rocks used in the compilation revolves round the use of the term trachyandesite. As it happens, all the trachyandesites in the tables of Carmichael *et al.* (1974) plot in what are here designated the mugearite and benmoreite fields. We prefer, however, to use the name trachybasalt loosely to express the idea of fairly basic rocks intermediate in mineralogical character between trachyte and basalt and this therefore includes our mugearite fields (benmoreites are better thought of as basic trachytes than trachybasalts). Trachyandesites, in contrast, should lie between trachytes and andesites. They are similar in many ways to mugearites but are more silica-rich. Such rocks do exist but by chance happen not to be in the Carmichael *et al.* compilation. We have therefore added some from other sources to aid in the construction of subsequent diagrams. The average trachyandesite of Le Maitre (1976, and see Appendix 2) falls approximately in the centre of the trachyandesite field of Figure 2.2.

A number of choices of name is also available for rocks similar to mugearites but more alkali-rich. We have chosen the term phonolitic tephrite to cover the field in which the tahitites of Carmichael *et al.* fall, and phonolitic nephelinite to cover a few rocks designated sanidine nephelinite and felsic nephelinite. Additionally, the nomenclature used in the diagram does not refer to ultrapotassic types since these were excluded from our sample. The nomenclature of the more potassic rocks is discussed in a later section.

Finally we note that the nomenclature discussed here and illustrated in Figure 2.2 should be regarded as a first step, a simplified approach, towards the complexity of nomenclature as actually used for research purposes. In the following sections we discuss the relationship of Mg, Ca, Fe and Al, as well as Na and K, separately rather than together. These relationships also contribute to the question of nomenclature, since for a particular rock type it is possible not only to consider the characteristic contents of alkalis and silica but also of these other elements. The system, however, does *not* contain enough dimensions to make some important classificatory distinctions, for example between rocks with differing Fe/Mg and Ca/Al ratios (though in a later section, as noted above, an attempt is made, albeit a first approximation, to distinguish rocks of varying K/Na ratio). Thus, disagreements will inevitably arise between names actually used and names suggested by the use of the figures; and in many respects our divisions are not fine enough (for example we do not distinguish between basalts of calc-alkaline, tholeiitic and alkalic types). There need, however, be no confusion if it is remembered that we are concerned only with illustrating the approximate ranges of chemical composition characteristic of various well-known broadly defined rock groups. Appendix 1 gives a commonly used mineralogical system of nomenclature (Streckeisen 1976) for comparison, while Appendix 2 gives some recently published average analyses of rock types. For readers already familiar with the CIPW normative calculation it would be of immediate value to make a careful comparison of Figures 2.2 and A1.1 (Appendix 1).

Relationships of Mg, Ca and Fe. In Figure 2.3, contoured alkali–silica diagrams are given for MgO, CaO and $FeO + Fe_2O_3$ which, given $Na_2O + K_2O$ and SiO_2, allow the predictions mentioned previously to be made for the contents of these oxides. The value of such predictions lies in their demonstration of the rational relationships between the various constituents. The accuracy is best judged empirically (see Exercise 3, p. 41). Too much should not be expected from this exercise but if errors are viewed in the light of the total variation of the parameter concerned then reasonable satisfaction should result. For example, MgO in lavas varies from 15 weight per cent (or more) down to less than 0.1%. If the diagram predicts 2% and the rock only actually contains 1.5%, this can, it is true, be regarded as a substantial error, but for present purposes it is a sufficiently good prediction.

The three constituents MgO, CaO and $FeO + Fe_2O_3$, behave as a coherent group, correlating strongly with each other over most of the field of variation. Rocks which combine *low* SiO_2 with *low* total alkalis are rich in these constituents while conversely rocks combining high

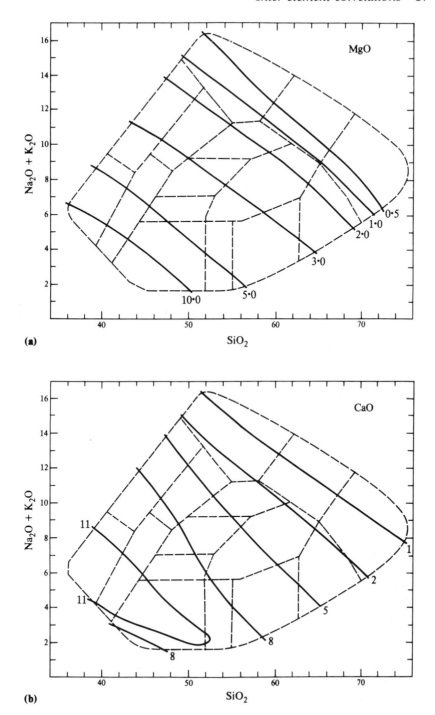

(a)

(b)

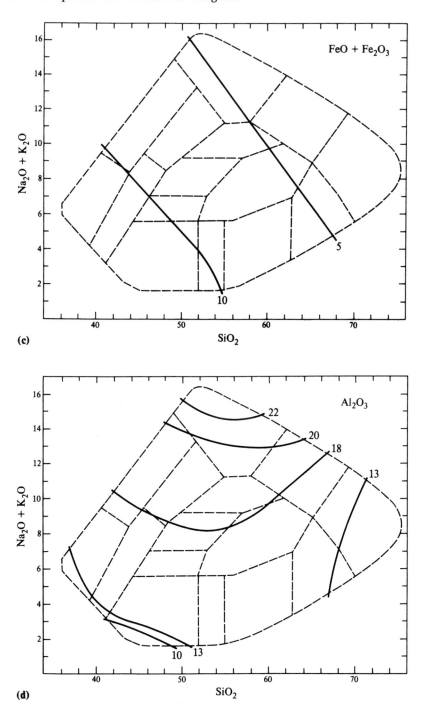

(c)

(d)

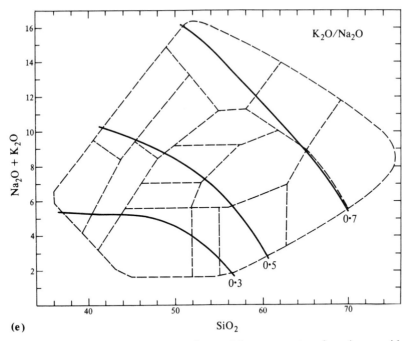

Figure 2.3 Approximate contours for weight per cents of various oxides superimposed on the alkali–silica diagram of Figure 2.2. (a) MgO, (b) CaO, (c) FeO + Fe$_2$O$_3$, (d) Al$_2$O$_3$, (e) K$_2$O/Na$_2$O.

SiO$_2$ and total alkalis are poor in them. Within the group, MgO shows the highest degree of fractionation (varying from about 0.1% to over 15% in normal magmatic liquids) and CaO shows only slightly less (normal range ca. 0.4–13%). Iron is less variable and the great majority of rocks show FeO + Fe$_2$O$_3$ in the range 2–14%. The concept of degree of fractionation employed here is of course expressed by the ratio of the highest normal content of the constituent to the lowest content.

Although, as mentioned above, there is a positive correlation between the elements of this group over most of the field, the relationship changes in the extremely alkali-poor, silica-poor rocks (picrite basalts). In these rocks the general tendency of CaO to correlate positively with MgO is reversed and the most highly magnesian basaltic rocks generally contain rather low CaO.

Finally, with regard to the elements of this group, we note that the total content of MgO, CaO and FeO + Fe$_2$O$_3$ has a dominating influence on the content of ferromagnesian minerals (and opaques) present in the rock after solidification and is therefore an approximate chemical expression of one of the oldest mineralogical classificatory concepts, that of the **colour index** (i.e. the total volume per cent of 'coloured silicates', that is, olivine, pyroxenes, amphibole, biotite). The contours for MgO, CaO and $FeO + Fe_2O_3$ in Figure 2.3 therefore express this fundamental classification and distinguish between groups such as phonolite–trachyte–rhyolite often termed **leucocratic** (pale-coloured) or **felsic** (rich in feldspars, feldspathoids or quartz) **salic** (rich in SiO_2 and/or Al_2O_3) on the one hand, and the nephelinite–basanite–basalt group termed **melanocratic** (dark-coloured), or **mafic** (rich in magnesium and iron minerals). Terms such as **mesocratic** are used for rocks of intermediate content of ferromagnesian minerals. The terms used above should not be confused with purely chemical expressions such as **acid, intermediate** and **basic** which are based *only* on SiO_2 content, and on the terms derived from the CIPW normative classification (Appendix 1) **salic** (rich in Al_2O_3 and/or SiO_2) and **femic** (rich in the chemical constituents of the ferromagnesian and other alumina-poor minerals.

Relationships of Al. In contrast to the considerable variations shown by the elements discussed above, Al_2O_3 contents of volcanic rocks are comparatively uniform, most containing between 12% and 20%. Highest values are generally found in phonolites while a restricted range of alkali-poor basic rocks (picrite basalts) shows low values (below 10%). From this it follows that Al_2O_3 shows a positive correlation with total alkalis, though it is not a strong one.

Relationships between Na and K. Up to this point it has been convenient to consider the two alkali metals, Na and K, together. They do indeed have many similarities of geochemical behaviour, for example they are both essential constituents of the alkali feldspars and feldspathoids. In a rather general way they correlate positively with each other, most rocks with $Na_2O + K_2O > 5\%$ containing substantial amounts of both elements. However, there are also significant geochemical differences resulting from their different ionic radii ($Na^+ = 1.10$ Å, $K^+ = 1.46$ Å) so that several important rock-forming minerals will accept one of them in strong preference to the other. Thus plagioclases contain far more Na than K while the reverse is true of biotites. At high pressures clinopyroxenes accept significant amounts of Na (as the jadeite endmember) while excluding K. Low pressure clinopyroxenes (containing the acmite end-member) exert a similar discrimination. It follows that crystallisation or melting involving such minerals is capable of fractionating Na and K relative to each other. However, it is also important to note that many minerals such as olivines and opaque phases contain *no* significant amounts of Na and K, and even minerals such as pyroxenes normally contain very little in total. Thus, working along with processes which have the potential to fractionate the alkali metals relative to each other, are additional processes tending to concentrate both of the alkali metals into the liquid during crystallisation and melting. This second factor gives rise to the crude positive correlation mentioned above, while the first factor impresses some degree of diversity on the ratio of K to Na.

The rocks selected for the study of variation so far demonstrate what may be thought of as normal K/Na relationships and we have deliberately excluded those rare rocks which are significantly more K-rich than normal. Figure 2.3e illustrates this 'normal' range of K/Na ratios. There is much scatter on this diagram and the predictive value of the contours is not so high as that of the earlier diagrams. Nevertheless there is a notable tendency for most basaltic rocks to have the ratio $K_2O/Na_2O \simeq 0.3$ (ocean floor basalts may be as low as 0.03) while trachytes and rhyolites have a ratio averaging about 0.7. In fact rhyolitic rocks with a ratio of 1.0 or more are comparatively common. It must be stressed that the concept of normality refers to the fact that most rocks conform approximately to this pattern. Much more rarely we find series with abnormally high K contents (the ultrapotassic rocks referred to earlier) which, in the more alkali-rich varieties, is expressed by the presence of leucite (or very rarely kalsilite) instead of the more usual nepheline. In alkali-poor rocks, in contrast, the mineralogical effects of high K/Na ratios are less obvious because they only involve increases in the ratio of K-feldspar to plagioclase or the presence of more K-rich alkali feldspars than normal. In these rocks no new phase is normally involved (though biotite may be more prominent than usual) and the nomenclature of K-rich varieties is therefore commonly derived from the normal nomenclature by the addition of a prefix such as 'K-' or 'potash-'. In the more alkalic rocks however a different system of nomenclature is normally employed for potassic as opposed to normal varieties (see Table 2.1). This table, like Figure 2.2, should not be regarded as a rigid classificatory scheme. It is intended to illustrate approximately the nomenclature of rock types which are similar to each other in most respects but differ in their K/Na ratios. As before a problem arises with the use of the word trachyandesite and its potassic equivalent given here, latite. The terms have frequently been used synonymously but the original latites (from the Latium region, Italy) are in fact distinctly potassic.

Summary of progress

The brief survey of variation in magmatic compositions given above has been largely descriptive but it should have the effect of creating a degree

Table 2.1 Approximate equivalents of some K-rich and normal rock types

Potassic variety	Normal variety
leucitophyre	phonolite
K-trachyte	trachyte
K-rhyolite	rhyolite
tristanite	benmoreite
latite	trachyandesite
leucitite	nephelinite
leucite basanite	basanite
leucite tephrite	tephrite
absarokite ⎫ shoshonite ⎭	basalt

of order in an abundance of data which otherwise might appear thoroughly confused. The exercises at the end of this chapter are designed to reinforce what has been described, but it must be emphasised that the generalisations which have been made are all subject to their exceptions. A frame of mind is required which can accept the generalisations while remaining appreciative of their vagueness.

In the most fundamental way, however, the explanation of the patterns of geochemical behaviour so far discussed constitutes one of the most important questions that the study of petrogenesis has to answer. Much of the rest of this book will be devoted to methods by which this can be attempted, and we continue now with an initial look at the topic of variation diagrams on the more specific level as opposed to the general approach adopted so far.

Variation diagrams – introduction

Sets of chemical analyses of igneous rocks from a particular igneous province, volcano, or intrusive complex, almost invariably show a considerable variation in the concentrations of individual elements. This is what may be termed 'within province' variation and expresses the idea that igneous rocks which are closely associated in time and space often show substantial variation. In most cases, however, such variation is confined within certain limits and it is exceeded by 'between province' variation, that is to say, gross differences which exist between the range of compositional types present in one province as opposed to another. One of the preliminary tasks in any research study is to devise a means of describing and displaying variation so that the numerous data for individual rocks become simplified, condensed, and rationally classified.

Table 2.2 Analyses of volcanic rocks from the Hanish-Zukur islands, Red Sea (data from Gass *et al.* 1973)

	1	2	3	4	5	6	7	8	9	10	11	12	13
SiO_2	45.51	45.74	45.83	47.36	48.37	49.00	49.82	50.09	52.45	56.17	56.87	61.01	61.22
TiO_2	3.52	3.93	3.92	3.30	2.82	2.73	2.98	3.08	2.29	1.61	1.40	0.68	1.00
Al_2O_3	15.24	15.08	15.47	16.32	16.01	16.33	16.90	16.83	16.09	17.13	16.96	17.14	17.10
Fe_2O_3	3.64	4.06	9.26	4.64	5.51	2.20	1.79	1.71	5.51	3.23	4.13	5.09	2.03
FeO	8.84	8.36	3.02	6.89	5.04	8.12	9.14	9.14	4.60	5.32	4.18	1.21	4.06
MnO	0.22	0.20	0.20	0.20	0.22	0.22	0.23	0.24	0.26	0.22	0.24	0.22	0.20
MgO	5.07	5.58	5.14	4.82	4.67	4.31	3.50	3.31	2.67	1.73	1.57	0.76	0.92
CaO	8.91	10.72	10.11	9.30	9.04	7.77	8.84	8.84	7.49	5.20	4.83	3.33	3.28
Na_2O	4.95	3.17	3.81	4.63	5.97	4.98	4.96	4.89	6.11	6.33	6.47	7.07	6.61
K_2O	1.19	1.23	1.46	1.49	1.28	1.66	1.35	1.28	1.64	2.22	2.43	2.87	3.05
H_2O+	0.97	1.15	0.98	0.89	0.39	1.98	0.23	0.26	0.38	0.52	0.46	0.24	0.21
H_2O-	0.00	0.00	0.00	0.00	0.68	0.00	0.26	0.33	0.51	0.32	0.46	0.38	0.32
P_2O_5	0.85	0.68	0.74	0.83	–	0.89	–	–	–	–	–	–	–
Total	98.91	99.90	99.94	100.67	100.00	100.19	100.00	100.00	100.00	100.00	100.00	100.00	100.00

The oldest method, still widely used, is a variation diagram in which oxides are plotted against SiO$_2$ (Harker 1909), often referred to as a Harker diagram (Fig. 2.1 is a type of Harker diagram). Table 2.2 gives analyses of individual lava specimens from a single volcano, arranged in order of increasing SiO$_2$. From the table a number of relationships can be noted, for example:

(a) TiO$_2$, FeO, MgO and CaO all show rather similar behaviour and fall together as SiO$_2$ rises.
(b) K$_2$O and Na$_2$O in contrast rise with SiO$_2$.
(c) Al$_2$O$_3$ does not show such strong variation as the other oxides, rising slightly as SiO$_2$ rises.

In Figure 2.4 the same data are given diagrammatically to show that the relationships are more readily digested than they are in the tabular form. However, as in all simplifying techniques, information has been lost in the process, in this case the precise analytical values for each element in the individual specimens.

The relationships discussed, not surprisingly in view of the general discussion given previously, indicate that the chemical parameters do not vary independently of each other. This sort of series is said to show geochemical 'coherence', that is to say, many pairs of elements show strong correlation coefficients, either positive or negative (e.g. in Figure 2.4 the correlation coefficient (r) between CaO and MgO is +0.95. A contrasting case where the correlation coefficient is low is shown by

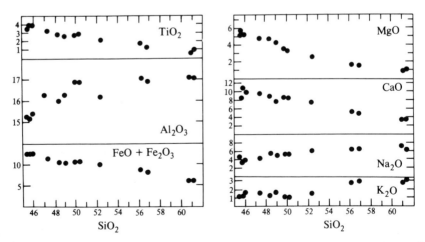

Figure 2.4 Weight per cents of oxides plotted against SiO$_2$ for the analyses of Hanish-Zukur lavas given in Table 2.2.

the relationships between SiO_2 and K_2O in Figure 2.11). We can now consider the reasons underlying the choice of SiO_2 as an index against which the other oxides were plotted, and it is apparent that any of the parameters TiO_2, $FeO + Fe_2O_3$, CaO, K_2O and Na_2O, all of which show moderate to good correlations (either positive or negative) with SiO_2, could alternatively have been chosen. In each case a diagram would have resulted showing essentially the same relationships, even though the individual analysed rocks would not appear in exactly the same sequence in each diagram. On the other hand it is clear that the choice of Al_2O_3 as an index would not have produced a particularly informative diagram because this element shows relatively little variation. As a first approximation it may be concluded that any element showing strong variation and showing high correlation coefficients with other elements is a potential choice for the abscissa of a useful variation diagram.

Before reading petrogenetic significance into such diagrams it is, however, necessary to consider the **constant sum effect** which results from the fact that the major oxides taken together make up very nearly 100% of a rock. In the case considered, since SiO_2 is by far the most abundant constituent and varies between 50% and 70%, the sum of all the other oxides must fall from 50% to only 30% as SiO_2 rises. At least some negative correlations with SiO_2 (though not necessarily the good correlations seen here) are therefore to be expected amongst the other oxides irrespective of petrogenetic considerations. Chayes (1964) gives an illuminating discussion of this topic. A related point is concerned with the scatter of data points on Harker-type diagrams which cover a wide range of SiO_2 variation. The trends for individual oxides are often observed to tighten up (i.e. show reduced scatter) towards the SiO_2-rich end of the diagram giving a misleading impression of greater coherence in the acid rocks than in the basic. The effect, of course, can be due to the constant sum factor and does not, as close inspection of such diagrams will show, necessarily imply that correlation coefficients between particular pairs of oxides are improving. It has sometimes been thought that the constant sum effect negates the usefulness of Harker diagrams for petrogenetic purposes but, as will be shown in Chapter 3, this is not so. The Harker diagram is simply one example of the whole range of possible two-element plots (rectangular diagrams in which one element or oxide is plotted against another) which is one of the most powerful data-handling techniques available to the petrologist.

The preceding discussion has considered the simplest possible type of variation diagram, which nevertheless may have a very useful function in displaying analytical data and focusing attention on correlations between the concentrations of different elements. The existence of coherent series of rocks and the existence of inter-element correlations within them are obviously amongst the most important facts upon which petrogenetic hypotheses can be founded. It is moreover a simple, obvious and tempting step, to interpret such series as evolutionary, and to begin to think in terms of a parental magma and its various derivatives. Before taking this rather important step it is useful to examine in more detail the 'fractionation index', that is to say the chemical parameter against which other parameters are plotted. So far, only simple diagrams involving the use of *one* of the oxides as a fractionation index have been discussed, but many more complicated functions have been devised.

Fractionation indices. Running through much of the older geological literature is the idea that analyses of individual rocks, if suitably plotted, can be arranged in an evolutionary sequence. However, considering the variety of possible fractionation processes (see Ch. 1), it is evident that evolutionary series can theoretically develop in a number of different ways. Firstly we may visualise the fractional crystallisation of a parental magma and the periodic eruption of the residual liquids to give rise to surface volcanic rocks. Alternatively the series, in the case of plutonic rocks, may be formed by successive accumulations of crystals from a magma body undergoing fractional crystallisation. Again, the series may be formed by progressive contamination of a magma by foreign material or by varying degrees of mixing of two magmas. These are not all the possibilities, but serve to illustrate that the idea of an evolutionary series is a complex one. In this situation it is clearly not reasonable to suppose that a single, all-purpose, fractionation index actually exists – an index which will automatically arrange analyses in their correct evolutionary order. However, as will become clear, the evolution of igneous rocks appears to be dominated by two processes which are closely related to each other. These are fractional crystallisation and partial melting (collectively termed **crystal–liquid fractionation processes**) and in this light it is worthwhile to enquire whether certain types of fractionation index can have a general value. A brief discussion of commonly used indices follows.

The **Harker index** (SiO_2 as abscissa), if it is to have *direct* evolutionary significance, depends on the commonly observed increase of SiO_2 in the successive liquids of fractional crystallisation, and conversely the decrease of SiO_2 in successive partial melts of reasonable source materials. With regard to fractional crystallisation, many studies of *basaltic* rocks show, however, that SiO_2 remains almost unchanged during the early stages of crystallisation. Wright and Fiske (1971) have provided a particularly good example with their study of the Alae lava lake, Hawaii, as it cooled. Their data are given in Figure 2.5 and

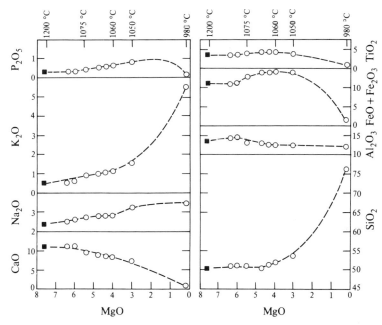

Figure 2.5 Variation diagram in which oxides are plotted against MgO for the Alae lava lake (data from Wright & Fiske 1971). The filled square represents the composition of the initial liquid. Open circles are analyses either of separated interstitial glasses or of residual liquid which oozed into the drill holes during sampling as the lava lake solidified. The temperature scale indicates the temperatures at which the different residual liquids were collected. Note the initially very slow increase in SiO_2, the sympathetic reduction of CaO and MgO, the sympathetic rise of Na_2O and K_2O (but with a strong rise in K_2O/Na_2O ratio), and the comparatively steady behaviour of Al_2O_3.

illustrate that SiO_2 does not begin to increase until the later stages of crystallisation. Other studies (e.g. Wager & Deer 1939, on the Skaergaard intrusion) have suggested that in some cases SiO_2 *decreases* in the residual liquids in the early stages of crystallisation. Crystallisation at this stage is dominated by pyroxenes and calcic plagioclase, both of which have SiO_2-contents similar to that of basaltic liquid. It is usually the onset of crystallisation of an SiO_2-free phase such as magnetite that eventually causes SiO_2 to become enriched in the residual liquid, and once this has happened the continued crystallisation of plagioclase and pyroxene will reinforce the SiO_2-enrichment trend. Irrespective of further detailed considerations it is already clear that though SiO_2 may be a useful index of evolution for intermediate and acid rocks its usefulness for basic rocks is limited. It is, as we shall see, rather typical of most simple fractionation indices. It can be used for some rock series but not for others. As a *universal* fractionation index it has no application.

Many authors, particularly those concerned with basaltic rocks, have preferred to use **MgO** as the abscissa for variation diagrams (e.g. Powers 1955). In the example of the Alae lava lake already quoted (Fig. 2.5) its use demonstrates the good positive correlation between K_2O, Na_2O and P_2O_5 (excluding the final stages of crystallisation), these elements being enriched in successive residual liquids, and the positive correlation between CaO and MgO as these are depleted. The use of MgO as an index depends on the fact that in most natural systems the solids crystallising from a magma contain collectively more MgO than the liquid or, conversely, during the melting the residual refractory solids at any stage contain more MgO than the partial melt liquid. The index can be effective in favourable circumstances in rocks far removed from basalt in composition, though it ultimately, in many trachytic and rhyolitic liquids, becomes unreliable because the concentration of MgO falls so low that analytical errors etc. become a serious factor.

However, as was the case with SiO_2, special circumstances may exist in which it is doubtful whether residual liquids during fractionation are actually impoverished in MgO, thus the applicability of MgO as a universal index is suspect. This arises because although Mg-bearing minerals such as olivine and the pyroxenes always, as far as is known, contain more MgO that the liquids from which they crystallise, their crystallisation is often accompanied by Mg-free minerals such as plagioclase and magnetite. The crystallisation of these latter minerals will of course tend to enrich the residual liquid in MgO and if they crystallise in sufficient quantity this may reverse the MgO-depletion of the ferromagnesian minerals.

Indices based on the **magnesium–iron ratio** have been widely and usefully employed (e.g. Wager & Deer 1939) and take several different forms (e.g. 100 $MgO/MgO + FeO$ or $Mg^{2+}/Mg^{2+} + Fe^{2+}$). Both the examples have the form $A/A + B$ since the index then varies between 0 and 1 or 0 and 100, which is preferable to the A/B form which can become infinite. The examples also illustrate two alternatives, the first in weight per cent of oxides, the second in atomic per cent of cations. Another variant has the form $MgO/MgO + FeO + Fe_2O_3$ which attempts to take account of the fact that the FeO value may be accidentally changed by late-stage oxidation (e.g. weathering).

The use of Mg/Fe indices avoids one of the principal defects of the MgO index because its value in residual liquids of fractionation is unaffected by the crystallisation of plagioclase. It is also useful for the study of cumulate rock sequences affected by crystal sorting (see Ch. 13). For example two associated olivine–clinopyroxene–plagioclase cumulates in which the mineral *compositions* are the same (and which therefore are assumed to have crystallised from virtually identical liquids) have almost identical magnesium–iron ratios irrespective of the *proportions* of the minerals present. This results from closely similar Mg/Fe ratios of coexisting olivines and pyroxenes.

In many basaltic liquids the Mg/Fe ratio falls steadily during the early stages of crystallisation because the common ferromagnesian minerals always have higher Mg/Fe ratios than the liquids from which they crystallise. Figure 2.6 gives simplified curves for a number of volcanic series and illustrates that strong iron enrichment relative to magnesium is characteristically restricted to the more basaltic (i.e. Mg-rich) liquids.

The Mg/Fe index has drawbacks, discussed in a later section, simply because it is a ratio. It is also, like the other indices discussed, fallible in certain circumstances, particularly when the crystallisation of ferromagnesian minerals is accompanied by magnetite. When this happens the index either falls much more slowly as crystallisation

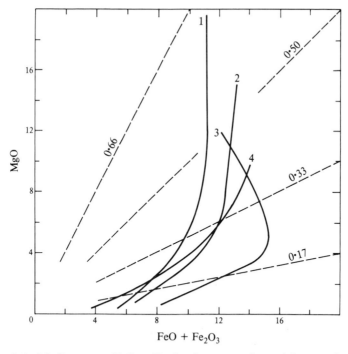

Figure 2.6 MgO versus FeO + Fe$_2$O$_3$ for a number of lava series (from Coombs & Wilkinson 1969). 1 – Gough Island, 2 – Hawaiian alkali series, 3 – Hebridean alkali series, 4 – Tristan da Cunha. Dashed lines have constant ratio of MgO/(FeO + Fe$_2$O$_3$ + MgO). Note the generally rapid change of this ratio in the more basic (MgO-rich) rocks.

continues (and as such becomes an insensitive index of crystallisation) or its behaviour may be slightly reversed.

The ratio **Ab/Ab + An** (normative albite/normative albite + anorthite) (normative calculations are discussed in Appendix 3) and other indices based more directly on the values of Na_2O and CaO in the rock analyses (e.g. the Felsic Index of Simpson (1954) which also includes K_2O) is similar in spirit to the Mg/Fe ratio and relies on the continual increase of the ratio in residual liquids dominated by plagioclase crystallisation. Its applicability to plagioclase-bearing cumulates is also evident because of its lack of sensitivity to mineral proportions. Its use, however, is restricted to plutonic series which are plagioclase-bearing and to volcanic rocks which crystallise plagioclase (i.e. have plagioclase phenocrysts) because it does not respond to the crystallisation ferromagnesian minerals alone.

In view of the way in which all the indices so far discussed can only usefully be applied to certain ranges of rock types it is not surprising that several attempts have been made to devise more complex indices intended to have a more comprehensive use. Two of these in particular are widely used, the **Solidification Index** of Kuno and the **Differentiation Index** (D.I.) of Thornton and Tuttle (1960). Solidification index is expressed as:

$$S.I. = \frac{100 \ MgO}{MgO + FeO + Fe_2O_3 + Na_2O + K_2O}$$

which, for basaltic rocks rich in MgO and FeO and relatively poor in alkalis, acts much like other forms of Mg/Fe index. In most magmatic series, however, residual liquids of crystallisation are enriched in alkalis so that the inclusion of Na_2O and K_2O in the index helps to offset the potentially poor performance of the Mg/Fe index in intermediate and acid suites showing little or no iron enrichment relative to magnesium (see Fig. 2.6). The Thornton and Tuttle index is expressed as:

$$D.I. = normative \ Q + Or + Ab + Ne + Ks + Lc$$

(See Appendix 3 for the discussion of the idea of the **norm**, a calculated theoretical mineral assemblage not to be confused with the **mode**, or actual mineral assemblage.) As opposed to the Kuno S.I., which is based on a blend of petrogenetic reasoning already discussed with relation to simple fractionation indices plus an admixture of empirical reasoning, the D.I. is based on the simple petrogenetic idea that in all crystallisation the constituents of 'petrogeny's residua system' (e.g. Schairer 1950, and see p. 133) will become concentrated in the residual liquid, since these are the minerals which begin to crystallise in general only at low magmatic temperatures. As generalisations about fractionation indices go, this is sound, and although the index is superficially complex it is in practice easy to comprehend because it is closely related to the old mineralogical concept of colour index, from which it differs mainly in the exclusion of calcic plagioclase. Like the other indices discussed, however, there are occasions, admittedly rare, when it serves no purpose, e.g. in rocks extremely depleted in ferromagnesian minerals and calcium (when D.I. tends to 100) and in rocks such as peridotitic cumulates where D.I. can remain close to zero despite considerable fractionation involving ferromagnesian constituents.

Indices based on the so-called **incompatible** or **dispersed** elements are of considerable theoretical interest and have occasionally been used (e.g. Barberi *et al.* 1975). An incompatible element is one which under given compositional conditions is excluded or largely excluded from the lattices of the minerals crystallising and therefore becomes concentrated in the liquid in a manner simply related to the amount of the original material crystallised. In the ideal case where the distribution coefficient K_D (expressed as concentration of the element in the solid divided by concentration in the liquid) approaches zero (see also pp. 345–6):

$$C = C_0 \frac{1}{F}$$

where C_0 is the concentration of the element in the original liquid, F is the amount of liquid remaining (expressed as a fraction of 1) and C is the concentration of the element in the remaining liquid. This is a simple hyperbolic curve in which for example C doubles as F is halved and is ideally a perfect evolutionary index which theoretically enables a complete description of the way the geochemistry of a liquid changes with progressive crystallisation to be made. Elements showing an approach to the required K_D over certain compositional ranges include K and Rb in basic liquids crystallising only olivines, pyroxenes and oxides, P in liquids not crystallising apatite, light rare earth elements such as Ce, and Th and U in considerable ranges of liquids. However, any incompatible element sooner or later becomes compatible as its concentration in the liquid increases and new phases appear. P forms a good example showing a rapid increase in the liquid in early stages of crystallisation, followed by a rapid fall as soon as apatite is stabilised (see Fig. 2.5). This type of index can thus never be used to monitor the *complete* crystallisation of a liquid, though in favourable circumstances it may be possible to study a considerable range of it. In practice C_0 is determined by analysing for the element in the most basic non-porphyritic rocks available, which are then assumed to represent the parental liquid. Then since $F = C_0/C$ a variation diagram can be made with F as abscissa, the value of f for each sample being calculated from the concentration of the element concerned and C_0. Although this type of variation diagram has a particularly sound theoretical foundation it does depend on the assumption that the rocks studied are all descended from a single definable parent. However, as is shown in a later part of this chapter, this assumption is not safe except in restricted circumstances. The lavas of the Roman region for example (see Fig. 2.11) show a large range in K at a *given* level of SiO_2. This means that it is not possible to define a *single* parent magma with a fixed content of K (the same is true of other elements) and an incompatible element type of fractionation index is not directly applicable. The case cited is extreme, but illustrates one of the principal difficulties of this type of study.

Fractionation indices in comparative use. Information which has a general petrogenetic usefulness is derived increasingly from individual rocks (e.g. see Nicholls *et al*. 1971), frequently from suites of associated rocks (e.g. studies of a particular volcano) and frequently also from comparative studies of different provinces. On the within-province scale petrologists frequently come to conclusions about the role of particular minerals in the fractionation of the suite and the mechanism of the fractionation process, and they may additionally be able to speculate profitably about the source material. Such conclusions are specific to the province. On the other hand, facts like the confinement of andesites to very specific tectonic environments, the geochemical similarities and differences between continental basalts and oceanic basalts etc. have an altogether wider nature and pose general petrogenetic problems. In most studies information of both a general and a specific nature can be drawn on with advantage, e.g. the SiO_2-poor nature of the highly potassic volcanic rocks of the Roman region has frequently been ascribed to limestone assimilation because the magmas pass through Mesozoic limestone sequences on their way to the surface. The fact, however, that extremely similar lavas are found in the Western Rift area of Uganda (Holmes & Harwood 1937) where limestone is virtually absent is an additional factor which might well be taken into account. Whatever the application, all such considerations involve comparative studies of different suites.

Much comparative work is concerned with the demonstration of similarities between suites, and the choice of variation diagram is thus of little consequence. Some of the more complex fractionation indices do, however, pose problems of comprehension when it is desired to

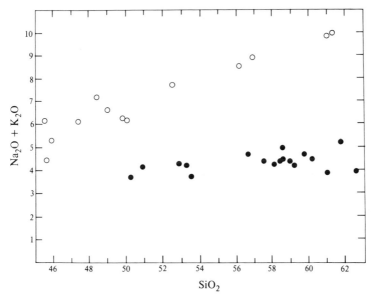

Figure 2.7 Alkali–silica diagram for two provinces. Open circles – Hanish–Zukur lavas (see Table 2.2) ranging from basalt to trachyte; filled circles – basalts and andesites of Mount Misery (data from Baker 1968).

comment on differences between provinces. An example of a simple comparative study is given in Figure 2.7 where $Na_2O + K_2O$ is plotted against SiO_2. For these suites, at a given value of SiO_2 one is consistently more alkali-rich than the other – a simple conclusion, but an entirely comprehensible one. Using a more complex index such as for example the Larsen Index ($= 1/3SiO_2 + K_2O - (FeO + MgO + CaO)$) (Larsen 1938)) it is possible to produce a very similar diagram (Fig. 2.8) from which one is strictly only entitled to conclude that one suite is more alkali-rich than the other at a given level of $1/3SiO_2 + K_2O - (FeO + MgO + CaO)$ – a simple conclusion, but the meaning is not immediately clear. What all such diagrams (providing obviously inappropriate indices are avoided) are hoped to demonstrate is that at comparable stages of evolution the suites differ in their alkali contents. We can at this stage only grasp rather intuitively what is meant by 'comparable stages of evolution'; but although a vague concept, the conclusion about these two suites can now be seen in more comprehensible terms. For present purposes, however, the main issue is whether there is any particular value in doing something in a complex way (e.g. Larsen Index) which can be done simply (e.g. SiO_2). A further example is afforded by the rocks illustrated in Figure 2.9. Here the

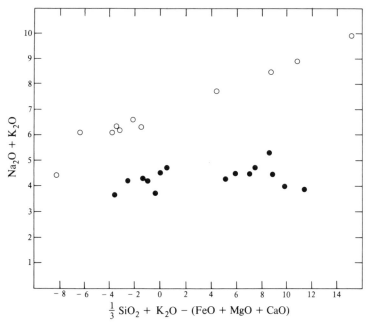

Figure 2.8 The same analyses as in Figure 2.7 plotted against the Larsen index.

variation diagram uses S.I. as the abscissa and the straight line given by the MgO trend will be noted. Clearly a variation diagram employing MgO as abscissa would convey almost exactly the same information.

Finally, on the subject of complex fractionation indices, the vague idea is sometimes held that a complex index has inherent value because it takes into consideration a larger number of the constituents of the rock than a simple index. However, since the effects of the individual constituents on the numerical value of the index cannot be separated, the argument is questionable. For example, if one rock has a higher value for the Larsen Index than another it may be because it contains more SiO_2, or more K_2O, or because it contains less MgO, CaO or FeO. It may be noted that the objection to simple indices is readily overcome if several different sorts of simple diagram are used in conjunction in the study of a single set of data (see Ch. 6).

Triangular variation diagrams

Diagrams showing the simultaneous relative variation of three chemical parameters are widely and usefully employed in comparative studies of

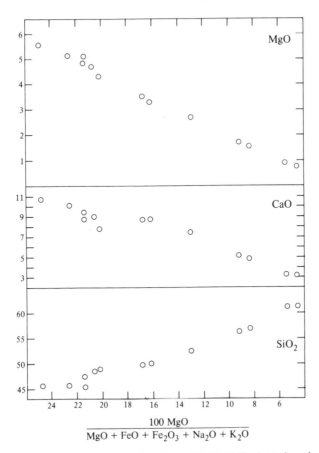

Figure 2.9 Analyses of Hanish–Zukur lavas (Table 2.2) plotted against Kuno's Solidification Index. Note the particularly good correlation between S.I. and MgO in this case.

rock suites. Commonly selected parameters are MgO, FeO (which may include recalculated Fe_2O_3), and $Na_2O + K_2O$, or, Na_2O, K_2O, and CaO. The values of the three selected parameters are summed and recalculated to total 100, then plotted on triangular graph paper as explained with reference to ternary phase diagrams at the beginning of Chapter 4. As an alternative to the use of weight per cent values of the oxides, the cations alone may be recalculated as atomic percentages.

Triangular diagrams have the disadvantage that they do not show absolute values of the parameters. They are, however, very useful in the illustration of features, for example the change of inter-element ratios, which do not necessarily show up clearly on two-oxide plots. The magnesium–iron–alkalis plot (so-called AFM diagram) is, for example, particularly useful in discriminating between tholeiitic and calc-alkaline suites, because iron enrichment relative to magnesium, as alkalis rise, is perhaps the most characteristic feature of the former group but is only very weakly demonstrated by the latter.

The liquid line-of-descent

In the previous discussion there has been a strong implication that variation diagrams for volcanic rock suites showing reasonable coherence illustrate the course of chemical evolution of liquids, whether formed by fractional crystallisation or by progressive partial fusion. Bowen (1928) discusses this in some detail, pointing out that only analyses of phenocryst-poor or aphyric volcanic rocks will give a true indication of this path, termed the **liquid line-of-descent.** Petrologists have generally embraced this idea and it has, with the rise of rapid methods of silicate analysis, led to enormous progress in the subject. The idea is, however, only rarely strictly true, often a mere approximation, and occasionally downright misleading.

In Figure 2.10 all the available analyses of lavas from the Aden volcano, South Yemen, are plotted in an alkali–silica diagram. At first sight this relatively continuous trend of composition points with comparatively little scatter represents a good example of a liquid line-of-descent. However, the different symbols refer to various lava groups distinguished on the basis of age relations and structure. Evidently then, although the chemical evolution of the lavas appears to be straightforward and simple, this is not an evolutionary process which can be directly related to a *time factor*. The individual members of the series are mainly erupted in a highly random time sequence. Clearly the simple idea that a parental magma fractionates progressively to give rise to the members of the series, may be replaced by the idea of a fractionation process which is *reproducible at different times*. The simplest hypothesis to explain the behaviour of this particular volcano is that batches of trachybasaltic (or possibly basaltic) magma were repeatedly supplied to high level chambers where they repeatedly underwent crystal fractionation of a specific type, being sampled from time to time by the eruptive process. The original 'liquid line-of-descent' is in fact a bundle of similar, overlapping, lines-of-descent. It is also perhaps too much to expect of a natural system of this complexity that the reproducibility of the fractionation process, or indeed the duplication of the parental magma supplied at different times, should be

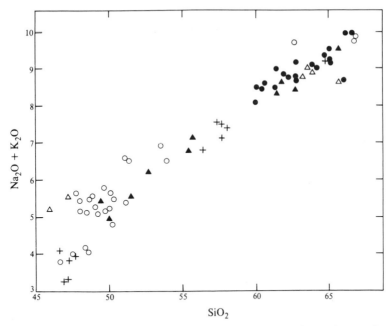

Figure 2.10　Alkali–silica diagram for lavas from the Aden volcano, South Yemen (data from Hill 1974). The rocks belong to successive eruptive cycles in the following sequence: Tawahi series (open triangles) (oldest series), Ma'alla series (filled triangles), Main Cone series (open circles), Shamsan Caldera series (filled circles), Amen Khal series (crosses). Note lack of correspondence between time and degree of chemical evolution.

exact. This accounts for some of the *scatter* of the variation diagram, though some of it is also due to the inclusion of porphyritic rocks in the analysed samples.

Even at this stage, the 'revolutionary' nature of the series acquires a new meaning, for it is evident that two lava flows of closely similar composition, such that one has the required chemical and mineralogical features to be 'parental' to the other, certainly cannot have this close relationship if they were erupted at quite different stages of the volcanic evolution. Finally let us note that there is nothing particularly unusual about Aden. Not all volcanoes behave so irregularly but nevertheless the literature contains many descriptions of volcanoes which show considerable chemical coherence in their eruptive products, but which produced them in a partially random time sequence.

On the other hand it is occasionally possible to identify smaller sequences of eruptive rocks in which it is more reasonable to postulate the existence of a single parental magma. On a relatively small scale such sequences are produced when magma from a chamber which is already compositionally stratified is forced rapidly to the surface and gives rise to products which overlie each other stratigraphically in an inverted fractionation sequence. That is to say, the liquids from the top of the chamber are erupted first to form ashes, individual lava flows, or the basal parts of larger ash flows, and are progressively covered by rocks formed from magmas derived from deeper parts of the chamber. Several apparent examples are known, e.g. various ash-flow deposits from the western U.S.A. (e.g. Smith & Bailey 1966), the ashes of Hekla in Iceland (Thorarinsson & Sigvaldason 1972), and the Shamsan Caldera sequence from Aden (Hill 1974). This topic will be returned to in Chapter 11 for such sequences give a rare opportunity to reconstruct the interior of a magma chamber at a specific moment during its solidification. As such they form an interesting corollary to the study of layered intrusions in which magma chambers can be studied only after their solidification.

In the preceding section we have seen that although it is sometimes possible to identify limited volcanic sequences which are probably derived from a single parent magma and whose aphyric members can therefore be expected to approximate to an ideal, single, liquid line-of-descent, sequences consisting of a large number of overlapping similar lines of descent are probably more common. However, there are many provinces and indeed individual volcanoes where even this approximation to the ideal does not hold. Such series do not show coherent trends for many oxide-pairs and it is convenient for the moment to think of them as consisting of many subparallel lines of descent, though to *prove* conclusively that this is so is a difficult exercise. One of the best examples is afforded by the Roman province of Quaternary volcanoes, an example from which is given in Figure 2.11. Here it will be observed that coherence is poor for some elements (e.g. SiO_2 and K_2O) but considerably better for others (e.g. CaO and MgO). The East Otago area (Coombs & Wilkinson 1969) also provides a good example of a restricted province in which several different lineages may be distinguished. Many series are found which show degrees of scatter intermediate between the ideal coherent case and the Roman province case and thus show varying degrees of what may be termed proliferation of the liquid lines-of-descent.

Having illustrated one aspect of the complex concept of the liquid line-of-descent, let us now examine in more detail some of the evolutionary aspects of volcanic series. In an ideal coherent series it is obviously permissible to think of successive magmas as having a parent–daughter relationship if they are evolving by fractional crystallisation. It is additionally quite possible to think in this way about

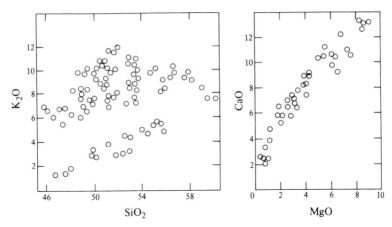

Figure 2.11 K_2O versus SiO_2 and CaO versus MgO for the lavas of Roccamonfina, Roman province, Italy (data from Appleton 1972). Most major elements in this volcano show the normal types of correlation (e.g. Ca and Mg) but incompatible elements such as K show widely variable behaviour relative to them (e.g. K relative to Si).

series of the Aden type, where a fractionation process has been reproduced at different times. In this case two similar magmas are more likely to be cousins than siblings, but since they all look very much the same this is a consideration which for many purposes need not deter us. However, the parent–daughter relationship is not in itself an *inherent* property of even ideal coherent magmatic series. It is for example a difficult concept to apply to the theoretically possible case of magmatic mixing, and equally difficult to apply to open system processes involving contamination or re-equilibration with wall rocks. Nor is it by any means easy to think of progressive partial melting in these terms.

In summary, if we examine the properties which have from time to time been ascribed to variation diagrams, or examine the assumptions which have sometimes been made in the course of their use, we see that most are suspect in specific circumstances and hence cannot be relied on to give *general* guidance in petrogenetic studies. It would in any case be quite wrong to make assumptions about the inherent petrogenetic properties of variation diagrams since this would be to prejudge the issue under investigation. Variation diagrams are central to many petrogenetic studies and they are properly used as a means of testing hypotheses, or as a means of erecting hypotheses to be tested elsewhere. In Chapter 6 this aspect of variation diagrams, that is, their role in the hypothesis-forming process, is considered in detail. However, it is helpful to have some knowledge of the ways in which magmas are likely to crystallise and the ways in which partial melts can form before proceeding to this stage. Chapters 3 to 5 which now follow therefore deal with aspects of experimental petrology.

Exercises

1. Total alkali and SiO_2 contents of all the rocks termed trachyte and hawaiite given in the compilation of Carmichael *et al.* (1974) are given in Table 2.3. Confirm that they each plot in a restricted field in Figure 2.2.

Table 2.3   Total alkalis and silica contents of rocks termed trachyte and hawaiite in compilation of Carmichael *et al.* (1974)

Trachytes

$Na_2O + K_2O$	SiO_2	$Na_2O + K_2O$	SiO_2
11.53	61.18	10.74	62.90
13.24	61.36	13.00	63.10
11.64	59.64	12.38	62.5
11.8	58.0	12.38	61.09
11.71	66.12	11.53	57.53
8.40	66.87	10.61	66.13
11.83	62.02		

Hawaiites

$Na_2O + K_2O$	SiO_2
8.18	54.04
4.9	47.26
6.10	50.4
5.27	47.49
8.53	47.56

Table 2.4   Partial analyses of volcanic rocks

	1	2	3	4	5	6	7	8
SiO_2	50.1	71.5	46.4	62.6	55.2	44.9	46.0	55.6
Al_2O_3	18.0	13.8	15.7	15.0	16.8	7.8	17.0	15.5
Fe_2O_3	4.4	1.5	5.3	3.3	2.1	2.0	3.8	3.6
FeO	2.1	0.5	3.4	3.4	8.3	9.8	7.5	5.0
MgO	0.4	0.8	6.0	0.4	1.7	21.1	4.8	6.5
CaO	7.3	1.6	11.6	1.2	4.5	9.1	9.5	7.0
Na_2O	6.9	3.7	1.6	6.4	5.0	1.6	4.0	3.2
K_2O	5.1	4.0	6.6	5.6	2.8	0.5	3.1	1.3

For sources of data see p. 419.

Table 2.5 Partial analyses of volcanic rocks (see Exercise 4)

SiO_2	50.5	51.0	50.9	50.8	50.5	51.3	52.0	53.9
Al_2O_3	13.5	14.1	14.5	14.0	13.0	12.5	12.5	12.2
MgO	7.6	6.4	6.0	5.4	4.8	4.5	4.0	3.0
CaO	11.0	11.2	11.1	10.0	9.2	8.7	8.8	7.5
Na_2O	2.40	2.53	2.62	2.68	2.79	2.84	2.91	3.20
K_2O	0.50	0.52	0.62	0.90	1.02	1.10	1.17	1.30
P_2O_5	0.29	0.32	0.31	0.42	0.51	0.59	0.65	0.80

2. Using the SiO_2 content and the total alkali content of the analyses given in Table 2.4, decide on a suitable name for each rock by plotting them on Figure 2.3. Check the K_2O/Na_2O ratios against Figure 2.3e to decide whether they are normal and whether the nomenclature should be derived from Figure 2.2 or alternatively from Table 2.1. Names and localities of these rocks are given on p. 419.

3. Using SiO_2 and $Na_2O + K_2O$ plot the rocks in Table 2.4 on Figures 2.3a–2.3e. Make predictions of the MgO, CaO, $FeO + Fe_2O_3$ and Al_2O_3 contents and compare these with the actual values.

4. Table 2.5 gives partial analyses of a series of volcanic rocks believed to be related to each other by fractional crystallisation. Assuming that no phosphorus is removed by the crystallising phases make a variation diagram by plotting oxides against f (the fraction of liquid remaining assuming the first analysis in the table represents the parent magma) using the relationship $C = C_0/f$.

   (a) What fraction of the original liquid must crystallise to produce the most evolved liquid?

   (b) Which other oxides would form suitable abscissae for producing a similar variation diagram (i.e. would arrange the rocks in a similar evolutionary sequence)?

   (c) Which of the other oxides enriches in the liquid almost as rapidly as P_2O_5 and hence must be almost excluded from the crystallising phases?

3
Phase diagrams – introduction

As indicated in earlier chapters, igneous fractionation is believed to be dominated by melting and crystallisation processes. Natural observable examples of these processes in operation are almost impossible to obtain, and while the study of the petrographic and chemical characters of natural rocks often allows inferences to be made about them, it is inevitable that experimental melting and crystallisation studies must be used to amplify the knowledge so gained. Natural rocks are, however, complex in terms of chemical composition, and, historically, experimental petrology has developed via the study of much simplified analogue systems. Because of the small number of chemical variables in these simple systems, graphical methods of representing results have been widely used and are known as **phase diagrams** or **equilibrium diagrams.** They portray the results of experiments designed to discover what phase assemblage is present under given conditions of pressure and temperature when a system is in equilibrium. The term **system** can be defined as that part of the universe which is arbitrarily or naturally isolated for purposes of consideration or experimentation. A **phase** is any part of the system which has a particular chemical composition and physical state but it need not be present as a single continuous entity. Thus a system consisting of a mixture of olivine and plagioclase crystals in equilibrium with each other contains two phases, olivine and plagioclase, irrespective of the number of crystals present. The science of experimental petrology has as its objectives the understanding of how magmas crystallise, how rocks melt, and how fractionation processes may operate within the known constraints of the experimental data. Experimental petrology says nothing directly about how geological materials behave in nature. Hypotheses derived from the study of natural rocks must, however, as far as possible be checked against experimental results to ensure that they are reasonable in a physico-chemical sense. Conversely hypotheses may evolve directly from experimentation and these must be applied to natural cases to find out if they have any basis in reality.

The experimental study of silicate systems was originated largely by

N. L. Bowen and co-workers at the Geophysical Laboratory in Washington from 1912 onwards. Most of the earlier studies were confined to comparatively simple synthetic systems but in recent years a growing amount of experimentation has been carried out on natural rock starting materials.

Phase relations may be studied in two essentially different ways. The approach adopted in this book is mainly empirical and may be contrasted with methods of study based on thermodynamics. Empirical methods take questions like, for example, how much does the addition of 10% of albite depress the melting point of anorthite? – and seeks to answer them by direct experiment. It has not traditionally been primarily concerned with the explanations of such phenomena in more fundamental physico-chemical terms. The thermodynamic approach in contrast attempts to predict the behaviour of systems from a relatively small amount of experimentally acquired thermochemical data such as heats of reaction and heat capacities. Such methods are widely applied in metamorphic petrology and are becoming increasingly important in the study of igneous rocks. At present, however, the empirical approach forms a convenient basis for a realistic introduction to the problems of igneous rocks evolution and for reasons of brevity we therefore omit detailed consideration of thermodynamic methods. The interested reader is referred to Wood and Fraser (1976) and Ernst (1976) whose treatments are complementary to ours.

Classification of systems

Systems may be classified initially in terms of the number of **components** which are required to describe the chemical compositions of all the phases which appear in them. If for example experiments are carried out on the chemical substance H_2O the only phases which appear over a considerable range of pressure and temperature conditions are ice, water and water vapour. Each of these phases has the composition H_2O, hence only one chemical parameter, H_2O, is required to describe their composition. The system is thus termed a **one-component** or **unary** system, even though two chemical constituents, hydrogen and oxygen, are present.

Experiments carried out at high temperatures and 1 atmosphere[1] pressure on mixtures of MgO and SiO_2 produce the solid phases periclase (MgO), forsterite (Mg_2SiO_4), enstatite ($MgSiO_3$), cristobalite (SiO_2) and tridymite (SiO_2), the latter two being high-temperature polymorphs of quartz. In this system three chemical constituents, Mg, Si and O, are present but the composition of all the phases is such that they may be expressed entirely in terms of the two oxides MgO and SiO_2 (forsterite consists of these two oxides combined in a $2:1$ molecular ratio, enstatite in a $1:1$ ratio). Hence this is a **two-component** or **binary** system. It must be emphasised that there is no simple relationship between the number of components and the number of elements present (the previous examples have both contained one less component than the number of elements) and, for example, mixtures of anorthite $(CaAl_2Si_2O_8)$ and albite $(NaAlSi_3O_8)$ also constitute a binary system despite the inclusion of five elements. Solid phases in this system consist only of plagioclase feldspars of which the compositions can always be expressed in terms of the two end-members albite and anorthite.

The system forsterite (Mg_2SiO_4) – diopside $(CaMgSi_2O_6)$ – silica (SiO_2) is an example of a **three-component** or **ternary** system. Phases produced by mixtures of these constituents at 1 atmosphere and high temperatures are forsterite, diopside, enstatite (see above), tridymite, and cristobalite. With this example it is possible to illustrate that components can be selected in more than one way. For example, components could be chosen as $CaMgSi_2O_6$, MgO, and SiO_2 or alternatively as $CaMgSi_2O_6$, $MgSiO_3$, and SiO_2. In the latter case it would be necessary to express the composition of forsterite as $2MgSiO_3$ *minus* SiO_2. Yet a third alternative set of components would be $CaSiO_3$, MgO and SiO_2. In this case diopside would be expressed as $CaSiO_3$ + MgO + SiO_2. There is in short no right way or wrong way of expressing components though some ways are more convenient than others. The number of components is a mathematical abstraction rather than a reference to real reactants and a direct analogy is afforded by the number of pieces of information it is necessary to give in order to define the position of a point in a plane. The answer is always two, though in one case x and y (Cartesian) co-ordinates might be used and in another the information might consist of a bearing and a distance from a reference point.

As a contrast to the above, consider experiments carried out on mixtures of anorthite $(CaAl_2Si_2O_8)$ and forsterite (Mg_2SiO_4). Solid phases appearing at high temperature and 1 atmosphere pressure are anorthite, forsterite, and spinel $(MgAl_2O_4)$. Despite the fact that all *bulk* compositions in the system are formed by mixing only two constituents, the solid phases cannot be expressed in any such simple way. Considering spinel, Mg and Al must be included in different components because they appear separately in the other two phases. Spinel contains no Si, but it is present in the other two phases so that Si must also be included in a separate component. Only one of the phases contains Ca, so this must be part of yet another component. Only oxygen, which is present in all the phases is a possible candidate for exclusion from the list of independent components. A trial and error examination of the compositions concerned shows that four components are needed, MgO, Al_2O_3, SiO_2 and CaO. The system is thus part of a **four-component** or **quaternary system.**

The above discussion is concerned with the determination of the number of components and it has been demonstrated that this is not simply related to the number of chemical elements present. It may also have become clear that neither is it simply related to the number of chemical constituents which are required to make up all the *bulk* compositions (as opposed to phase compositions) in the system. In the last example only *two* starting materials were needed (anorthite and forsterite) to make up a series of phases which could only be described in quaternary terms and the expression 'part of a four-component system' was used. The term **end-member** is used to describe a material which may be mixed with other end-members to make up the compositions required for experiments, and thus a system, apart from being classified in terms of the number of components may also be classified according to its number of end-members. When the number of end-members is less than the number of components the system studied is always part of a larger system. Thus the system anorthite–forsterite may be described as a two end-member join (the term 'join' is derived from the graphical representation of compositions in phase diagrams) in the quaternary system $CaO–MgO–Al_2O_3–SiO_2$ and is an example of what is known as a **pseudobinary** system, because it has two end-members but is not itself a binary. The system forsterite–anorthite–diopside, a three end-member join, provides an example of a **pseudoternary** because the phases produced include forsterite, anorthite, diopside, and spinel which, as in the previous example can only be described by the four components CaO, MgO, Al_2O_3 and SiO_2.

Coincidence of phases and components. Confusion sometimes arises when a phase and a component appear to be the same thing, that is when for example a system with SiO_2 as a component contains a mineral such as quartz. To avoid possible ambiguity phase diagrams should ideally be presented with phases and components labelled differently. Hence in the discussion of theoretical systems which follows, if a component Y crystallises as a pure phase the phase should ideally be designated *crystals of Y*. This procedure is, however, cumbersome and at the risk of causing slight confusion the notation illustrated in Figure 3.4 has been adopted throughout this book. Here the symbols X and Y have a dual meaning. When applied to compositional axes they stand for chemical components, when used to label fields within the diagram they stand for pure phases. Hence X + Y signifies crystals of composition X co-existing with crystals of composition Y.

Experimental method

The technique used for the determination of most silicate phase diagrams is termed the quenching method. A charge of known composition is loaded into a small capsule, usually of platinum or some other refractory and non-reactive metal, held in a furnace at a known temperature and pressure, and frequently under controlled conditions of oxygen fugacity, until equilibrium is reached. It is then rapidly quenched by switching off the furnace, or dropping the capsule into a container of cold water or mercury. Any liquid present in the charge will ideally quench to a glass and phases present will not have time to react with each other and will be preserved metastably in the final product. Identification of the phases is then carried out by microscopy, X-ray diffraction or by the electron microprobe. Experimental difficulties are considerable, some of the problems being failure of equilibrium to be reached, reaction of the charge with the capsule, difficulty in the identification of phases, control of temperature and pressure, failure of proper quenching (e.g. the liquid may crystallise), and leaking of the capsule. However, the interpretation of phase diagrams is often substantially aided if the principles underlying the experiments are remembered. This point will be returned to frequently in the ensuing pages.

Representation of results

Three main variables are involved in the representation of phase relations, pressure, temperature and composition (P, T and X). In one-component systems no graphical representation of composition is required and two-dimensional diagrams having P and T axes may be constructed. For two-component systems a compositional axis is needed so that full graphical representation requires three dimensions. Commonly however binary diagrams show X and T axes only, and represent **isobaric** (single pressure) sections across the full P–T–X diagram. It is of course necessary to specify the pressure in such diagrams. Note that in binary diagrams composition is expressed in terms of the percentage mixing of the two end-members. Pseudobinary diagrams are usually presented in a similar way.

Three-component systems require two dimensions for the representation of the *X* parameter and are thus mainly based on a compositional triangle in which the three components are plotted at the corners with temperature as the remaining axis. Such diagrams are three-dimensional triangular prisms (sometimes known as *T–X* prisms) and can only be fully represented by models. They are, however, usually constructed as projections along the *T*-axis onto the *X*-triangle and temperature information is conveyed by means of contours. They are thus analogous to topographic maps and the inconvenience involved in projection is minimal. **Polybaric** (many pressures) representation of a three-component system is only possible by means of a series of separate diagrams, for example a number of *T–X* prisms given for different pressures, or a number of *P–T* diagrams given for different compositions. The representation of quaternary systems is complex, three dimensions being required for the compositional data alone.

The phase rule

Before discussing the actual construction of phase diagrams from experimental data it is helpful to introduce an equation which relates numbers of phases and components in systems which are at equilibrium. This is the phase rule, first enunciated by the physical chemist Willard Gibbs and stated as:

$$P + F = C + 2$$

where *P* is the number of phases co-existing, *F* the number of degrees of freedom, and *C* the number of components. The number of **degrees of freedom** possessed by a system is conveniently defined as the number of parameters (chosen from *P*, *T* and *X*) which it is possible arbitrarily to fix (i.e. put a value on) before the system is completely defined. Note that *X* is a single parameter because at equilibrium it is only necessary to define the composition of one phase in order to fix the compositions of all the others. If, for example, a system involving two solid phases shows solid solution, then under given conditions a particular composition of one phase (e.g. plagioclase) is in equilibrium with a specific composition of the other (e.g. olivine).

Figure 3.1 shows a one-component phase diagram for a substance such as H_2O. It consists of three fields in each of which a single phase exists. Each pair of fields meets in a curve along which two phases co-exist, and the three fields meet in a point U (so-called triple point) at which all three phases co-exist. If *p* represents 1 atmosphere pressure

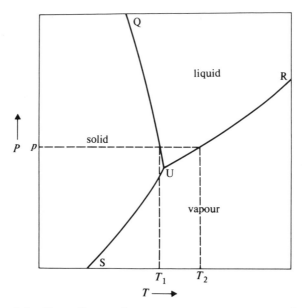

Figure 3.1   Phase diagram for a one-component system such as H_2O.

then T_1 is the melting point at 1 atmosphere and T_2 is the boiling point. Curve UQ represents the variation of melting point with pressure, UR is the boiling point curve, and US is a sublimation curve. Note that in this diagram UQ has a negative slope, which is typical of water and a few other substances though most materials show positive slopes. Consideration of Le Châtelier's principle indicates that the negative slope of the melting point curve arises from the fact that water is denser than ice. Hence increasing pressure promotes melting.

For the construction of a phase diagram such as that for water (ignoring the various higher pressure modifications of ice), it is not necessary to use the quenching method, it is only necessary to determine melting, boiling and sublimation points at a variety of pressures. However, it is relatively easy to determine such points because of latent heat effects. As heat is supplied to the system or removed from it, each phase change is accompanied by a readily identifiable period when the temperature does not change. Applying the phase rule, when two phases co-exist in a one-component system $F = 1$, and the system is said to be **univariant.** Since any given experiment is carried out at a fixed pressure an arbitrary value has in effect been chosen for one of the three parameters (in this case P), and no further arbitrary choice (for example of T) can be made. The two phases must co-exist under these conditions at one temperature and one temperature only. Thus, for example, during heating of an ice–water mixture no temperature change will take place until all the ice has melted. The number of phases has now dropped to one so that F rises to two and the temperature is free to vary even though the pressure is still fixed.

When three phases co-exist in a one-component system, the system is said to be **invariant** as it has no degrees of freedom. Pressure, temperature and, in a one-component system, composition, are all fixed and no choice of parameter values is available. It is obviously experimentally almost impossible to reproduce the conditions of an invariant equilibrium, but it may be located by the extrapolation of the other curves to their intersection. Ice, water and water vapour which often co-exist in nature do not of course normally do so in equilibrium.

An arbitrary choice of values can be made for both T and P in divariant equilibria in one-component systems which consist of one phase only. If the system consists for example of water at 50°C and 1 atmosphere pressure then both T and P can be altered at will through certain limits without changing the nature of the phase present.

Construction of a phase diagram by the quenching method

At the high temperatures and pressures used in the study of many silicate systems it is not easy to determine a melting point or other phase change by an essentially calorimetric method as could be used for water. Hence it is normal to make up several charges of the same composition and run them at a number of different temperatures and/or pressures, and then quench them. In a one-component system the determination of a melting point curve could be carried out as illustrated in Figure 3.2. Here each charge is represented on the diagram in a position depending on the P and T at which it was equilibrated, and it is recorded as consisting of either liquid (glass) or solid (no glass). The melting point curve is then interpolated as shown. The accuracy with which this may be done depends on the spacing of the runs in terms of P and T. One of the most important features of this type of experimentation is the way in which a diagram can be made with a relatively small number of runs. Having located a boundary by the two sets of runs at p_1 and p_2 the experimenter might well be able to fix it with only two more runs (x and y) at a higher P and T. Furthermore the phase rule enables predictions to be made about the way fields must fit together. For example, in the present case the experimenter knows that the melting point curve must be a single curve on the diagram and not a broad transition zone, as would be required in a binary system (where the assemblage of one solid

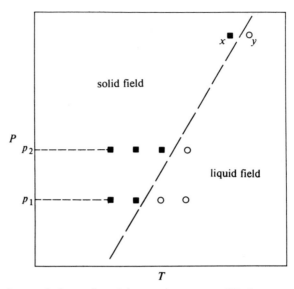

Figure 3.2 Interpolation of melting point curve. Black squares – charges consisting only of solid material; open circles – charges consisting of glass.

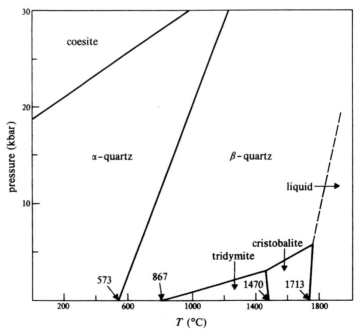

Figure 3.3 Stability fields of polymorphs of SiO_2 (after Boyd & England 1960).

phase + liquid has an additional degree of freedom). Many other examples of the usefulness of the phase rule in the geometrical treatment of phase diagrams will appear in the succeeding chapters. As an example of a geologically-useful one-component system the diagram for the various polymorphs of SiO_2 is given in Figure 3.3.

The reading of phase diagrams

Before proceeding to the detailed study of different types of phase diagram, it is worth pausing to consider the sort of information that they can give and the way in which this information can be applied to petrological problems. Firstly a diagram can give complete information about the way liquids within the system under consideration crystallise, that is the sequence and compositions of solid phases which appear during cooling, and the way the liquid composition changes during this process. Crystallisation paths may be read on the assumption of complete equilibrium between all phases present (perfect equilibrium crystallisation) or alternatively with complete absence of equilibrium (perfect fractional crystallisation). The more complex the system in terms of the number of components the more closely such paths may be expected to coincide with fractionation sequences observed in natural magmas *if* natural sequences do indeed develop by means of crystal–liquid fractionation. Even simple systems, however, are capable of giving much qualitative information about the way natural magmas behave, though it is incumbent on the petrologist to demonstrate this statement to be true, and it should not without reason be assumed to be true.

As a complement to the 'down-temperature' way of considering phase diagrams, they can be used with equal facility to model melting processes both of equilibrium and fractional type. The reading of a phase diagram is the process of extracting all these types of information and in the succeeding pages various types of diagram will be considered with this in mind. Facility in reading diagrams is of great benefit to the petrologist but requires a certain amount of practice. Exercises are therefore appended at several stages in the text.

Binary diagrams

In systems with two or more components, solid solutions can be formed as well as the more familiar solutions normally expected in liquid or vapour phases. Since the thermodynamic definition of equilibrium requires that the chemical potential of a particular component must be the same in all co-existing phases it follows that all components must be present in all phases. In practice solid solution between pairs of phases varies from complete (e.g. albite and anorthite) to virtually nil (e.g. albite and quartz). In the latter example the amount of albite which can be dissolved in quartz is so small as to be negligible, and is indeed below the limits of detection for many analytical techniques. However, it is important to remember that some solution, however small, does exist as some of the applications of the phase rule in the following sections show.

Figure 3.4 shows a typical binary diagram in which there is negligible solid solution between two end-members. This type of diagram is typical of many mineral pairs when one end-member is a ferromagnesian mineral (e.g. forsterite or diopside) and the other a feldspar or feldspathoid.

The diagram consists of a number of different elements which may be described as follows:

(a) *The liquid field* – all charges plotting in this field (by which is meant for a point such as *a* that it was held at the temperature T_1 before quenching, and has composition 74% Y) were found by the experimenter to consist only of glass after quenching.

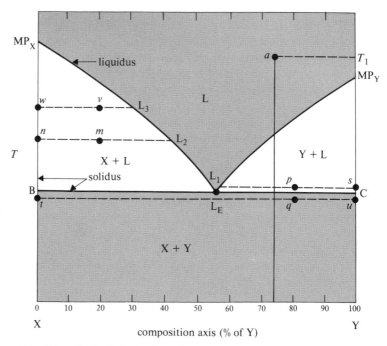

Figure 3.4 Hypothetical isobaric diagram for a binary system showing negligible solid solution.

(b) Two fields representing *a solid phase + liquid*. Charges plotting in these fields were found to consist of crystals either of the phase X or the phase Y, plus glass. The texture of such charges when examined microscopically is essentially that of a porphyritic volcanic rock with a glassy groundmass.

(c) A field of *solid X plus solid Y*.

The descriptions of the fields given above are complete in terms of the nature of the phases but do not tell us how much of each phase is present. To answer this question we consider one of the most fundamental geometrical properties of phase diagrams, the lever rule.

The lever rule. Consider a charge m which lies in the X + L field of Figure 3.4 (where X means crystals of phase X, and L means liquid). The crystals of X must lie at the point n in the diagram (they consist of the pure X component in compositional terms, and they have the same temperature as the whole system). Since the crystals lie to the left (i.e. on the X-rich side) of the bulk composition of the system, m, it follows that the liquid must lie somewhere to the right and must be relatively X-poor. The liquid in fact lies at L_2, on the boundary curve separating the X + L field from the L field. This is a logical necessity because any liquid more Y-rich than L_2 would lie in the all-liquid field and therefore *could not be in equilibrium with solid X* (the experimenter would find that if he made up a charge having a composition a little more Y-rich than L_2 and then equilibrated it at the temperature of L_2, the crystals would promptly dissolve). Similarly any liquid more X-rich than L_2 would partially crystallise if equilibrated at the temperature of L_2. The appearance of crystals of pure X would cause the liquid composition to change to the right until it reached L_2.

The principles of this argument may be applied to any two-phase field in a binary system. An isothermal line crosses the two-phase field and intersects its boundaries at both ends. The two phases concerned lie at the ends of the isothermal line and the bulk composition of the system lies on the line somewhere *between* them. To demonstrate the latter, it is arithmetically obvious that two phases containing say 5% and 20% of CaO can only mix together to give bulk compositions with between 5% and 20% CaO. The bulk composition clearly depends on the proportions of the two phases present, for example, equal parts of the two will give a bulk CaO content of 12.5%. The graphical generalisation of this calculation is the lever rule and referring to bulk composition m consisting of the two phases n and L_2 is stated as follows:

amount of n is proportional to distance mL_2

and:

amount of L_2 is proportional to distance *mn*

hence:

$$\text{percentage of } n = \frac{mL_2}{nL_2} \times 100$$

$$\text{percentage of } L_2 = \frac{nm}{nL_2} \times 100$$

Equations of reactions. Throughout this work the term **reaction** will be used to refer to any change in the system which involves either a change of phase proportions or of phase assemblage as either P or T varies. Considering the points p and q in Figure 3.4, both of which refer to the same bulk composition, since p lies in the Y + L field and q in the X + Y field, it follows that cooling the system from the temperature of p to the temperature of q involves a reaction. The change of phase assemblage must take place when the system has a temperature corresponding with the horizontal line separating the two fields. To investigate the nature of this change imagine that p lies infinitesimally above the line and q infinitesimally below. At the temperature of p the phases present are liquid L_1 and solid Y, labelled s. The relative proportions of these are given by the lengths sp and L_1p, in this case about 55% of solid and 45% of liquid. At the temperature of q the phases present are solid X (labelled t) and solid Y (u). The proportions here are 20% X and 80% Y (derived from the relative lengths of qu and qt). Thus the down-temperature reaction taking place at the field boundary is:

$$55Y + 45L \rightarrow 20X + 80Y$$

Simplifying, by the subtraction of 55Y from each side (see below), this gives:

$$45L \rightarrow 20X + 25Y$$

which on dividing through becomes:

$$L \rightarrow 0.45X + 0.55Y$$

While this reaction is taking place, three phases co-exist. Application of the phase rule ($P + F = C + 2$) shows that $F = 1$ and this is thus an example of a univariant reaction. Since the pressure is already fixed, the temperature cannot vary while the reaction takes place, the graphical expression of this being the horizontal (isothermal) nature of the line separating the two fields.[2] *All* bulk compositions that, like *p*, lie in the Y + L field at a temperature fractionally above that of the boundary line will undergo the same reaction at the same temperature as they cool. The significance of the 55% of Y which was cancelled from the equation above is now clarified. It represents that amount of Y crystals which was present both before and after the reaction and therefore did not participate in the change, i.e. for practical purposes acted as an inert excess of Y throughout. If the calculation is repeated for another bulk composition, e.g. 70% Y (written for convenience as $X_{30}Y_{70}$) we shall derive exactly the same equation as the above but the amount of Y to be subtracted from both sides will be less, for consideration of the lever rule shows that $X_{30}Y_{70}$ contains less solid than *p* does at a temperature just above that of the univariant reaction.

Compositions lying in the X + L field at a temperature immediately above its isothermal lower boundary clearly undergo an analogous univariant reaction on cooling into the X + Y field. This time the change is from the assemblage X + L to X + Y and at first sight would appear to differ from the reaction described above. However, the only difference is that here excess (inert) crystals of X are present throughout, and after cancellation the same equation as before is obtained. Clearly if temperatures only infinitesimally above the temperature of the univariant reaction are considered the liquid composition L_E is the same whether the bulk composition initially lies in the X + L or the Y + L field. This liquid is known as a binary **eutectic** liquid and lies at the intersection of the univariant reaction line and the two curves bounding the liquid field. The equation derived above expresses what happens to the eutectic liquid as it solidifies. A bulk composition having the eutectic composition changes at the temperature of the univariant reaction from the all-liquid state to the all-solid state and then consists of 45% X and 55% Y, the proportions derived in the equation. Measurement of the distances $L_E C$ and $L_E B$ will confirm this. Finally it follows inevitably that the horizontal line BC representing the univariant reaction on the diagram is itself a special case of a 'field'. Any charge which lies exactly on the line must consist of the three phases, L_E + X + Y. However, for such charges the lever rule cannot be applied to determine the phase proportions, which depend on the energy content of the system, e.g. during cooling they depend on how much heat has been extracted from the system since it first entered the univariant condition.

Divariant reactions. Divariant reactions in binary systems involve only two phases and take place over a temperature range even at constant pressure. In the system shown in Figure 3.4 only one type of divariant reaction is possible, that is one involving a solid phase and liquid. During the cooling of bulk composition $X_{80}Y_{20}$ from the temperature of v to the temperature of m the proportion of the solid phase (represented by the lengths of the lines vL_3 and mL_2) increases continuously. Specifically, a high-temperature liquid L_3 which co-exists with a smaller amount of solid gives rise to a low temperature liquid L_2 plus a larger amount of solid. The liquid composition changes continuously during the reaction since pure X is extracted from it. The down-temperature reaction can be expressed by the general equation:

$$L \rightarrow X$$

since liquid can be cancelled from both sides of the more specific equation $L_1 \rightarrow X + L_2$ (because L_1 can be expressed as $L_2 + X$).

Liquidus and solidus. These terms are frequently useful in the discussion of phase diagrams and are defined as follows: The **liquidus** is the curve or set of curves above which (in the thermal sense) the system is completely liquid. In Figure 3.4 this is the line $MP_X-L_E-MP_Y$. The **solidus** conversely is the line or set of lines below which the system is entirely solid. In Figure 3.4 this is the line $MP_X-B-C-MP_Y$, i.e. it is the isothermal line representing the univariant reaction $L = X + Y$, plus a vertical extension at either end rising to the melting points (MP) of the pure compounds X and Y. Thus the vertical lines bounding the diagram as far up as the two melting points are further examples of special (one-dimensional) 'fields', each representing one solid phase only. These are one-component compositions within the binary and as such are divariant. Even at fixed pressure, pure X and pure Y are capable of stable existence over a range of temperatures. On melting at MP_X and MP_Y these unary compositions lose a degree of freedom and become univariant, i.e. the temperature remains constant until all the solid has disappeared.

In Figure 3.4 the liquidus and the solidus touch at three points, that is the two melting points of the pure end-members and the eutectic point. Such points have a general significance in phase-diagram reading because they are the only places where liquids can crystallise to solids having the same bulk composition as themselves. This is an important aspect of the treatment of fractional crystallisation and will be covered in a later section.

Sequence of equilibrium crystallisation. In Figure 3.5 we follow the crystallisation behaviour of a liquid a which is cooled from an initial temperature T_1. This acts as a⸱ model for crystallisation of all compositions in the system.

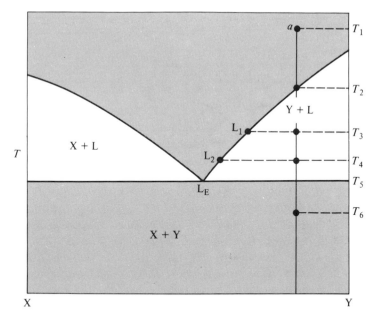

Figure 3.5 System as in Figure 3.4.

As heat is extracted from the system the liquid cools until it reaches the liquidus at T_2. Solid Y begins to crystallise at this temperature and increases in amount with further cooling, the liquid meanwhile becoming more X-rich. At T_3 for example the solid is 50% of the charge and the liquid has composition L_1. At a lower temperature T_4 the solid has increased to 60% of the total and the liquid has composition L_2. Shortly below this the liquid reaches L_E, the eutectic composition, and the system becomes univariant as X begins to crystallise. With further heat extraction the temperature remains fixed at T_5 while X and Y continue to crystallise and the remaining liquid disappears. When all the liquid has solidified the solid assemblage X + Y cools further, e.g. to T_6.

Figure 3.6 shows qualitatively a curve of temperature plotted against time in such a crystallisation sequence on the assumption that heat is extracted at a uniform rate. Notice the rapid initial fall of temperature due to the relatively low specific heat of the liquid. Then follows a much slower cooling period because it is necessary to remove the latent heat of crystallisation of Y as well as the specific heat of the system. Next follows the univariant condition when temperature is fixed, and finally there is a renewal of rapid cooling below the solidus.

Other compositions would show similar late stages of crystallisation history but would initially crystallise either Y or X at a variety of

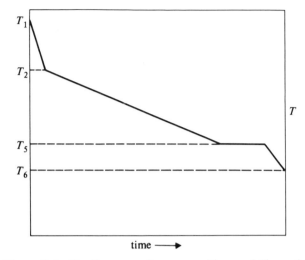

Figure 3.6   Cooling curve for composition *a* of Figure 3.5.

temperatures depending on composition. Only a liquid having the eutectic composition would show the simultaneous initial crystallisation of both X and Y. All compositions would of course produce the same final liquid as they entered the univariant condition.

Sequence of equilibrium melting. Equilibrium melting paths are the exact reverse of those of equilibrium crystallisation. In Figure 3.5 if heat is supplied to the bulk composition represented by *a* held at an initial temperature of T_6 the solid assemblage of X + Y warms up to T_5 where it begins to melt isothermally with the production of liquid L_E. The two solids melt together contributing to the liquid in the eutectic proportions (the eutectic liquid is slightly richer in Y than in X) until all the X has melted. The temperature begins to rise again and the system proceeds up-temperature in the Y + L condition, Y melting continuously until it disappears at T_2. The system is now all liquid. Note that *all* mixtures of X + Y will melt at the same temperature with the isothermal production of the eutectic liquid. Bulk compositions lying on the Y-rich side of L_E will lose X first and enter the Y + L equilibrium while those more X-rich than L_E will lose Y first.

Intermediate compounds. Several geologically important binary systems show the appearance of a solid phase which has a composition intermediate between the two end-members. The system nepheline ($NaAlSiO_4$) – silica (SiO_2) for example contains albite ($NaAlSi_3O_8$) as an intermediate compound, while periclase (MgO) – silica (SiO_2) shows the appearance of forsterite (Mg_2SiO_4) and enstatite ($MgSiO_3$). Intermediate compounds are found if the components represented by the end-members can be combined to produce them. Thus:

$$NaAlSiO_4 + 2SiO_2 = NaAlSi_3O_8$$

     nepheline   quartz   albite

and:

$$MgO + SiO_2 = MgSiO_3$$

     periclase   quartz   enstatite

$$2MgO + SiO_2 = Mg_2SiO_4$$

     periclase   quartz   forsterite

The satisfying of an equation as above is not, however, the sole criterion necessary because the intermediate compound may not be stable under reasonable conditions of P and T. Thus for example although enstatite is stable under 1 atmosphere conditions in a dry system its iron analogue ferrosilite ($FeSiO_3$) is not stable under similar conditions in the system wüstite (FeO) – silica. The existence of a stable intermediate compound means that two compounds lying compositionally on either side of it *cannot co-exist*. This is the basis of the recognition of antipathetic mineral pairs which have been traditionally important in the classification of igneous rocks, e.g. forsteritic olivine and quartz, nepheline and quartz, leucite and quartz, do not co-exist in equilibrium in normal rocks. Fayalitic olivine in contrast is stable in the presence of quartz because of the non-stability of ferrosilite. In general, however, olivines and quartz react to give pyroxenes, while feldspathoids and quartz react to give feldspars.

Congruent melting. Binary intermediate compounds are said to melt **congruently** if they melt to a liquid of their own composition. The phase diagram for such a system is illustrated in Figure 3.7 and is similar to that of nepheline–silica. The diagram is the same as two phase diagrams of the type illustrated in Figure 3.4 placed end to end. There are two different sub-solidus assemblages, A + AB (where AB is the intermediate compound) and AB + B. The AB composition creates a thermal maximum on the liquidus since its melting point is depressed by the addition of either A or B. The AB + L field is divided into two parts by the vertical line extending up to the liquidus and representing the solid assemblage AB. Pure AB melts to a liquid of its own composition at MP_{AB} but the addition of even a small amount of A to the system will cause melting to occur at the much lower temperature of T_{E1} with the

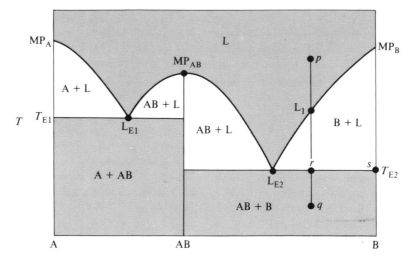

Figure 3.7   Hypothetical isobaric diagram for a binary system containing a congruently melting intermediate compound.

production of the eutectic liquid L_{E1}. Similarly the addition of B to pure AB will induce melting at T_{E2} and the formation of eutectic liquid L_{E2}. Under equilibrium conditions all bulk compositions to the left of AB will solidify to the assemblage A + AB at temperature T_{E1} and all bulk compositions to the right will solidify to the assemblage AB + B at temperature T_{E2}.

Fractional crystallisation. The equilibrium crystallisation discussed does not automatically find an analogy in natural magmatic processes since it has been assumed that crystals and liquids remain in contact throughout solidification. However, from what has been discussed in earlier chapters it is evident that in nature there is a strong tendency for crystals and liquids to become separated by gravitative and other means. Fractional crystallisation is the term used to imply a lack of equilibrium between crystals and liquids and we shall first discuss the case in which crystals are instantaneously removed from contact with the liquid as soon as they are formed. This is clearly not completely plausible in a mechanical sense but it forms an 'end-member' amongst crystallisation mechanisms. Neither perfect equilibrium nor perfect fractional crystallisation is to be expected in nature but under varying circumstances natural systems may approach one or the other more or less closely.

   In the discussion of fractional crystallisation the important factors are the path of liquid evolution and the sequence of solid phases fractionating from the liquid. If the mechanism of crystal removal is assumed to be crystal settling the fractionating solids give rise to cumulate rocks in the lower part of the system while periodic tapping off of the liquid fraction forms an analogy to lava eruptions. Thus in Figure 3.7 a liquid p cooled under fractional conditions begins to precipitate crystals of B at L_1. With falling temperature the precipitation of B continues and the liquid moves down the boundary curve from L_1 to L_{E2}. This liquid then precipitates crystals of AB and B without changing composition until it is all consumed. During this process the sequence of solids precipitated is first B and then AB + B. An imagined layered intrusion would consist of a lower layer of B (a monomineralic rock) overlain by a layer consisting of AB and B. Liquids varying in composition from L_1 to L_{E2} could have been erupted during the crystallisation process.

Considering all possible liquid compositions in Figure 3.7, four essentially different trends of liquid evolution are possible:

(a) Liquids lying initially between A and L_{E1} evolve by fractionation of A towards more B-rich compositions (using B in the sense of a component), the final residual liquid being L_{E1}.
(b) Liquids lying between L_{E1} and AB evolve towards more A-rich compositions by fractionation of AB. Like the preceding group the final liquid is L_{E1}.
(c) Liquids lying between AB and L_{E2} fractionate AB and become more B-rich, giving L_{E2} as the final liquid.
(d) Liquids more B-rich than L_{E2} fractionate to a final liquid at L_{E2} by crystallisation of B.

This classificatory exercise has two important general features. It illustrates firstly how liquids which are originally rather variable may fractionate towards a very limited number of end-points. Secondly, certain liquids which are initially similar in composition, e.g. liquids immediately either side of AB, may fractionate to quite different end-points. Both of these are important features of natural rock series.

Fractional melting. Fractional melting involves the removal of liquid instantaneously during the melting process, and while it may not be mechanically possible to remove very small amounts of melt from a source rock in the natural environment, like fractional crystallisation it forms a convenient theoretical end-member mechanism. However, it differs in several important respects from fractional crystallisation, and liquid evolution paths are *not* precisely the reverse of those of fractional crystallisation.

Consider the fractional melting of a solid mixture of AB and B represented by q in Figure 3.7. On heating, the system begins to melt at temperature T_{E2} to a liquid of composition L_{E2}. Since the liquid formed is immediately removed and is more A-rich than the original bulk composition, the bulk composition of the residual solids must become progressively more B-rich as the liquid is extracted. Ultimately all the AB must disappear leaving a solid residue consisting only of B. Pure B, however, does not melt until the temperature of MP_B is reached so that continued heating of the system does not immediately result in the formation of more liquid. However, at MP_B the system becomes univariant and all the B melts to a liquid of its own composition with no further rise of temperature. In this example the liquid is produced in *two* discrete batches at two different temperatures and thus contrasts strongly with the way a liquid of the same initial bulk composition evolves during fractional crystallisation.

Incongruent melting. An intermediate compound is said to melt **incongruently** if it melts to a liquid *not* of its own composition, plus a new solid phase. Expressed as an up-temperature reaction this is:

$$\text{solid A} \rightarrow \text{solid B} + L$$

In a binary system the new solid and the liquid formed must lie compositionally on either side of the original compound to maintain a constant bulk composition.

Figure 3.8 is an example of a system in which the intermediate compound AB melts incongruently to A + L. Characteristic of such a system is the way the field of A + L extends over the composition of the intermediate compound AB. As we have seen the sequence of phase assemblages during equilibrium melting is read directly from the diagram by running the eye up the vertical line representing the bulk composition. A charge of composition pure AB remains solid until the temperature MP_{AB} is reached. Above this temperature the same bulk composition lies in the A + L field. Thus the reaction on melting is:

$$AB \rightarrow A + L$$

and at the temperature of MP_{AB} the system is univariant as AB, A and L must co-exist. On heating, as the system enters the univariant condition, only solid AB is initially present but as more heat is supplied AB must diminish in amount, and liquid and A must appear and increase in amount. Thus melting involves the decomposition of AB into pure A and a more B-rich liquid. The liquid concerned, by analogy with earlier arguments, lies at L_P, and is known as a **peritectic** liquid.

Consider now melting of a mixture of A and AB represented by o in Figure 3.8. Below the solidus it consists of equal parts of A and AB,

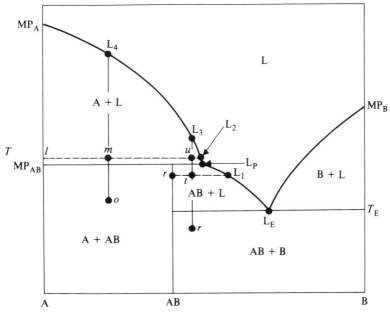

Figure 3.8 Hypothetical isobaric diagram for a binary system containing an incongruently melting intermediate compound.

since it lies compositionally midway between them. At a temperature slightly above MP_{AB} however it consists of A + L in the approximate proportions 60:40 (see relative lengths of mL_2 and $m1$). Thus during the univariant reaction the amount of A has *increased*, while AB has disappeared. The 50% of AB which has melted is now represented by 40% of liquid L_2 (ideally L_p) plus 10% of newly formed A. Further heating sees the progressive melting of A, the system becoming completely molten when it reaches L_4. This argument should demonstrate that *all* mixtures starting as A + AB show behaviour similar to that of pure AB. The initial liquid is L_p in all cases, and during the univariant reaction the amount of A increases as heat is supplied.

More complex melting behaviour is shown by compositions lying between AB and L_p. A bulk composition consisting of solid AB and B represented by r melts at T_E with the formation of the eutectic liquid L_E and passes up-temperature into the AB + L equilibrium with the loss of solid B. AB continues to melt, its proportion at a higher temperature being represented by the line tL_1 (here it is about 35% liquid, 65% AB). When the point representing the bulk composition reaches the top of the AB + L field just above t the system is about 60% liquid and contains 40% of crystals of AB. Above the temperature of the invariant reaction, however, the system consists of A + L (bulk composition represented by u) and only contains about 7% of crystals. The system now consists of liquid plus a new phase which was *not* present in the starting assemblage. Further heating sees the melting of A until the liquid reaches L_3 when all A has disappeared. The last liquid of melting has of course the bulk composition of the starting material.

Mixtures of AB and B which have bulk compositions lying to the right of L_P are the only compositions to show simple melting behaviour not involving the appearance of A in the univariant reaction $AB \rightarrow A + L$. In these charges all the AB melts before the temperature of MP_{AB} is reached and the melting behaviour is thus the same as that in the simple eutectic binaries discussed previously.

Equilibrium crystallisation in a system with an incongruently melting compound. Equilibrium crystallisation in the system described above is now considered. It is the reverse of the melting process but is sufficiently important to justify separate discussion. In Figure 3.9 the same system is shown as in Figure 3.8 and the liquid field is divided up into four classes (I–IV) each of which shows a different crystallisation behaviour.

Down-temperature any bulk composition which crosses the line $P–MP_{AB}$ must become involved in the reaction:

$$L + A \rightarrow AB$$

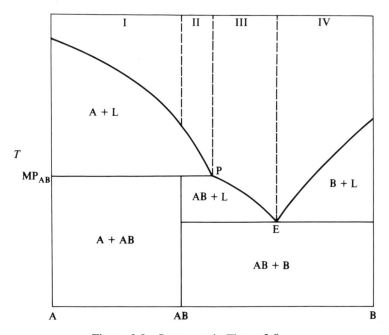

Figure 3.9   System as in Figure 3.8.

that is, crystals of A will react with the liquid to form AB, and the amount of A present must diminish. The other general consideration is that, irrespective of the complexities of crystallisation behaviour encountered *between* the liquidus and the solidus, the diagram can be read immediately to determine the final solid assemblage. Thus all Class I liquids solidify eventually to A + AB, and all other classes solidify to AB + B. These considerations allow a rapid and logical analysis of the crystallisation behaviour as follows:

Class I. Liquids start by crystallising A until the temperature of MP_{AB} is reached. AB begins to crystallise as A reacts with the liquid. Since A is present in the final assemblage the liquid must disappear before all the A has become resorbed. This class shows *partial* resorption of early formed A at the temperature of the univariant reaction. The last liquid present is the peritectic liquid P.

Class II. In this case since A is not present in the final assemblage the resorption of A must go to completion. The system then passes into the AB + L field and the liquid proceeds down-temperature to E. This is the final (eutectic) liquid and the system solidifies to AB + B. This class shows *complete* resorption of early formed A.

Classes III and IV. Neither of these shows resorptional phenomena as the A field is not encountered. The crystallisation of both classes is that of the simple eutectic system, the difference being only that Class III liquids show AB as the first crystals whereas Class IV liquids produce crystals of B first.

Special cases. The liquid forming the boundary between Classes I and II has the special composition AB. It shows the simultaneous disappearance of A and liquid, leaving only solid AB below the temperature of MP'_{AB}.

Fractional crystallisation. Fractional crystallisation in a system with an incongruently melting intermediate compound presents a new feature, the petrological importance of which will be discussed with relation to the system $MgO–SiO_2$ in a later section.

Referring to Figure 3.9, by analogy with earlier discussions liquids of Classes III and IV will show the familiar type of fractional crystallisation leading in both cases to residual liquids at E. The fractionation of A from liquids in Classes I and II leads in all cases to a liquid at P which under *equilibrium* circumstances would react with A to form AB. However, in the fractional case there are no crystals of A present to take part in the reaction. Careful consideration of this reaction is evidently required to decide what will happen under fractional conditions.

Under *equilibrium* conditions when A is present, the liquid P is subjected to two contrasting influences as heat is extracted. On the one hand A is dissolving in it, thus tending to make it more A-rich, on the other hand AB is growing from it thus tending to make it more B-rich. These effects balance out so that P remains fixed in composition while A dissolves and AB grows. In the absence of A the only possible result is that AB crystallises and the liquid becomes richer in B. An alternative way of looking at this is to observe that P lies on the upper boundary of the AB + L field. Considering P as a starting composition it is clear that cooling must make it pass immediately into this field, i.e. it precipitates AB and the liquid evolves down-temperature towards E.

Thus, although under equilibrium conditions it can act as a final liquid of crystallisation or a first liquid of melting (Class I compositions), the liquid P *cannot* act as a final liquid of fractional crystallisation. The point P on the diagram is often referred to as a 'reaction point' and it is a characteristic of such points that fractionating liquids pass down-temperature through them. 'Reaction points' do not lie in thermal lows on the liquidus and are not places where solidus and liquidus touch. Hence the liquid can under no circumstances solidify to a solid of its own composition and under fractional conditions there is no means of disposing of such a liquid. It must always crystallise a solid phase having a composition other than its own and therefore evolves down-temperature changing composition. Under equilibrium conditions, in contrast, such a liquid does disappear as heat is extracted by combining with an existing solid phase to make a new one. The term 'reaction point' is unfortunate because all changes in systems are conveniently thought of as reactions, and here we are dealing with one particular type, that involving the down-temperature resorption of a solid by a liquid.

The important general conclusion to be drawn is that all liquids in Classes I and II give rise to final liquids of fractional crystallisation lying at E. Thus the system only contains one final liquid of fractional crystallisation, and compared with a system (Fig. 3.7) in which the intermediate compound melts congruently there is no thermal high on the liquidus. Whether the presence of an intermediate compound acts as a barrier to fractionating liquids thus depends on whether it melts congruently or incongruently.

The system $MgO–SiO_2$. Phase relations for this system, which has much petrological interest are shown in Figure 3.10. It includes the congruently melting compound forsterite (Mg_2SiO_4) and the incongruently melting compound clinoenstatite ($MgSiO_3$) as well as a field of liquid immiscibility in silica-rich compositions. Experimentally

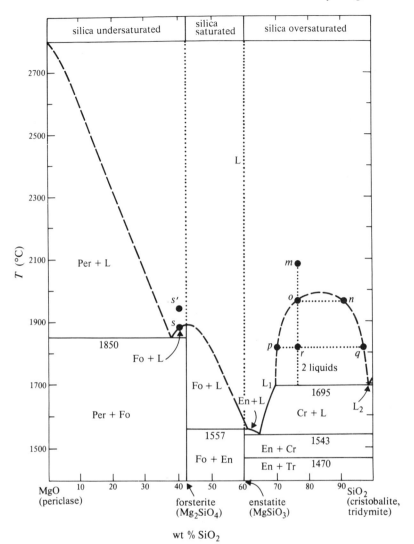

Figure 3.10 The system MgO–SiO$_2$ at 1 atmosphere pressure (after Bowen & Anderson 1914 and Greig 1927). Abbreviations: Per – periclase; Fo – forsterite; En – enstatite; Cr – cristobalite; Tr – tridymite; L – liquid. Dashed curves are of uncertain location.

liquid immiscibility is detected by the presence of two types of glass in the quenched charge, one dispersed as 'droplets' in the other. If two immiscible liquids show, for example, a significant density difference, then mechanical separation is possible and a potential fractionation mechanism exists. However, although liquid immiscibility is well established between sulphide and silicate melts, and may well account for the formation of certain magmatic sulphide bodies, immiscibility between different silicate melts appears to be restricted to compositions, as in this case, substantially different from those of common natural magmas. Hence fractionation by the separation of liquids from each other does not appear to be a commonly occurring natural process.

In Figure 3.10 a homogeneous liquid such as m begins to exsolve droplets of a more silica-rich liquid n when the temperature drops to that of the boundary curve at o. The continued exsolution of silica-rich liquid with falling temperature causes the initial liquid to become more magnesian so that it moves down the left-hand boundary curve while the exsolved droplets change composition towards silica enrichment. At a lower temperature the co-existing liquids are represented by p and q, the bulk composition being r. Further cooling leads to the univariant reaction at 1695 °C where $L_1 + L_2$ (infinitesimally above the temperature of the reaction) changes to L_1 + cristobalite. Using the lever rule to assign quantities to these phases we find that:

$$77L_1 + 23L_2 \rightarrow 78L_1 + 22 \text{ cristobalite}$$

and simplifying by subtracting 77% of L_1 from each side this gives:

$$23L_2 \rightarrow 1L_1 + 22 \text{ cristobalite}$$

Hence the reaction consists of the break-up of the more siliceous liquid into a silica mineral plus a small additional amount of the more magnesian liquid.

The upper boundary curve of the two-liquid field is a typical example of a **solvus** curve. It represents the limits of solubility of one phase in the other and in common with most solvi becomes narrower up-temperature indicating that solubility increases. This solvus closes completely at its upper limit signifying complete solubility of the two phases. Some solvus curves which will be discussed in later sections encounter other equilibria up-temperature and solution does not become complete at any stage.

The two different polymorphs of silica, cristobalite and tridymite (see Fig. 3.3) present a second feature of interest in this system. The inversion temperature between them is at approximately 1470 °C and it is important to note that it is for practical purposes in the binary system independent of the bulk composition. This is a characteristic of polymorphic transitions of phases which do not show significant solid solutions towards other components (for contrasting behaviour see the up-temperature conversion of nepheline to carnegieite in Figure 3.19).

Like cristobalite and tridymite, which are high-temperature modifications of the much more common naturally occurring mineral quartz, the enstatite in this system is a high-temperature monoclinic phase encountered only rarely in natural rocks where it is normally represented by the orthorhombic form. High-temperature phases are commonly encountered in simple systems with few components. In the systems considered so far the addition of a second component has a strong effect on reducing the temperature at which the original phase crystallises from the liquid. In natural magmas consisting of a large number of components the effect is magnified so that the temperatures at which natural magmas crystallise are generally far lower than those of the simple systems. Lower crystallisation temperatures result in the appearance of quartz as the common silica mineral of igneous rocks and orthopyroxene as the common calcium-poor pyroxene.

The system $MgO-SiO_2$ provides an excellent illustration of the concept of silica-saturation and the basis of the **CIPW norm** which is widely used in petrology and discussed in detail in Appendix 3. Magmas are regarded as **silica-saturated** if their compositions are such that they crystallise to mixtures of olivine, pyroxenes, and feldspars under anhydrous equilibrium conditions. If, however, excess silica is present this results in the appearance of a silica mineral, usually quartz, in addition to the assemblage mentioned. Such magmas are termed **silica-oversaturated**. Conversely if insufficient silica is present, minerals of the feldspathoid group (usually nepheline or leucite) appear and the magmas are termed **silica-undersaturated**. Magmas and the igneous rocks they give rise to can be classified into three major groups on this basis. The significance of the classification will be discussed more fully at a later stage but we note that in the system under discussion the three groups are represented by the compositional ranges given in Figure 3.10 though here the silica-undersaturated group is represented by compositions crystallising to a periclase-bearing assemblage rather than the feldspathoid-bearing assemblages of natural rocks. From the previous discussion of incongruently melting intermediate compounds it follows that in $MgO-SiO_2$ magmas under equilibrium conditions will crystallise to one of the three assemblages Per + Fo, Fo + En, or En + Cr. However, under conditions of fractional crystallisation, because enstatite melts incongruently, liquids which are initially silica-saturated may give rise to silica-oversaturated residual liquids. Similarly, fractional melting or low degrees of fusion under equilibrium conditions may produce silica-oversaturated initial liquids from silica-saturated source materials. Hence there is no barrier in this system (at 1 atmosphere pressure) to liquids crossing from the silica-saturated to the silica-oversaturated condition during down-temperature fractionation or changing up-temperature in the opposite direction. In contrast, because forsterite melts congruently and produces a thermal high on the liquidus, no fractionation paths enable a change from silica-saturation to silica-undersaturation (or the converse) to take place. Similar considerations apply in more complex systems and the division of magmas on the basis of their silica-saturation thus has important genetic overtones. Thermal highs create barriers which restrict possible fractionation paths and place convenient constraints on petrogenetic hypotheses. Also restricted are the possibilities inherent in contamination. A liquid such as *s* in Figure 3.10 which lies on the silica-poor side of the forsterite composition and is in equilibrium with forsterite crystals, cannot for example be converted to a silica-saturated or -oversaturated liquid by the addition of SiO_2 as a contaminant. Adding SiO_2 will move the bulk composition to the right and cause more forsterite to crystallise so that the liquid remains silica-poor. A silica-undersaturated liquid such as *s* could only become silica-saturated by contamination if its initial temperature were higher than that of the divide (e.g. *s'*). Then if the added contaminant were sufficiently hot the liquid composition could move to the right without encountering the forsterite + liquid field.

The effect of pressure. Phase relations in $MgO–SiO_2$ so far discussed refer to 1 atmosphere pressure. However, under increased pressure and anhydrous conditions the liquidus field of forsterite contracts relative to that of enstatite with the consequence that at about 5 kbar enstatite begins to melt congruently. At pressures above 5 kbar the enstatite composition becomes a thermal divide as shown in Figure 3.11. This has two important consequences for the petrological discussion of mantle processes elaborated in later sections. First it implies that initial melts in equilibrium with magnesian olivine and magnesian orthopyroxene become progressively more olivine-rich with increasing depth; secondly it implies that the free passage of liquids from silica-saturated

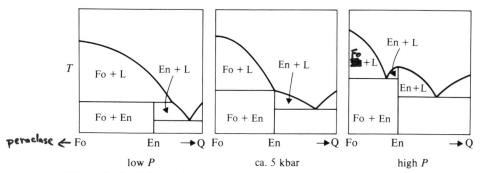

Figure 3.11 Schematic illustration of the effect of pressure on forsterite–enstatite equilibria.

(olivine-bearing) compositions to silica-oversaturated (quartz-bearing) compositions possible at low pressures is restricted at higher pressures.

Solid solution. Figure 3.12 illustrates a hypothetical binary eutectic system in which there is partial solid solution between the two solid phases. Maximum solid solution takes place at the temperature of the eutectic T_3 and in this case X is capable of dissolving 20% of Y, and Y is capable of dissolving 15% of X. If these solubility limits were to be reduced the diagram would pass by transition into that shown in Figure 3.5 where solid solution is negligible. In most systems, as in this case, the solvus curves limiting the solid solution fields below the solidus converge up-temperature indicating that increased temperature favours more extensive solution. A composition such as p which consists of a single homogeneous phase at temperature T_3 begins to exsolve the other phase (Y_{ss} of composition r) if it is cooled to T_2. Further cooling to T_1 sees the X_{ss} becoming more X-rich until it reaches composition t while the exsolved Y_{ss} becomes more Y-rich (composition s). The natural analogy of this cooling process results in the formation of exsolution lamellae in minerals such as alkali feldspars (perthites) and pyroxenes. Reheating of exsolved structures will of course tend to rehomogenise them.

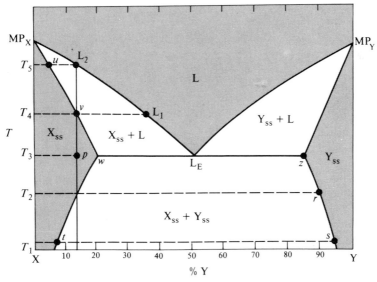

Figure 3.12 Hypothetical isobaric diagram for a binary system showing solid solutions (indicated by subscript 'ss').

Above the solidus the curves limiting solid solution fields fall back towards the melting points of the pure components at MP_X and MP_Y. This results from the fact that the more extensive solid solutions begin to melt at lower temperatures than the less extensive ones. For example the single phase composition p if heated begins to melt at T_4 to the liquid L_1. It is completely melted at T_5 to a liquid L_2 having the same composition as the starting material. The last solid to disappear has composition u. A more Y-rich solid solution than p would begin to melt at a lower temperature to a liquid lying between L_1 and L_E.

Melting, as illustrated, involves the evolution of a Y-rich liquid and the composition of the solid solution must become concomitantly richer in X. Conversely, on cooling composition p from a temperature above T_5, the first crystals to appear have composition u and these must be progressively made over to the composition v as crystallisation proceeds. When the crystals reach v the last drop of liquid L_1 disappears. Only starting liquids containing between 20% and 85% of Y (i.e. outside the maximum limits of mutual solubility of the two solid phases) produce final liquids of eutectic composition and solidify to a mixture of the two maximum solid solutions w and z. Further cooling results in each exsolving the other component and giving rise to two parallel sets of exsolved crystals. Conversely, only compositions lying between w and z are capable of melting to give the eutectic liquid as an initial product.

Fractional crystallisation in systems involving solid solutions can take place in two ways which differ essentially in scale. Firstly crystals may be removed continuously, perhaps to form cumulate sequences elsewhere, or alternatively crystals may remain within the system but fail to equilibrate with the liquid. In the latter case compositions are not made over to the appropriate new equilibrium composition as temperature changes, and during cooling **zoned crystals** result. These have central compositions corresponding with the initial temperature of crystallisation (for example u) and are overgrown progressively by lower temperature compositions (for example v and w). As in previous examples of fractional crystallisation there is no means of disposing of the liquid during cooling until a point where it can crystallise to a solid or solids of its own composition is reached. In this system all perfectly fractionating liquids thus evolve to the composition L_E and the accompanying zoned crystals or cumulates evolve to the maximum limits of solid solution, w and z. The difference between the two types of fractional crystallisation are important petrologically. In the first case, when crystals are removed, the liquid changes along a well-defined evolutionary path and may be represented in natural cases by periodic eruptions of lavas or changing bulk composition. When crystals are not removed and zoning takes place, the bulk composition of the system does not change but there is a strong fractionation of compositions on the hand-specimen scale within the resultant rock. Since rapid cooling, especially when coupled with high viscosity in the melt, is ideal for the production of zoned crystals this second type of fractionation in natural cases typically takes place after the emplacement of small intrusions and after the eruption of lavas. It shows up as strong marginal zoning in phenocrysts, particularly plagioclase, and in the appearance of interstitial patches of micropegmatite, analcite, and various quenched glasses of more or less extremely alkali- or silica-enriched composition.

Fractional melting in the system considered also differs significantly from the pattern found in binaries without solid solution. In this case as in the other systems two phase assemblages which consist of the two maximum solid solutions *w* and *z* melt initially to give a finite batch of eutectic liquid. However, as soon as one of the solid phases has disappeared there is not in this case a pause in the evolution of liquid as heat is supplied. The remaining solid solution itself must melt at a temperature only infinitesimally above that of the eutectic and leaves a solid residue infinitesimally enriched in the more refractory component. Thus continuous evolution of liquid of changing composition is to be expected, and the last drop of liquid must have the composition of the pure component since this is the only solid phase capable of melting to a liquid of its own composition. Quantitative assessment of the fractional melting of solid solutions is complex because it involves the integration of infinitesimal increments of melting, using solidus and liquidus curves which are not necessarily capable of simple mathematical expression. However, perfect fractional melting is probably not geologically realistic and a better approximation is perhaps afforded by the model known as **incremental batch melting** in which successive batches of equilibrium melt are removed from the source material. If many small batches of equilibrium melt are removed the effects approach those of fractional melting.

In Figure 3.13 we consider the incremental batch melting of a solid solution equivalent to *w* of Figure 3.12 and we suppose that each melting episode involves the creation of 20% of liquid. Under natural conditions it may be realistic to suppose that a certain degree of melting is necessary before a batch of liquid will be extractable. Here it is assumed that all the liquid is extracted each time and the remaining solid is then subjected to a further 20% of melting.

Considering *w* as the initial solid, trial and error is used to find the tie line *ac* in such a position that *ab* is 20% of the total length. At the temperature of the tie line 20% of liquid *c* is in equilibrium with 80% of the solid *a*. The first batch of liquid removed is thus *c* and the new bulk composition of the solid is *a*. In the present case (Fig. 3.13) the process is repeated twice more to give two more batches of liquid *f* and *i* and the

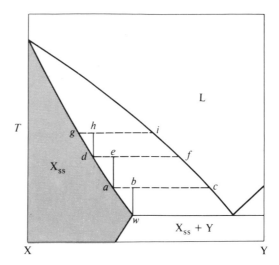

Figure 3.13 Batch melting in a system similar to that of Figure 3.12.

residual solid *g*. The amount of the original solid left is now 51.2% of the original, the three batches of liquid having represented 20, 16 and 12.8% respectively. This is, of course, an infinite series and after any finite number of steps there is always some residual solid remaining. Hence a finite amount of liquid having the composition of the pure component cannot be produced. In the limit this argument applies to perfect fractionation and demonstrates that fractional melting, although its end-product is the highest-temperature liquid available, cannot produce this liquid in a finite amount.

Partial and complete solid solution. In Figures 3.12 and 3.13, the system considered showed partial solid solution between the two solid phases and although solution increased with temperature the solids began to melt before it was complete. Figure 3.14 shows the form of binary diagrams where solid solution is (a) almost complete, and (b) complete before the onset of melting. Figure 3.14b is similar in form to that for orthoclase–albite. In this type of system a minimum is found on the solidus and liquidus, which touch tangentially. In Figure 3.15a the phase diagram for a binary showing complete solid solution below the solidus but no minimum in the liquidus is illustrated. The plagioclase feldspars (the system albite–anorthite, Fig. 3.16) form a geological example. Figure 3.15b shows what this type of diagram changes into when solid solution is incomplete at solidus temperatures.

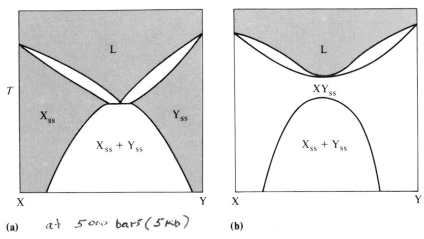

(a) at 5000 bars (5Kb)   **(b)**

Figure 3.14 (a) Binary diagram showing almost complete solid solution just below the solidus. (b) As above, but solution is complete below the solidus and there is a liquidus minimum.

The system albite–anorthite. This system illustrated in Figure 3.16 is an example of complete solid solubility at magmatic temperatures as CaAl in the feldspar lattice is replaced by NaSi. Equilibrium crystallisation involves the early precipitation of calcic plagioclases which are made over with falling temperature to more sodic varieties. Fractional crystallisation leads to residual liquids rich in the albite component and final residual liquids which are theoretically of pure albite composition.

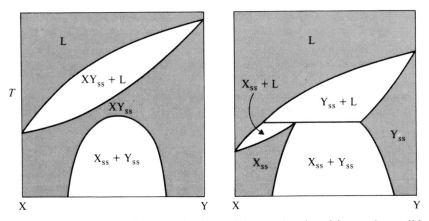

Figure 3.15 Systems without a liquidus minimum showing (a) complete solid solution just below the solidus, (b) incomplete solid solution. This system shows down-temperature resorption of early formed Y_{ss}.

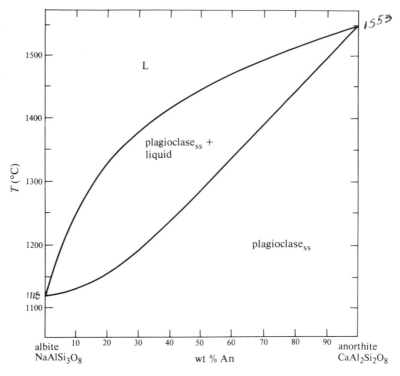

Figure 3.16 The system albite–anorthite at 1 atmosphere pressure (after Bowen 1913). A revision of the system is given by Kushiro (1973b).

The effect of the addition of extra components to this system is considered later but the essential features of the diagram are certainly applicable to most more complex systems. Zoning in natural plagioclases is, for example, an almost ubiquitous phenomenon and, as we would predict from the diagram calcic cores with sodic overgrowths, are very common as a result of imperfect equilibrium during cooling. Furthermore if a particular magma undergoes fractionation by crystal settling we would expect early formed plagioclase cumulates to be more calcic than later ones. This sequence of cryptic layering in plagioclases of layered intrusions is also very commonly observed.

These observations lead to the conclusion that in complex systems the general form of the melting loop is retained, and that fractionation of plagioclase will lead to liquids having higher albite/anorthite ratios. However, the temperatures at which particular compositions of plagioclase will precipitate and the precise relations between the albite/anorthite ratio of liquids and co-existing crystals cannot be predicted for more complex systems from this diagram. It is generally true that simple systems have higher crystallisation temperatures than complex systems, and the distribution coefficient which relates compositions of co-existing solids and liquids is likely to be both temperature and composition dependent. Nevertheless in series of rocks which are obviously closely related, e.g. fractionates erupted periodically from a single fractioning magma chamber, the composition of plagioclase is likely to form a good guide to relative temperatures of formation. Even on a more general level it is probable that rocks containing a calcic plagioclase were formed at higher temperatures than rocks carrying sodic plagioclase. These considerations highlight the traditional importance of plagioclase composition in the classification of igneous rocks. A classification based on plagioclase composition is a first step, albeit crude, towards a system based on temperatures of formation.

The system forsterite–fayalite. Phase relations in this system (Fig. 3.17) are similar to those in albite–anorthite. Here the iron-olivine fayalite forms the low-temperature end-member and the magnesian-olivine forsterite the high. Down-temperature fractionation of liquids is towards iron enrichment relative to magnesium. Since Mg^{2+} is readily replaced by Fe^{2+} in all Mg-bearing silicate lattices and since distribution coefficients are such that Mg is preferentially incorporated in the solid, Fe/Mg ratios have the same general classificatory significance as the plagioclase compositions discussed above. In series of rocks which contain both plagioclase and olivine there is generally a fairly good positive correlation between the fayalite content of the olivine and the albite content of the plagioclase.

The olivine series is almost unique amongst natural solid solutions in behaving with a close approach to thermodynamic ideality, and having a distribution coefficient relating the Mg and Fe^{2+} contents of co-existing solids and liquids which is virtually temperature independent. This has been established for a fairly wide range of natural basaltic liquids by Roeder and Emslie (1970). Their predictions are presented graphically in Figure 3.18 where the mole per cents of MgO and FeO in the liquid are plotted to predict the composition of the equilibrium olivine and the temperature at which it will crystallise at 1 atmosphere. The radial lines representing olivine composition indicate that olivine composition is controlled only by the *ratio* of the MgO and FeO contents of the liquid. At a given MgO/FeO ratio the *total content* of these oxides determines the temperature at which that olivine will crystallise. (If the temperature of crystallisation is not required then the molecular *ratio* of FeO/MgO may be determined directly without recalculating the whole analysis.) Compositions with high total MgO + FeO approach the synthetic system forsterite–fayalite and lie off the top and the right-hand side of

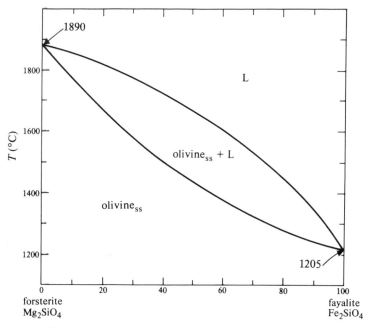

Figure 3.17   The system forsterite–fayalite at 1 atmosphere pressure (after Bowen & Schairer 1935).

the diagram. The diagram is most conveniently used for whole-rock or groundmass compositions in which olivine is the first phase to crystallise on cooling. The weight per cent analysis is first converted to mole per cent by dividing each oxide by its molecular weight and recalculating to 100%. The diagram can then be used directly to determine the temperature of crystallisation and the composition of the equilibrium olivine. This information can then be compared with actual olivine compositions to determine, for example, whether olivines present in a rock are in equilibrium with their groundmass or alternatively whether they might be xenocrystal. Practical difficulties, however, arise from the ready oxidation of Fe^{2+} to Fe^{3+} during alteration of natural rocks so that Fe^{2+} content of a rock is not necessarily a good guide to the Fe^{2+} content of the original magma.

It is not at present certain whether the distribution coefficient is independent of pressure as well as temperature though Roeder and Emslie suggest that it may be. If this is so, the calculations can be extended to decide whether basaltic liquids could be in equilibrium with mantle olivines of specified composition and hence could be regarded as primary magmas. For example, Figure 3.18 shows that a magma

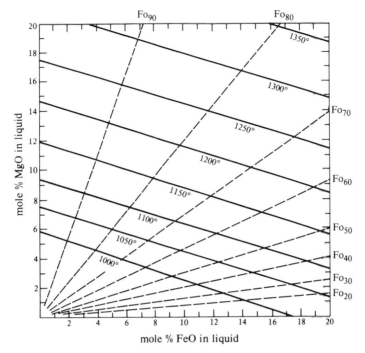

Figure 3.18   Relationships, after Roeder & Emslie (1970), between mole per cents of MgO and FeO in liquid to composition of olivine and temperature of olivine crystallisation at 1 atmosphere pressure. Note that the diagram does not predict whether olivine will crystallise or not (since this depends on numerous other compositional factors) and is applicable only to liquids known or assumed to be in equilibrium with olivine.

equilibrating with mantle olivine of composition Fo_{90} must have a MgO/FeO ratio in the liquid of about 2.7.

Notes

1. Pressure in petrological experiments is normally measured in bars, kilobars (10^3 bars), or occasionally kg cm^{-2}. The bar and the atmosphere are almost the same size and approximately equal to 1 kg cm^{-2}. The kilobar (kbar) is useful for high-pressure experiments because a pressure in kilobars multiplied by 3 (the approximate density of the Upper Mantle) gives the depth in kilometres at which that hydrostatic pressure would be expected, e.g. experiments at 20 kbar approximately simulate the pressure condition at 60 km depth.

    SI units (pascal or newton m^{-2}) of pressure are not in common use in experimental petrology because of their small size.

Conversion factors are as follows:

$$1 \text{ atmosphere} = 1.013 \text{ bars} = 1.033 \text{ kg cm}^{-2}$$
$$1 \text{ bar} \qquad = 0.987 \text{ atmospheres} = 1.020 \text{ kg cm}^{-2}$$
$$= 10^5 \text{ N m}^{-2} \text{ (or Pa)}$$

2. Practice varies in the classification of equilibria in terms of their number of degrees of freedom. Throughout this work equilibria are classified according to the phase rule $P + F = C + 2$. However in diagrams determined at a fixed pressure some authors use the equation $P + F = C + 1$ because one degree of freedom has been taken up. Hence a eutectic such as that just discussed is referred to as invariant rather than divariant. This is strictly incorrect, although the term *isobarically* invariant is more acceptable. The incorrect practice has as its attraction the coincidence between the apparent number of degrees of freedom and the number of dimensions needed to represent the loci of the corresponding liquid compositions in the isobaric diagram (points are invariant, areas are divariant etc.). However, the use of this condensed version of the phase rule diverts attention from the essentially polybaric nature of igneous processes, apart from the immense confusion which can result from forgetting to state in each case whether the polybaric or isobaric case is being referred to.

Exercises

1. The following phases are present in an experimentally investigated system: forsterite (Mg_2SiO_4), diopside ($CaMgSi_2O_6$), and solid solutions between

Table 3.1

Composition	Temperature (°C)	Phases present
Di_{100}	1392	gl
	1385	Di
Di_{95}	1385	gl
	1365	Di + gl
	1355	Di + Tr
Di_{85}	1370	gl
	1365	Di + gl
	1350	Di + Tr
Di_{80}	1410	gl
	1390	Tr + gl
	1370	Tr + gl
Di_{75}	1500	gl
	1480	Cr + gl
	1355	Di + Tr

Abbreviations: Di – diopside; Tr – tridymite; Cr – cristobalite; gl – glass.

pyrope $(Mg_3Al_2Si_3O_{12})$ and grossular $(Ca_3Al_2Si_3O_{12})$. How many components has the system, and could the above assemblage melt isothermally?

2. Table 3.1 gives phases present in charges made up in the system diopside–silica. Use this information to make a phase diagram for the diopside-rich half of the system and determine the temperature of the eutectic between diopside and tridymite. What is the composition of the eutectic liquid?

3. Using the phase diagram for albite–anorthite (Fig. 3.16), determine:
    (a) The composition of the first plagioclase to crystallise on cooling of a liquid of composition An_{50}.

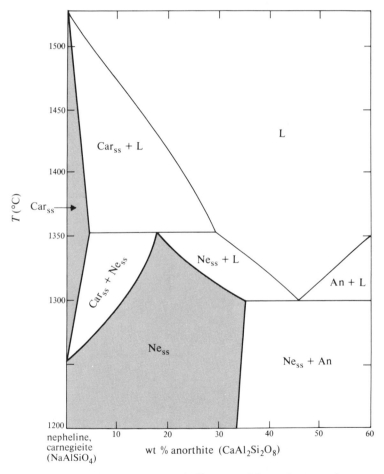

Figure 3.19 Part of the system nepheline–anorthite at 1 atmosphere pressure (after Bowen 1912).

(b) The composition of the last drop of liquid to disappear during equilibrium crystallisation of the above liquid.
(c) The composition of the liquid and the remaining solid when a bulk composition of An_{50} is 20% molten.
(d) The composition and temperature of the final liquid of fractional crystallisation of liquid An_{50}.

4. Figure 3.19 gives part of the diagram for the system nepheline–anorthite at 1 atmosphere pressure. Nepheline shows extensive solid solutions towards anorthite, while its high-temperature polymorph, carnegieite, shows limited solid solutions.

(a) Describe with full details of temperature, phase proportions and phase compositions the melting of a nepheline solid solution of composition $Ne_{90}An_{10}$.
(b) Write an equation to represent the univariant reaction affecting this composition at 1353 °C.
(c) If another univariant reaction is represented in the diagram, what is its temperature and what equation would you write?
(d) What is the composition of the last liquid to disappear on equilibrium crystallisation of a liquid of composition $Ne_{78}An_{22}$?
(e) What is the composition of the last liquid of perfect fractional crystallisation of $Ne_{90}An_{10}$?

5. The analysis of a picrite basalt is given below. Calculate the olivine composition and the temperature at which olivine will begin to crystallise at 1 atmosphere pressure (see Fig. 3.18). Assume firstly that FeO as reported represents FeO of the liquid, and then alternatively that all the reported Fe_2O_3 was originally present as additional FeO in the liquid. This gives a bracket on both composition and liquidus temperature. Small amounts ($< 1\%$) of olivine (Fo_{85}) are present in the rock. Are these demonstrably xenocrystic?

SiO_2	TiO_2	Al_2O_3	Fe_2O_3	FeO	MnO	MgO	CaO
45.70	2.62	14.26	6.68	4.30	0.21	7.93	9.86

Na_2O	K_2O	P_2O_5
2.12	1.80	0.43

4
Ternary systems – I

The extra component in ternary systems compared with binaries represents a substantial step towards petrological reality, and there are many ternaries which form a relatively close analogue to natural magmas. The study of ternaries provides a useful basis for the formulation of petrological ideas, but is nevertheless a foundation which is far from complete. The question of how additional components might affect phase relations should not be allowed too far out of sight during the following discussions of ternary systems.

Representation of composition

Two dimensions are needed for the representation of ternary compositions and triangular plots are normally employed. In Figure 4.1a, three components X, Y and Z are placed at the corners of an equilateral triangle on which a 10% triangular grid has been drawn. The amount of X at any point within the triangle is represented by the grid lines parallel to the YZ edge, thus for example all points on the line *ab* contain 30% of X. A point *c* which represents the composition $X_{30}Y_{30}Z_{40}$ illustrates how the three co-ordinates total to 100%. Figure 4.1b demonstrates the same geometrical relationships transferred to a scalene triangle.

Phase proportions in ternary diagrams are also determined by the method above. If X, Y and Z in Figure 4.1b are taken to represent the compositional points of three phases (solid or liquid) the point *c* is a bulk composition consisting of 30% of phase X, 30% of Y and 40% of Z. Furthermore the lever rule as used in binaries applies in exactly the same fashion in ternaries. Hence, in Figure 4.1a, if *c* is mixed with *e* all compositions produced lie on the straight line between them, and it can readily, for example, be verified that a point *d* which lies so that the lengths *de* and *cd* are in the ratio 1 : 3 can be expressed as a mixture of

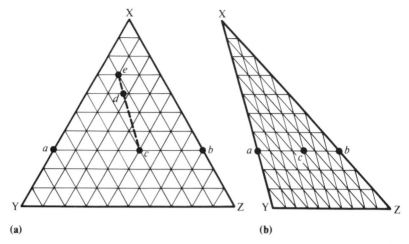

Figure 4.1   Plotting compositions in ternary diagrams. Grid lines are at 10% intervals.

one part of c to three parts of e. The calculation is carried out as follows:

$$
\begin{aligned}
e &= 70X + 20Y + 10Z \\
c &= 30X + 30Y + 40Z \\
c + 3e &= 240X + 90Y + 70Z
\end{aligned}
$$

dividing by 4

$$
\frac{c + 3e}{4} = 60X + 22.5Y + 17.5Z = d
$$

This is the same as the composition of d found graphically in Figure 4.1a. The same construction would clearly also apply to the scalene triangle of Figure 4.1b.

In practice during the reading of ternary diagrams it is frequently useful to determine the composition of points within scalene triangles, but it is not necessary to construct a complete grid. Comparison of Figure 4.2 with Figure 4.1b shows that the ratio of the lengths cx to Xx if expressed as a percentage is the X content of c. Similar arguments allow the Y and Z contents to be found. To plot the point c, given its composition, it is probably easiest to calculate the ratio $X : Y$ (in this case $1 : 1$) and thus find the point z. Then divide the line zZ in the ratio of $(X + Y) : Z$ (in this case $3 : 2$). The principle here is the lever rule which demands that c is made up of a mixture of Z and X + Y (the last two having between them a bulk composition of z) in the proportions determined.

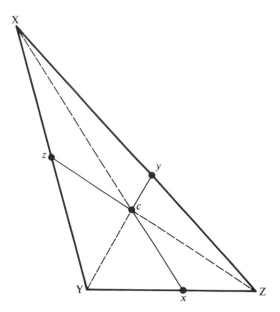

Figure 4.2 Plotting a ternary composition without the use of a grid.

Liquidus projections

Ternary phase relations are most commonly presented as so-called liquidus projections in which the plane of the diagram is used for the representation of composition and features of the liquidus surface are displayed. Such diagrams are projections of triangular $T-X$ prisms down the temperature axis onto the base. Thermal information is conveyed by contours or by the labelling of specific points in the same way that topographic height is indicated on maps. Figure 4.3 illustrates the relationship between the $T-X$ prism and the liquidus projection. This is an example of the simplest type of ternary system bounded by the three eutectic binaries XY, YZ and XZ. Binary fields in the visible front faces of the $T-X$ prism are labelled. The liquidus projection consists of the three fields $X + L$, $Y + L$, and $Z + L$ (these are termed **primary phase fields**) which are extensions into the ternary of similar binary fields. They represent the compositional ranges of liquids which show the appearance of a particular solid phase as the first crystals to appear on cooling. The primary phase fields are separated by boundary curves which are the loci of liquids in equilibrium with two solid phases, that is, those of the neighbouring primary phase fields. These curves are the ternary extensions of the binary eutectic points. The three boundary

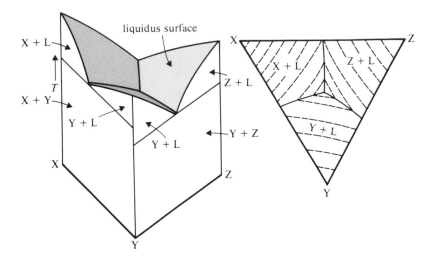

Figure 4.3 Perspective view of ternary $T–X$ prism (left) and liquidus projection with thermal contours (right).

curves meet in a point which is the unique liquid capable of equilibrating with all three solid phases. In the diagram illustrated this point is a **ternary eutectic**, the liquid concerned crystallising isothermally to all three phases as heat is extracted (but note that not all such points in ternary diagrams are eutectics. Some show down temperature resorption of solid phases and are analogous to the binary peritectic points).

The solidus

The arrangements of sub-solidus fields is shown in Figure 4.4 for the system illustrated in Figure 4.3. It consists of:

(a) 3 one-dimensional single-phase fields each representing a pure component, in the absence of solid solution, and terminating upwards, at the melting point (MP) of the phase.
(b) 3 two-dimensional two-phase fields representing the two-phase sub-solidus assemblages of the bounding binaries and terminating upwards at the temperature of the appropriate binary eutectic, e.
(c) 1 three-phase field (the assemblage X + Y + Z) forming a triangular prism (upper surface shaded in the figure) terminating upwards at the temperature of the ternary eutectic, E.

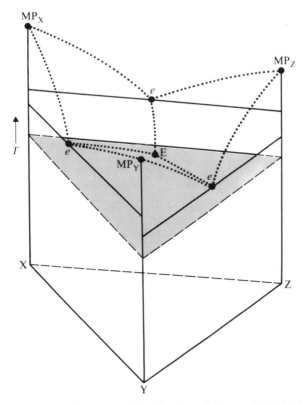

Figure 4.4 Solidus of the ternary $T-X$ prism of Figure 4.3. Melting points of pure components – MP; binary eutectics – e; ternary eutectic – E. Outline of the liquidus surface shown by dotted line; shaded area is upper surface of X + Y + Z sub-solidus volume.

Within the ternary the liquidus surface touches the solidus only at E. Between the liquidus and solidus lies a volume of complex geometrical shape which consists of fields of one solid phase + liquid, and two solid phases + liquid. This is discussed in a later section.

Primary phase fields

We now consider crystallisation paths in equilibria involving one solid phase + liquid. Such equilibria are trivariant and represented by the primary phase fields of the liquidus projection. Figure 4.5 shows a field with its bounding divariant curves (two solid phases + liquid) and thermal contours. A liquid a cools until it meets the liquidus at 1100 °C

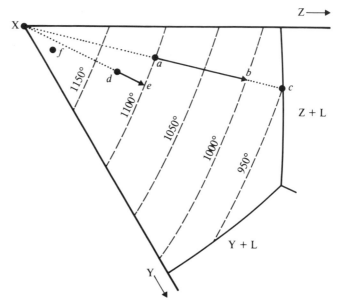

Figure 4.5   Crystallisation paths in a primary phase field (cf. X + L field of Fig. 4.3).

when crystals of X begin to separate. The residual liquid becomes impoverished in the component X and must migrate directly away from the X composition point. By the time the temperature has fallen to 1000 °C the residual liquid has changed to composition *b*, the amount of X that has crystallised by this stage being represented by the ratio of the length of *ab* to X*b* (lever rule). As heat is extracted X will continue to crystallise until the liquid reaches *c* on the boundary curve and begins to precipitate crystals of Z at 950 °C. From the above it follows that liquid paths in both equilibrium and fractional crystallisation follow a set of straight lines radiating across the primary phase field from the composition point of the phase concerned. Thermal contours do not necessarily lie at right angles to liquid paths and the precise form of the liquidus surface must normally be experimentally determined. However, liquid paths must always have a down-temperature component of movement as they radiate away from the composition point of the primary phase.

Isothermal sections

Isothermal sections are of prime importance in the reading of ternary diagrams and are slices across the $T-X$ prism (see Fig. 4.3) parallel to the base. An isothermal section is therefore a representation of what the experimenter would see in charges of different compositions all of which were quenched from the same temperature. It is actually experimentally convenient to run several charges simultaneously in a furnace at a given temperature so that much experimental information originates in the form of sets of isothermal data from which the liquidus diagram is subsequently constructed. However, as an introduction we shall initially derive an isothermal section from the liquidus diagram. By so doing we are investigating the nature of the complex area of the $T-X$ prism lying between the solidus and the liquidus.

Referring to Figure 4.5, suppose that a number of charges with different compositions lying in the primary phase field of X are simultaneously equilibrated at 1100 °C. All charges on the X-poor side of the 1100 °C isotherm will remain completely liquid because none of them is capable of crystallising X until a lower temperature is reached. A charge such as d on the other hand if cooled to 1100 °C would start to crystallise X at about 1130 °C and by the time the temperature had fallen to 1100 °C would consist of the liquid e plus crystals of X represented in proportion by the ratio of de to Xe. Thus, all charges lying on the X-rich side of the 1100 °C isotherm must by analogy consist of X + liquid, and all the liquid compositions must lie on the thermal contour. Charges in this range will only vary in their proportions of X to liquid. A composition such as a will contain only an infinitesimally small amount of X whereas a charge such as f, which is very rich in the X component, will be about 75% crystalline at 1100 °C. The isothermal section for 1100 °C is shown in its normal form of representation in Figure 4.6. The radial

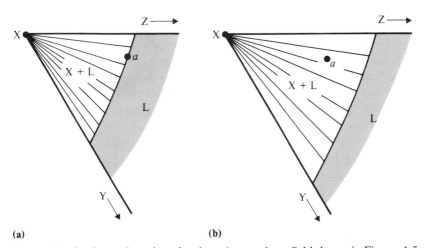

(a)   (b)

Figure 4.6 Isothermal sections for the primary phase field shown in Figure 4.5.

lines across the X + L field signify that various liquids along the isotherm co-exist with X. Figure 4.6b shows the isothermal section for the lower temperature of 1050 °C. The juxtaposition of two isothermal sections for neighbouring temperatures provides a very useful tool for the logical solution of phase diagram reading problems. There is an area lying between the 1100 °C and 1050 °C contours which belongs to the L field at the higher temperature and the X + L field at the lower. Compositions lying in this range have therefore begun to crystallise X within the temperature interval concerned. This may seem an obvious conclusion but it is an example of the reasoning which was previously employed to solve problems in binaries when isothermal tie-lines at temperatures immediately above and immediately below reactions of interest were considered. The isothermal sections discussed here are simply ternary versions of the isothermal tie lines used in the previous chapter and will be found to have the same logical power.

Equilibria involving two solids + liquid

The boundary curves separating primary phase fields in the liquidus projection are the loci of liquids equilibrating with both solid phases. Such equilibria must involve the precipitation of at least one of the solids with cooling but the other solid phase may either crystallise or be resorbed. The two types of equilibrium are thus analogous to the eutectic and peritectic equilibria of binary systems but in the presence of an extra component they are divariant, not univariant.

The Alkemade theorem. This theorem enables the determination of the direction of temperature change along a boundary curve to be made in the absence of experimental information on the temperatures concerned. At any point on the boundary curve a tangent is constructed to project back until it intersects the line joining the composition points of the two solid phases or its extension. The boundary curve then goes down-temperature away from the point of intersection. A complex ternary with an intermediate binary compound YZ is illustrated in Figure 4.7. The liquidus surface is not contoured but the Alkemade theorem makes possible the placing of temperature arrows on all the divariant equilibria. For example the boundary curve PT represents equilibrium of liquid with X and Y. The X–Y join is located and the temperature arrow inserted pointing away from it. Curve PR illustrates a case where no back tangent to the curve cuts Y–YZ, the appropriate join, but does cut the join extended to R. The temperature arrow points down the curve away from the extension of the join. Curve PQ is an

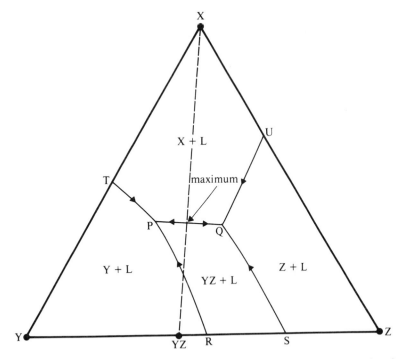

Figure 4.7 Ternary system with an intermediate compound showing incongruent melting. Directions of falling temperature in divariant equilibria are shown by arrows. Broken line is the stable sub-solidus join X–YZ.

example which has a thermal maximum on it, which results from the fact that the boundary curve crosses its own join, X–YZ. It therefore has two temperature arrows pointing in opposite directions.

Isothermal sections. The nature of different types of divariant equilibria is best investigated by the isothermal section method. Figure 4.8 shows a partial isothermal section of a ternary for a boundary curve such as QU in Figure 4.7. The section contains a triangular area where all charges consist of X + Z + L at a temperature of 1000 °C. This is flanked on either side by areas of X + L and Z + L which are themselves limited by the 1000 °C isothermal contour where they come into contact with the L field. L_1 is the special liquid at 1000 °C which equilibrates with both X and Z, and lies on the boundary curve QU, shown as a broken line. A composition a which lies within the X + Z + L field crystallises X at a temperature above 1000 °C and its residual liquid evolves to meet the boundary curve at L_2. Subsequent cooling sees the

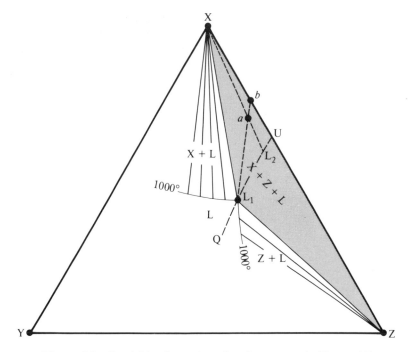

Figure 4.8   Partial isothermal section for system in Figure 4.7.

precipitation of both X and Z so that the liquid moves down the boundary curve away from the X–Z join, upon which the bulk composition of
the crystallising material must lie. By the time 1000 °C is reached the
liquid has reached L_1 where the boundary curve and the 1000 °C
liquidus contour intersect. By analogy *all* bulk compositions lying within
the X + Z + L triangle must at 1000 °C consist of the liquid L_1 plus
varying amounts of X and Z. Compositions lying on the Z-rich side of
QU will have reached this condition via the early precipitation of Z.
Phase proportions at 1000 °C are determined by the construction of the
line from L_1 through the bulk composition until it cuts the X–Z join.
The charge *a* for example consists of L_1 plus a mixture of X and Z
represented by *b*. In this case the ratio X*b* : XZ is 24 : 100 so that the
solid present is 24% Z and 76% X. The amount of liquid is derived from
the length *ab* and is in this case 18%. Thus the charge consists in total of
18% liquid with 20% of Z crystals and 62% of X.

In constructing isothermal sections it is in fact *not* necessary to
consider *how* each composition might have achieved its particular
condition by cooling. The nature of equilibrium is such that a liquid in
equilibrium with one crystal of a solid phase will remain in equilibrium

with many. Thus the area of the section which consists of X + Z + L is readily derived by considering all possible bulk compositions that may be made by adding X and Z to the liquid. This is a useful general rule for the construction of isothermal sections.

In Figure 4.9a two isothermal sections for an equilibrium of the same type as discussed above are superimposed. Broken lines refer to the higher temperature where L_1 is the liquid; solid lines to a slightly lower temperature involving liquid L_2. Compositions *a* and *b* are such that the residual liquids have only just reached the boundary curve at the temperature of the higher isothermal section and they thus contain only an infinitesimally small amount of the second solid phase. However, both are firmly inside the X + Z + L triangle at the lower temperature of L_2. Thus the transition down-temperature from L_1 to L_2 must involve the crystallisation of both phases. This type of equilibrium is termed **co-precipitational** (or sometimes **cotectic**) and is readily identified in a ternary diagram because the back-tangent to the boundary curve cuts the join linking the two solids *between* the two composition points (in the case discussed this is approximately the point U because the boundary curve is almost linear).

Figure 4.9b shows the contrasting case for the boundary curve PR of Figure 4.7 which, as already noted is characterised by back-tangents which cut the *extension* of the Y–YZ join rather than the join itself. Successive isothermal sections of the Y + YZ + L field do not nest inside each other as in the former case but cross over as illustrated. A bulk composition such as *a* which clearly contains all three phases at the higher temperature of L_1 lies *outside* the triangle at the lower temperature of L_2 and is in fact in the field of YZ + L. Hence this composition has shown complete resorbtion of Y in the temperature interval between L_1 and L_2. Resorption, viewed geometrically, is caused by the rotation of the tie line L–YZ which sweeps towards the compositioin point *a* and crosses it at some temperature between L_1 and L_2. Resorption of Y is complete at this moment. Composition *c* is such that it shows complete resorption when the liquid reaches L_2. A composition such as *b* demonstrates that during the fall in temperature the other phase YZ is precipitating, since *b* contains no significant YZ at the higher temperature but a relatively large amount at the lower. However, *b* also demonstrates the *partial* resorption of Y because the percentage of Y present at the higher temperature is represented by the ratio $bL_1 : YL_1$ and at the lower temperature by $cb : Yc$ which is clearly lower. This equilibrium is thus the precise ternary equivalent of a binary peritectic reaction such as that between forsterite and liquid to give enstatite. As we saw in the preceding chapter all compositions which encounter the reaction undergo at least some down-temperature

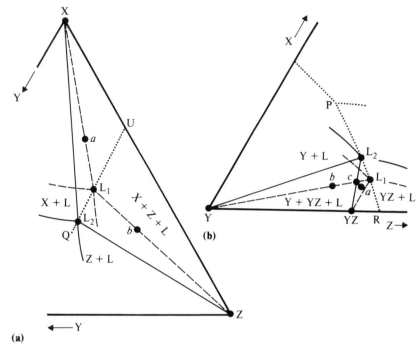

Figure 4.9 Use of successive isothermal sections to determine nature of divariant reactions in system illustrated in Figure 4.7. (a) Co-precipitational, (b) resorptional.

resorption of one phase, and many compositions show complete resorption. Considering all the possibilities in the present case it is evident that the sweeping tie-line YZ–L can cover the whole of the area between the line YZ–P and the boundary curve PR. All bulk compositions lying in this area will therefore show complete resorption of early formed Y as they undergo equilibrium crystallisation. Conversely all compositions lying in the triangle Y–P–YZ will behave like composition *b* and show partial resorption of Y. No other bulk compositions in the system can encounter this equilibrium. Liquids, for example, starting in the field of YZ + L evolve on radial lines away from the YZ composition point and thus do not reach the boundary curve PR.

Equations of divariant reactions. The specific equation for the co-precipitational type of divariant reaction is of the form:

$$L_1 = L_2 + X + Z$$

However, since L_1 can be thought of (see Fig. 4.9a) as a mixture of L_2 and another liquid L_j lying on the join X–Z where the line L_1–L_2 back-projects to it, this equation reduces to:

$$L_j = X + Z$$

The liquid L_j is that part of the original liquid which goes to form the crystals of X and Y and its composition expresses the proportions in which the two phases separate during cooling. This proportion, unlike that of the eutectic liquid of a binary system, e.g. U, is not, however, constant because L_j moves in position as a result of any curvature of the boundary curve. Many boundary curves are however approximately straight so that it is useful to think of them as characterised by particular proportions in which they precipitate two phases. Such proportions are termed 'cotectic' to distinguish them from eutectic proportions.

The resorptional type of boundary curve as exemplified by the composition *c* of Figure 4.9b represents a reaction of the type:

$$L_1 + Y = L_2 + YZ$$

for these are the assemblages *c* consists of at the two temperatures. However, as argued above, L_1 can be regarded as a mixture of L_2 and L_j, the latter in this case lying approximately at R on the extension of the Y–YZ join. Hence the general equation is:

$$L_j + Y = YZ$$

As in the binary case we see the phase undergoing down-temperature resorption on the same side of the equation as the liquid. As a rule of thumb it is useful to remember that the resorbed phase lies at the far end of the tie-line from the extension where L_j lies.

Equilibria involving three solids + liquid

In ternary systems equilibria involving three solid phases + liquid are univariant. They are represented in liquidus projections by the intersection of three boundary curves of the two solid phases + liquid type. The point formed by the intersection of these curves is the unique liquid capable of equilibrating with the three primary phases of the adjacent fields. In isobaric diagrams this equilibrium takes place at a fixed temperature.

Figure 4.10 shows the liquidus projection for the system $MgO–Al_2O_3–SiO_2$, a ternary showing relatively little solid solution (which will be ignored in the present treatment) and illustrating all the possible types of equilibria involving three solids and a liquid. We

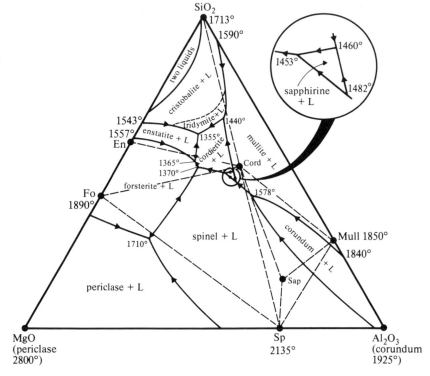

Figure 4.10 Liquidus projection for the system MgO–Al₂O₃–SiO₂ at 1 atmosphere pressure (after Osborn & Muan 1960 and Eitel 1965). Abbreviations are: En – enstatite; Fo – forsterite; Cord – cordierite; Mull – mullite; Sap – sapphirine; Sp – spinel.

introduce a convenient shorthand notation where M is used to signify MgO, A is Al₂O₃, and S is SiO₂. Four binary compounds are present – forsterite (M₂S), enstatite (MS), mullite (A₃S₂), and spinel (MA). Ternary compounds are represented by cordierite (M₂A₂S₅) and sapphirine (M₄A₅S₂). The diagram is plotted in weight per cent so that compounds like MA and MS which would fall in the middle of their respective joins in a molecular per cent plot do not actually do so (enstatite for example contains 60 weight per cent SiO₂, but 50 mole per cent).

Straight broken lines on the diagram (**tie-lines**) represent a highly important feature of liquidus projections and link pairs of phases which are capable of equilibrating with each other in the presence of liquid. Thus the cordierite and forsterite composition points are shown joined because the forsterite + liquid and cordierite + liquid primary phase fields come into contact. Mullite–forsterite is not joined because these two fields do not come into contact. The joins shown are best termed **stable binary sub-solidus assemblages**. Cordierite–forsterite is a binary assemblage which may be raised to the solidus and melted to give a liquid in equilibrium with both solids. There is no liquid anywhere in the system, however, which can equilibrate with forsterite and mullite. The latter therefore is not a stable assemblage at melting temperatures. It is important to realise that the term 'sub-solidus assemblage' here refers to stability *immediately* below the solidus. The diagram does not tell us whether such assemblages retain their stability or instability at lower temperatures. If changes in stability relationships do occur at lower temperatures these would be classified as metamorphic.

The stable binary joins are of course those lines which were used in the preceding section in the application of the Alkemade theorem to the determination of directions of falling temperature along boundary curves of the two solids + liquid type. They are therefore also known as Alkemade lines. Careful inspection of the diagram will show that every divariant boundary curve in the system is matched by a stable binary join somewhere in the system (those in the bounding binary systems being shown by solid lines).

The stable binary joins divide the whole diagram into a series of triangles, each of which represents an assemblage of three solid phases which is capable of melting and is stable immediately below the solidus temperature. This follows from the argument that if phase X is capable of equilibrating with liquid and phase Y simultaneously, and is also capable of equilibration with phase Z and liquid *at the same temperature*, then Y and Z must also be in equilibrium. The only liquid which satisfies the requirement of fixed temperature is that which lies at the intersection of the three appropriate boundary curves.

Inspection of the diagram shows that there are nine different ternary sub-solidus assemblages. These are:

Tr + En + Cord,	Fo + En + Cord,
Fo + Sp + Cord,	Fo + Sp + Per,
Tr + Cord + Mull,	Cord + Sap + Mull,
Cord + Sap + Sp,	Sap + Mull + Sp,

and   Sp + Mull + Corundum.

Each of these is represented by a point on the liquidus surface where the three appropriate primary phase fields meet at a specific temperature. These are the compositions of the first liquids to appear on melting of each of the three-phase solid assemblages or the compositions of the liquids which solidify to give those assemblages.

In the section on divariant equilibria it was demonstrated that the geometrical relationships between liquid composition and the sub-solidus join representing the appropriate pair of solid phases determines the nature of the reaction, i.e. whether co-precipitational or resorptional. In univariant equilibria the same applies but here there are three different geometrical possibilities and hence three types of reaction, which can be termed **co-precipitational**, **mono-resorptional**, and **bi-resorptional**. Where X, Y, Z represent solid phases, the three types by analogy with binary reactions previously discussed are:

$$\text{co-precipitational} \quad L = X + Y + Z$$

$$\text{mono-resorptional} \quad L + X = Y + Z$$

$$\text{bi-resorptional} \quad L + X + Y = Z$$

The geometrical relationship which allows ready identification of reaction types concerns the position of the liquid relative to the sub-solidus triangle as illustrated in Figure 4.11 with examples from $MgO–Al_2O_3–SiO_2$. For co-precipitational equilibria, the liquid lies *inside* the sub-solidus triangle; mono-resorptional equilibria have their liquids lying *outside* the triangle in a position where the liquid can 'see' *one* edge of the triangle; while in bi-resorptional equilibria the liquid is *outside* the triangle and can 'see' *two* edges of it.

Reference back to the system $MgO–SiO_2$ illustrates that this is an extension of binary relationships. The liquid equilibrating with tridymite and enstatite is co-precipitational and lies between the composition points of these two phases, the binary equivalent of lying inside the triangle. The liquid equilibrating with forsterite and enstatite which shows down-temperature resorption of forsterite, conversely lies outside the forsterite–enstatite join.

Isothermal sections can be used to demonstrate the nature of the ternary equilibria and involve drawing the four possible three-phase triangles for temperatures immediately above and immediately below that of the univariant reaction. Since three boundary curves meet at the point representing the composition of the liquid in equilibrium with all three solid phases, three of the isothermal triangles concerned must be L + X + Y, L + Y + Z and L + X + Z, where the liquid in each case is on the boundary curve, infinitesimally displaced from the temperature of the univariant reaction, and may be either above or below the temperature of the univariant reaction itself. The fourth three-phase triangle represents the sub-solidus assemblage X + Y + Z which *must* lie on the down-temperature side of the reaction. A rule is then applied which states that three-phase triangles which cross each other or overlap cannot refer to the same temperature. Thus we derive Figure 4.12 which

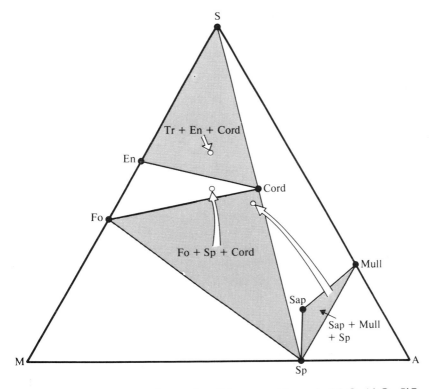

Figure 4.11 Three three-phase sub-solidus assemblages in MgO–Al$_2$O$_3$–SiO$_2$ showing spatial relationships to their equilibrium liquids (small circles). Large arrows link sub-solidus triangles to the appropriate liquid.

shows the way in which the three-phase triangles may be arranged in non-overlapping sets for the three equilibria considered in Figure 4.11. The three-phase triangles of L + X + Y type have been drawn according to the principles illustrated in Figure 4.8 and 4.9, the intervening fields of L + X type having been reduced to infinitesimal width because of the convergence of the boundary curves on their point of intersection.

Considering the bi-resorptional equilibrium between liquid, sapphirine, spinel, and mullite as an example we can now write out a number of reactions which take place in the univariant condition. Each reaction below refers to a different bulk composition. All the compositions lie in the L + Sp + Mull triangle at a temperature infinitesimally above that of the reaction but in L + Sap + Sp, L + Sap + Mull, or Sap + Mull + Sp at an infinitesimally lower temperature. For

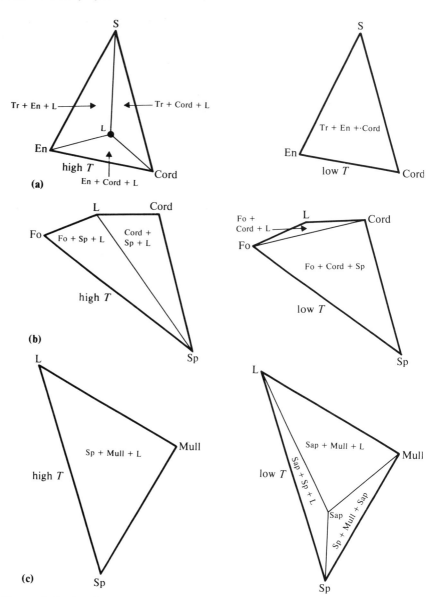

Figure 4.12 Isothermal sections for the three assemblages of Figure 4.11 drawn immediately above (left-hand figures) and immediately below (right-hand figures) the temperatures at which three solid phases equilibrate with liquid.   (a) Co-precipitational equilibrium of tridymite, enstatite, cordierite and liquid;   (b) equilibrium of forsterite, cordierite, spinel and liquid, in which spinel is resorbed down-temperature;   (c) equilibrium between sapphirine, spinel, mullite and liquid in which spinel and mullite are resorbed down-temperature.

the three classes of bulk composition the reactions are respectively:

$$L + Sp + Mull = L + Sap + Sp \qquad (1)$$
$$L + Sp + Mull = L + Sap + Mull \qquad (2)$$
$$L + Sp + Mull = Sap + Mull + Sp \qquad (3)$$

From this it is immediately clear that sapphirine is precipitated in all the reactions irrespective of bulk composition. The first reaction demonstrates the complete resorption of mullite, the second the complete resorption of spinel. However, from what we have seen previously in binary systems and in the consideration of ternary divariant equilibria it might be suspected that *all* bulk compositions should show at least partial resorption of mullite and spinel. A single example should suffice to show that this is in fact so. Figure 4.13 considers a bulk composition *c* which lies inside the Sap + Mull + Sp triangle and is thus a representative of the behaviour shown by reaction (3) above. At a temperature above that of the univariant reaction this composition consists of L + Sp + Mull having a proportion of mullite represented by *ac/af* and of spinel by *ec/eg*. However, after solidification the

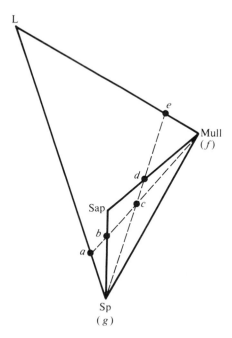

Figure 4.13 Phase proportions during the down-temperature resorption of mullite and spinel.

proportions of mullite and spinel are represented by bc/bf and dc/dg both of which are substantially lower. Hence, in general the three types of bulk composition which become involved in the univariant reaction concerned can be classified as those which show (a) complete resorption of mullite with partial resorption of spinel (Equation 1), (b) complete resorption of spinel and partial resorption of mullite (Equation 2), and (c) partial resorption of both spinel and mullite.

In Equations (1)–(3) if a cancelling procedure is used to eliminate what we have previously termed inert excesses of reactants, the generalised equation below is derived:

$$L + Sp + Mull = Sap$$

The specific bulk composition which shows this behaviour is the sapphirine composition itself. Equilibrium crystallisation of this composition sees spinel and mullite reacting with liquid and sapphirine crystallising as heat is extracted. The last crystal of mullite and spinel and the last drop of liquid all disappear simultaneously.

Having established which assemblages lie respectively up-temperature and down-temperature of a univariant reaction, in-spection of Figure 4.10 will confirm that the application of the Alkemade theorem to the associated boundary curve of divariant equilibria gives temperature arrows which confirm this. Thus the three types of univariant equilibria may also be recognised as follows:

(a) Co-precipitational reactions show three boundary curves coming down-temperature into the intersection.
(b) Mono-resorptional reactions show two boundary curves coming down-temperature to the intersection and one proceeding down-temperature from it.
(c) Bi-resorptional reactions show one boundary curve coming down-temperature into the intersection and two proceeding down-temperature out of it.

Co-precipitational equilibria therefore represent thermal lows in the liquidus surface and are termed **ternary eutectics**. The two types of resorptional equilibrium do not represent thermal lows and hence never involve final liquids of fractional crystallisation, though they do include the last liquids of equilibrium crystallisation for bulk compositions lying inside their appropriate sub-solidus triangle.

Equilibrium crystallisation and melting paths in ternary systems

Figure 4.14 gives the liquidus projection of a hypothetical ternary which will be used as an example in the reading of equilibrium crystallisation

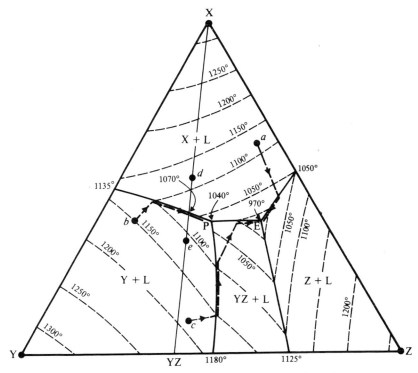

Figure 4.14 Equilibrium crystallisation paths in a hypothetical ternary system X–Y–Z.

and melting paths. An intermediate compound YZ has composition $Y_{60}Z_{40}$ in weight per cent. Preliminary examination of the diagram shows the following features:

(a) The intermediate compound YZ melts incongruently to a Z-rich liquid at 1180 °C. All the other binary reactions are eutectic.

(b) The divariant equilibria are all co-precipitational except for the equilibrium Y + YZ + L which is resorptional down-temperature for Y. (In referring to equilibria, as opposed to reactions, the + sign is used only to signify 'in equilibrium with'.)

(c) One of the univariant equilibria (X + Y + YZ + L at 1040 °C) is resorptional down-temperature (for Y). The other, X + YZ + Z + L is co-precipitational at 970 °C (liquid E).

(d) Considering the three bulk compositions *a*, *b* and *c* it is clear that *a* and *c*, since they lie inside the sub-solidus triangle X–YZ–Z, must crystallise eventually to this assemblage and their final liquid must be E. Conversely *b* crystallises to the assemblage X + Y + YZ, the final liquid being P.

Knowing the nature of all the equilibria concerned and knowing the final products of crystallisation it is now possible to read crystallisation paths directly from the diagram.

Composition *a* shows simple behaviour. It begins to crystallise X on cooling to 1125 °C. The liquid evolves directly away from the X composition point and reaches the boundary curve against the Z field at about 1020 °C. Then X and Z precipitate together while the liquid moves down the boundary curve to the eutectic E at 970 °C. YZ begins to crystallise at this point and the system becomes univariant. X, Z and YZ crystallise simultaneously without temperature change until all the liquid disappears leaving the solid assemblage X + YZ + Z.

The behaviour of *b* is similar until the liquid reaches the point P. This is a resorptional equilibrium for Y so further heat extraction is accompanied by the resorption of Y, the continued precipitation of X, and the crystallisation of YZ. The liquid disappears before all the Y has been resorbed, leaving the solid assemblage X + Y + YZ.

Compositon *c* shows a more complex behaviour because it encounters a resorptional equilibrium at an earlier stage. Having crystallised Y from a temperature of about 1185 °C the liquid encounters the boundary

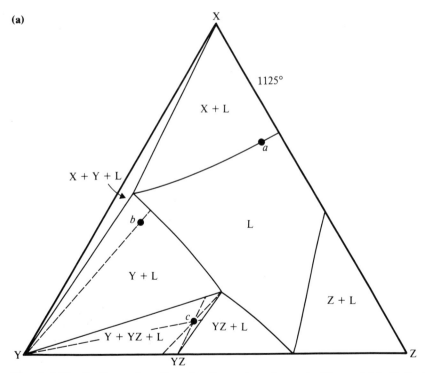

Figure 4.15   Isothermal sections for the system shown in Figure 4.14. Dotted lines are used for the determination of phase proportions.

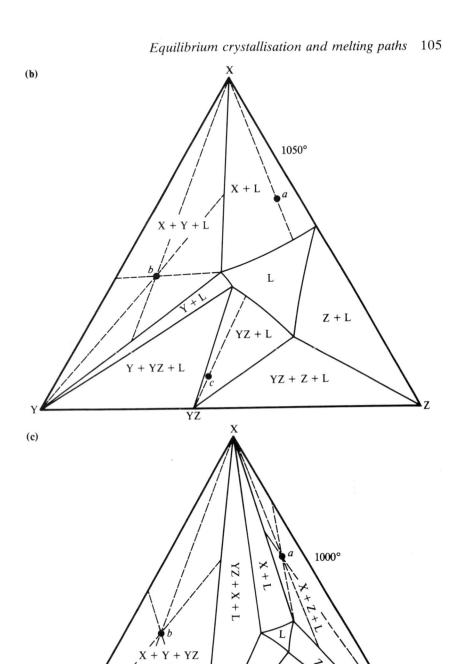

curve at 1150 °C. The liquid then moves down the boundary curve resorbing Y and precipitating YZ. Resorption of Y is complete at 1100 °C when L, c and the YZ composition point are co-linear. The liquid then enters the YZ field and evolves directly away from YZ until it reaches the boundary of the X field at 1000 °C. The two phases X and YZ then precipitate together until the liquid reaches E. Solidification takes place with the co-precipitation of X, Z and YZ.

The method of reading used above is specific for each composition. However, if the problem is alternatively approached via the drawing of complete isothermal sections it becomes possible to read any crystallisation path at will and moreover it makes the determination of phase proportions at any stage relatively simple. The closeness of spacing of the sections (in a temperature sense) determines the degree of detail which can be derived. In Figure 4.15 isothermal sections for the above system are given for 1125 °C, 1050 °C and 1000 °C. The compositions of a, b and c are marked for reference. In the 1125 °C section we note that a is just beginning to crystallise X, c consists of Y, YZ and L in the approximate proportions $Y_5YZ_{45}L_{50}$ (derived from a consideration of its position within the three-phase triangle using the dotted construction lines shown), while b consists of Y + L in the approximate proportions Y_8L_{92}. At 1050 °C, a consists of X + L in the proportions $X_{26}L_{74}$, c has finished resorbing Y and now consists of YZ + L in the proportions $YZ_{70}L_{30}$, and b consists of X + Y + L in the proportions $X_{25}Y_{37}L_{38}$. At 1000 °C a has now picked up the second phase Z and consists of $X_{45}Z_{10}L_{45}$, while the liquid of c has just reached the boundary curve with the X field. At this temperature c consists of YZ + L in the proportions $YZ_{76}L_{24}$. The remaining composition b is now completely solid and consists of X + Y + YZ in the proportions $X_{40}Y_{35}YZ_{25}$. With the experience gained in the preceding examples it is possible to make the completely generalised diagram representing all possible crystallisation sequences in this system, there being fourteen essentially different patterns. Areas in Figure 4.16 represent compositional ranges of initial liquids which will encounter the equilibria (expressed as reaction equations so that resorption is evident) in the varying sequences as listed:

A. L = X, L = X + Z, L = X + YZ + Z

B. L = X, L = X + YZ, L = X + YZ + Z

C. L = X, L = X + Y, L + Y = X + YZ, L = X + YZ, L = X + YZ + Z

D. L = X, L = X + Y, L + Y = X + YZ

E. L = Y, L = X + Y, L + Y = X + YZ

F. L = Y, L + Y = YZ, L + Y = X + YZ

G. $L = Y, L = X + Y, L + Y = X + YZ, L = X + YZ, L = X + YZ + Z$

H. $L - Y, L + Y = YZ, L + Y = X + YZ, L = X + YZ, L = X + YZ + Z$

I. $L = Y, L + Y = YZ, L = YZ, L = X + YZ, L = X + YZ + Z$

J. $L = Y, L + Y = YZ, L = YZ, L = YZ + Z, L = X + YZ + Z$

K. $L = YZ, L = X + YZ, L = X + YZ + Z$

L. $L = YZ, L = YZ + Z, L = X + YZ + Z$

M. $L = Z, L = YZ + Z, L = X + YZ + Z$

N. $L = X, L = X + Z, L = X + YZ + Z$

With the construction of this list, the reading of equilibrium crystallisation paths for the system can be regarded as complete with the exception of the special case of binary compositions lying on the X–YZ join. These show the same behaviour as groups D and F but the last drop of liquid disappears just as the resorption of Y is complete in the reaction $L + Y = X + YZ$. Hence they solidify to two-phase assemblages not three-phase. Compositions in the binary joins bounding the system at no stage give rise to ternary liquids and in all cases solidify to two-phase mixtures with final liquids either at one of the three binary eutectics or at the peritectic at 1180 °C (compositions on the Y–YZ join).

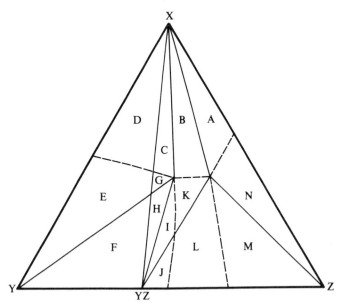

Figure 4.16   Classification of initial liquid compositions of Figure 4.14. Each labelled area shows a different sequence of reactions.

Equilibrium melting paths show the sequence of equilibrium crystallisation reactions in reverse. In any problem involving melting it is essential to start by determining the nature of the assemblage which is being melted. Thus for example we know that composition *a* of Figure 4.14 consists of X + YZ +Z in the sub-solidus. The only liquid in the system which equilibrates with these phases is the eutectic liquid E. Therefore E is the liquid produced on initial melting in the univariant reaction. Next, we know that one solid phase must disappear before melting can proceed and that the liquid will then migrate up one of the boundary curves. A simple test using an isothermal section for a temperature a little above that of the univariant reaction is then applied (see Fig. 4.15c). Since the bulk composition lies in the X + Z + L triangle it is evident that YZ must have been the first solid phase to disappear, and that the liquid must migrate up the boundary curve between the X and Z fields. The liquid will move along the boundary curve until the liquid composition L, the bulk composition, and one of the solid phase composition points become co-linear. This is the point at which the other solid has become completely melted, and the point at which the liquid leaves the boundary curve to enter the primary phase field of the remaining phase. The liquid then tracks towards the bulk composition point and when it reaches it the charge is completely melted.

The composition *c* shows more complex behaviour which involves the *up-temperature* crystallisation of Y while the liquid is on the boundary curve between the YZ and Y fields. Whatever the complexities encountered, they can be solved by the careful inspection of partial isothermal sections for a succession of temperatures.

A binary composition such as *d* (Fig. 4.14), lying on the X–YZ join presents a special problem as a melting study. Since it consists of X + YZ in the solidus it follows that the first liquid of melting must lie somewhere between P and E, for these are the only liquids capable of equilibrating with the solids concerned. However, it is not immediately apparent whether the correct liquid is P or E or something in between. One solution to this sort of problem is work out the equilibrium crystallisation sequence, covered above, and then reverse it. However, a more interesting argument goes as follows: a *finite* amount of liquid lying somewhere between P and E can only be produced *if* a phase on the X-poor side of the X–YZ join is simultaneously formed. Then the bulk composition can lie within a triangle with liquid as one apex. Y is the only phase which satisfies this requirement and thus the only possible three-phase triangle is L–Y–X which represents the condition of the system as soon as one phase (YZ) has melted. However, from this it follows that X, YZ and Y must co-exist during initial melting. Therefore the initial liquid must be P.

Fractional crystallisation paths

Once the nature of the various divariant and univariant equilibria in a ternary *without solid solution* is understood the construction of liquid fractionation paths is simple and a unique set of paths may be constructed for the whole system. The rules to be applied are as follows:

(a) In trivariant equilibria (primary phase fields), fractionation paths form a set of straight lines radiating from the composition point of the phase concerned.
(b) Fractionation paths follow divariant boundary curves if they are co-precipitational. Fractionation paths cross resorptional boundary curves and change direction to move away from the composition point of the new primary phase. Occasionally boundary curves show sufficient curvature to change their nature from co-precipitational to resorptional or vice versa at some stage along their length. Fractionation paths change accordingly.
(c) Fractionation paths terminate at co-precipitational univariant points (eutectics) but pass through mono-resorption points and continue down the down-temperature boundary curve. At bi-resorption points fractionating liquids pass directly into the primary phase field of the phase not undergoing down-temperature resorption.

These rules are all based on a consideration of the phases which are precipitated from the liquid with cooling. Phases which undergo down-temperature resorption during equilibrium crystallisation have no influence on fractionation paths. Fractional and equilibrium liquid paths only coincide in the absence of such resorption.

Figure 4.17 shows fractionation paths for the system discussed in Figures 4.14–4.16. Note that all fractionation paths lead to the eutectic E and any that reach P pass through it and continue along PE. The boundary curve between the Y + L and YZ + L fields is shown by a dotted line. This is not a fractionation path but marks the curve along which fractionation paths emanating from the Y + L field change direction to move away from YZ.

Figure 4.18 shows the pattern of fractionation paths in the region of the sapphirine field of $MgO–Al_2O_3–SiO_2$ (Fig. 4.10) if solid solution in the phases is ignored. The divariant equilibria Sap + Sp + L, Sap + Mull + L and Mull + Cord + L (*bd* and its extension) are all resorptional, for spinel, mullite, and mullite, respectively. Fractionating liquids approaching across the spinel and mullite fields therefore cross these curves, changing direction to move away from either the sapphirine or the cordierite point. Incidentally there is a thermal maximum

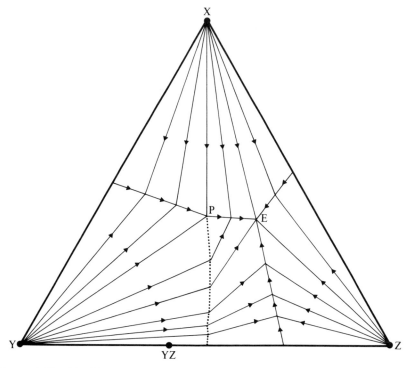

Figure 4.17   Liquid paths of fractional crystallisation in the system illustrated in Figures 4.14–4.16.

on Mull + Cord + L at *d* where the extension of the Mull–Cord join crosses the boundary curve. Inspection of Figure 4.10 shows that all liquids lying on the silica-rich side of the join must ultimately fractionate to the Tr + En + Cord eutectic whereas those on the silica-poor side will eventually fractionate towards Fo + En + Cord.

A special fractionation path *ae* across the sapphirine field is the route taken by *all* liquids which have previously reached the co-precipitational Mull + Sp + L boundary by earlier fractionation of spinel, corundum, and mullite.

Pseudobinary systems

The join X–YZ in the system illustrated in Figures 4.14–4.17 is an example of a pseudobinary system. If mixtures of X and Y were made up and their phase relations presented as a *T*−*X* diagram the result would be as shown in Figure 4.19 which is constructed from Figure 4.14

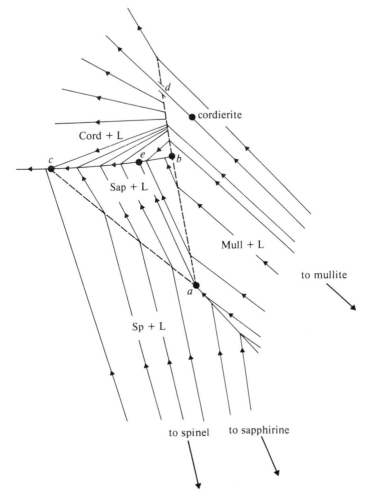

Figure 4.18 Fractionation paths in the vicinity of the sapphirine + L field of MgO–Al$_2$O$_3$–SiO$_2$. Solid solution is neglected.

by reading the equilibrium crystallisation history of every composition on the join. For example a liquid of pure YZ composition begins to crystallise at 1225 °C and continues to precipitate crystals of Y until it reaches the binary peritectic at 1180 °C. It then becomes univariant, YZ begins to appear and Y reacts with the liquid. The last drop of liquid and the last Y disappear simultaneously and the charge solidifies to YZ alone. As another example the critical composition *e* at 34% X begins to crystallise Y at 1110 °C and the residual liquid moves directly to P where it solidifies to X + YZ as the reaction L + Y = X + YZ takes

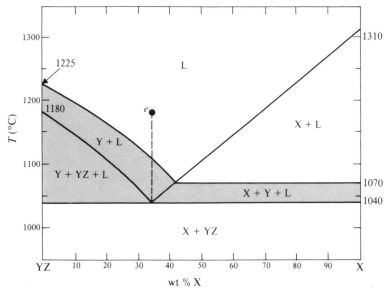

Figure 4.19 The pseudobinary $T-X$ diagram for the join YZ–X of Figure 4.14. Shaded areas are non-binary.

place. Slightly more X-rich compositions than *e* have residual liquids which reach P via the X + Y + L boundary curve, while less X-rich compositions reach P via the Y + YZ + L boundary. These do not resorb all their Y in this reaction and the residual Y is resorbed at P. Compositions starting in the X field produce residual liquid which all reach P having met the X + Y + L curve at a temperature of 1070 °C.

This system is called a pseudobinary because of the presence of phase Y which cannot be expressed in terms of the components X and YZ and clearly lies off the join. This particular pseudobinary is part of a ternary system, though this is not a necessary part of the definition. Pseudobinaries are two-end-member joins across any system of more than two components.

In all fields in the pseudobinary which include the phase Y the liquid must lie off the join on the side opposing Y (such areas are shaded in Fig. 4.19). Hence a pseudobinary cannot be read like a binary for these areas, though it may be truly binary in other parts (non-shaded areas). However, the experimental determination of pseudobinary joins is of immense practical importance because it allows the intelligent exploration of more complex systems. Suppose in the present case that the only information about the ternary X–Y–Z consisted of the pseudobinary discussed plus an experimental determination of the binary X–Y. The boundary curve X–Y–L would then be located at the

1135 °C point of Figure 4.14 and at the point (1070 °C) where it crosses the X–YZ join. Moreover the pseudobinary shows that this curve leads to a resorptional equilibrium for Y at 1040 °C. The assumption that the boundary curve was linear, and the extrapolation of the two known liquidus temperatures on the curve would give an approximate location for P. However, we also know that a line drawn from the Y composition point through the critical composition *e* on the pseudobinary (Fig. 4.19) also goes directly to P. Hence the position of P is quite closely established. An extrapolation of the two known liquidus temperatures (melting point of Y, and 1225 °C liquidus of YZ) on the Y–Z join would also lead to an approximate location of the 1180 °C peritectic. Hence the position of the X–YZ–L boundary curve could be reasonably fixed at either end, as well as likely liquidus temperatures along it. This is in total quite an impressive amount of information about the ternary derived from experiments in only two joins and is a small illustration of the inherent elegance of experimental petrology.

Some ternary systems of geological interest

Liquidus projections for 1 atmosphere pressure are given for the systems forsterite–anorthite–silica, forsterite–diopside–anorthite, diopside–leucite–silica, and anorthite–leucite–silica in Figures 4.20–4.23. The system forsterite–anorthite–silica is divided into a silica-saturated part (forsterite–enstatite–anorthite) and a silica-oversaturated part (enstatite–anorthite–silica). The incongruent melting of enstatite in $MgO–SiO_2$ persists into anorthite-bearing compositions so that solid silica-saturated assemblages of forsterite + enstatite + anorthite melt to silica-oversaturated liquids at P. Likewise, since the boundary curve between the forsterite and enstatite fields is resorptional for olivine throughout its length, liquids originating in the forsterite field fractionate to silica-over-saturated compositions. The system extends beyond the forsterite–anorthite join into silica-undersaturated space (part of the larger $CaO–MgO–Al_2O_3–SiO_2$ system) but there are no fractionation paths through this join from the silica-rich side. Since liquids in this system lying fairly close to the forsterite–anorthite join bear some compositional similarity to natural basalts this is an indication that natural magmas may behave in the same way, that is, silica-saturated basaltic liquids may fractionate towards silica-oversaturated products at 1 atmosphere but are separated from silica-undersaturated liquids by a thermal divide. As we shall see in a later section, however, the effect of pressure on such systems is strong and the divide ultimately breaks down.

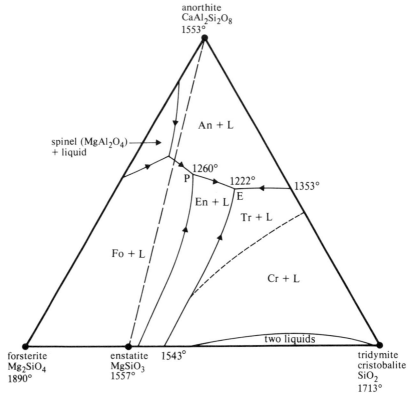

Figure 4.20   Liquidus projection for the system forsterite–anorthite–silica at 1 atmosphere pressure (after Andersen 1915).

The leucite-bearing systems of Figures 4.22 and 4.23 show divisions into silica-undersaturated and silica-oversaturated compositions. Silica saturation in these cases is shown only by the one-dimensional diopside–K-feldspar and anorthite–K-feldspar joins. Passage from silica-undersaturated to silica-oversaturated liquids is possible because of the incongruent melting of K-feldspar and the penetration of the leucite field across the K-feldspar–diopside and K-feldspar–anorthite joins. However, this is of little petrological importance because only modest water pressures are needed to cause a substantial contraction of the leucite field, and natural magmas rich in potential leucite are in any case rare.

The system forsterite–anorthite–diopside simulates basaltic compositions reasonably closely and represents the plane separating silica saturated and undersaturated compositions. It is sometimes known as the plane of critical undersaturation within the larger system $CaO–MgO–Al_2O_3–SiO_2$.

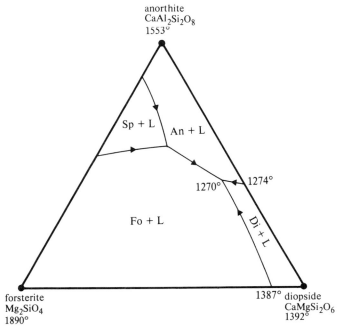

Figure 4.21 Liquidus projection for the system forsterite–diopside–anorthite at 1 atmosphere pressure (after Osborn & Tait 1952). The spinel + L field is non-ternary.

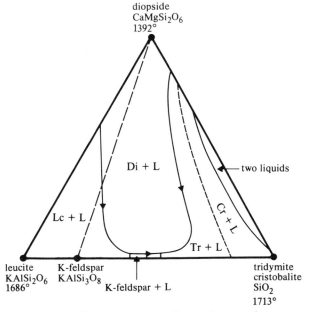

Figure 4.22 The system diopside–leucite–silica at 1 atmosphere pressure (after Schairer, 1938).

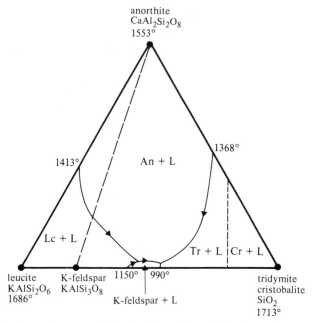

Figure 4.23 The system anorthite–leucite–silica at 1 atmosphere pressure (after Schairer & Bowen 1947).

Relative solubilities of minerals. The size of a primary phase field is an indication of the solubility of the phase in melts within the system. Large fields generally signify insolubility because in parts of the field remote from the composition point of the phase only relatively small concentrations of the necessary components are sufficient to cause precipitation. Conversely small fields generally signify that a phase does not precipitate until its concentration in the melt is high. In Figures 4.20 and 4.21 we see that forsterite is relatively insoluble compared with anorthite and diopside while the latter two have approximately equal solubilities. However, Figures 4.22 and 4.23 show that both diopside and anorthite are very insoluble relative to K-feldspar. Fractional crystallisation and low degrees of melting tend to produce liquids with high concentrations of soluble constituents, of which K-feldspar is one of the most important, accompanied in natural magmas by albite, its sodium analogue. Many high-temperature magmas rich in potential olivine, pyroxenes, and anorthite will tend to evolve by fractionation of these phases towards relatively alkali feldspar-rich compositions.

The effect of additional components. Ternary diagrams provide a good opportunity to study the effects of the addition of a third component to a binary. In ideal solutions the factor which determines whether a phase will crystallise is the concentration in the melt of the constituents needed. In a previous section we saw that the crystallisation of a given olivine composition depends on the molecular ratio of MgO + FeO in the melt, while the temperature at which it crystallises depends upon the concentration of these constituents. Most magmatic solutions are, however, far from ideal and instead of concentrations it is necessary to speak of the *activities* of components, which depend on the structure of the melt and the extent to which components are polymerised into more complex molecular arrangements within it. Thus in non-ideal solutions two factors come into play. On the one hand the addition of the extra component reduces the concentrations of the original components so that temperatures of crystallisation will tend to be reduced. On the other hand the activities of the original components may be altered and relative solubilities may become reversed.

To elaborate this argument, consider a composition lying on the mid-point of the diopside–silica join in Figure 4.22. In the binary, at temperatures below that of the immiscibility gap in the liquid field, this composition crystallises cristobalite. Adding about 15% of leucite, however, has the effect, because of a drop in the crystallisation temperature, of bringing the bulk composition into the tridymite field. When 20% or more of leucite is added, however, diopside comes onto the liquidus. Thus, although the diopside–silica ratio is exactly the same as in the original charge, crystallisation now produces diopside instead of a silica mineral. In other words, adding potassium and aluminium to the melt has reduced the activity of silica in the melt, thus suppressing the crystallisation of the silica mineral. Perhaps not surprisingly the exactly equivalent effect of adding leucite to anorthite–silica mixtures can be seen in Figure 4.23. The result is not quite so striking but again the crystallisation of the silica mineral is suppressed.

The significance of the *directions* of divariant boundary curves in ternary liquidus projections can now be elaborated. If the additional component has *no* effect on the activities of existing components or affects them equally, then a field boundary in the binary should extend into the ternary in a direction running straight towards the composition point of the third component. This is the line which retains a constant ratio of the existing components. The boundary between the forsterite and diopside fields in forsterite–diopside–anorthite (Fig. 4.21) provides a good example of behaviour approaching this ideal. The forsterite–enstatite field boundary in Figure 4.20 exhibits comparable behaviour. The boundaries of the diopside field with tridymite and leucite in Figure 4.22 and of anorthite with tridymite and leucite in Figure 4.23 provide fine examples of the converse behaviour, where the effects of the extra component are strong.

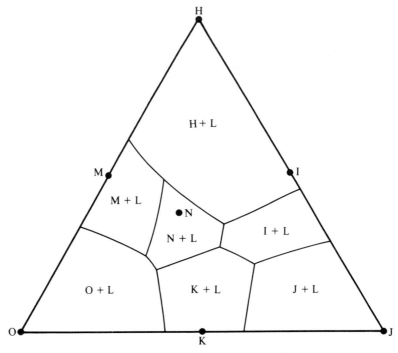

Figure 4.24   The system H–O–J.

Exercises

1. Figure 4.24 is a liquidus projection of a ternary system H–O–J which includes the binary compounds M, I and K, and the ternary compound N. Insert all the stable two-phase sub-solidus assemblages (Alkemade lines) and hence mark temperature arrows on all divariant equilibria including those in the bounding binaries. Show thermal maxima on all equilibria. Determine which of the divariant equilibria are resorptional down-temperature and identify the resorbed phase. Carry out the same exercise for the univariant equilibria and identify the resorbed phase or phases.

2. A ternary liquidus projection for the system A–B–C is given in Figure 4.25.
   (a) Draw a complete isothermal section for 1050 °C.
   (b) Identify the range of initial liquid compositions which shows complete resorption of early-formed A followed by re-precipitation of A.
   (c) Plot the composition $A_{56}B_{34}C_{10}$. What is the mineralogy of this charge immediately below the solidus? What is the liquid composition, the composition of the residual solid material, and the temperature, when this charge is 50% molten?

3. Sketch the pseudobinary $T–X$ diagram for the join enstatite–spinel in $MgO–Al_2O_3–SiO_2$ (Fig. 4.10).

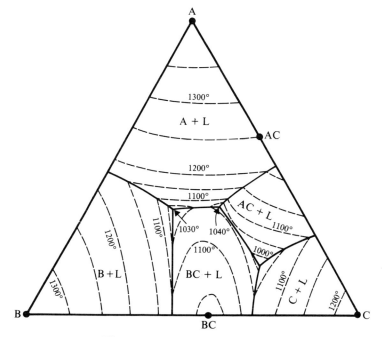

Figure 4.25 The system A–B–C.

5

Ternary systems with solid solutions

The existence of solid solutions in ternary systems complicates diagram reading substantially for two reasons. Firstly liquid paths in primary phase fields are no longer rectilinear, and secondly the range of compositions capable of crystallising to two-phase assemblages is much extended. Ternaries of this sort cannot be read precisely unless a considerable amount of experimental information is available concerning the compositions of co-existing phases. As a hypothetical example we consider in Figure 5.1 a system which shows complete solid solution between two components X and Y, while the third component Z shows negligible solid solution. As a result the liquidus projection consists of only two primary phase fields, one for the pure component Z and the other for XY solid solutions (XY_{ss}). Systems which show this general relationship include the geologically important diopside–albite–anorthite and orthoclase–albite–silica. Note that systems of this type contain solid solutions involving two end-members, such solid solutions being termed binary. Ternary (three end-member) solid solutions are considered in a later section.

The diagram shows four experimentally determined tie-lines linking liquids L_1–L_4 to their equilibrium solids a–d. The X/Y ratio of the liquid is always, in this example, higher than that of the solid with which it is in equilibrium. Assuming this is true for all liquids in the XY_{ss} field it follows that the binary system XY is of the albite–anorthite type with X as the low melting component. Liquid L_3 is shown as the apex of a three-phase triangle linking it to the XY_{ss} of composition c and to pure Z, the two phases with which it is in equilibrium. Applying the Alkemade theorem, the boundary curve between the Z + L and XY_{ss} + L field must go down-temperature to the left, and furthermore the equilibrium between XY_{ss}, Z and liquid must be of the co-precipitational type. If the X/Y ratio of the liquid is always higher than that of

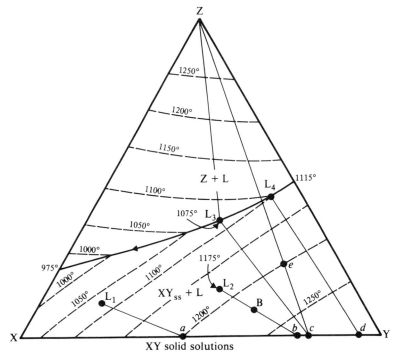

Figure 5.1   Liquidus projection of the hypothetical system X–Y–Z where solid solution between X and Y is complete. This system is closely analogous to diopside–albite–anorthite.

the accompanying XY_{ss} then the entire boundary curve must go down-temperature to the left, and since Z and XY_{ss} always lie on opposite sides of the curve it must be co-precipitational throughout its length.

Infinitesimal cooling of liquid L_3 results in the crystallisation of Z and the XY_{ss} of composition c. Hence a new liquid with a higher X/Y ratio is formed which must in turn equilibrate with an XY_{ss} of increased X content. Thus cooling involves the migration of a three-phase triangle with L moving down the boundary curve, c moving towards b, and the upper corner pivoting about Z. The line Z–XY_{ss} is thus capable of sweeping across the entire diagram and when a bulk composition is intersected by this line it becomes solid. Bulk composition e is an example of one which has just reached this condition and solidified to a mixture of Z and c, the final liquid of crystallisation being L_3. Figure 5.2 gives a complete isothermal section for the temperature of L_3 (1075 °C) and shows tie-lines across fields of one solid + L linking each liquid to its appropriate equilibrium solid. Liquids L_1 and L_3 with their equilibrium solids a and c are shown for reference purposes.

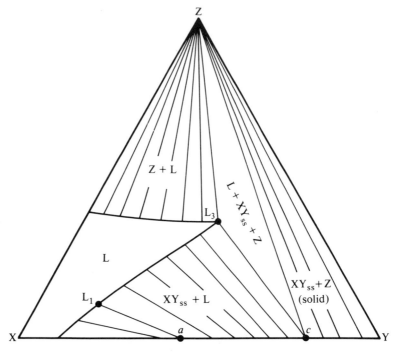

Figure 5.2   An isothermal section for the system shown in Figure 5.1.

Experimental determination of tie-lines. Direct chemical analysis using the electron microprobe of co-existing crystals and glasses is possible in some cases. However, it is often convenient to determine tie-lines from a knowledge of the composition of only one of the phases. If the liquid for example does not quench to a glass completely then a direct analysis may be difficult because of the presence of quench crystals. Hence the tie-line must be determined from a knowledge of the composition of the primary phase crystals only. The additional information needed in this case is a knowledge of the temperature behaviour of the liquidus surface.

In the system illustrated in Figure 5.1 suppose that a sufficient number of different bulk compositions has been studied so that the thermal contours for the liquidus surface in the $XY_{ss} + L$ field are known. Then suppose that a charge of bulk composition B quenched from 1175 °C is examined and the crystal composition found to be *b*, either by electron microprobe analysis or by X-ray diffraction if the dependence of lattice parameters on composition is known. Then since B, *b* and the liquid composition must be co-linear and since L must lie on the 1175 °C liquidus contour, the liquid composition, in this case L_2, can be found.

It must be emphasised that it is rarely possible to determine the phase *proportions* in experimental charges with any degree of reliability. Otherwise a knowledge of the bulk composition, the composition of the XY_{ss} crystals, and their proportion in the charge, would be sufficient to determine the liquid composition.

For the determination of three-phase triangles such as L_3–c–Z in Figure 5.1, the same liquidus temperature method can be used as for two-phase tie-lines. The composition L_3 is determined by knowing in advance the position of the boundary curve and the distribution of temperatures along it. In this case any charge with bulk composition inside the triangle quenched from 1075 °C will contain the liquid L_3 co-existing with Z and XY_{ss} of composition c.

Fractionation paths. If sufficient tie-line information is available for the primary phase field of a solid solution it becomes possible to construct a set of fractionation paths across the field, which by analogy with systems discussed earlier are unique and do not cross each other. A knowledge of the pattern of fractionation paths then allows equilibrium paths to be read (see next section). In practice few systems are known adequately, but the principle is important. Hamilton and MacKenzie (1965) discuss fractionation paths in some detail.

The tie-lines given in Figure 5.1 tell us the composition of the solids in equilibrium with each of the liquids. When a liquid such as L_1 is cooled from above the liquidus temperature the first crystals to appear have composition a. Therefore an infinitesimal degree of cooling below the liquidus will result in the liquid composition evolving directly away from a. For the compositional point L_1 the direction of the tie-line thus gives an infinitesimally short part of a fractionation path.

In Figure 5.3a data are given in the form of short arrows for a number of different compositions within the $XY_{ss} + L$ field. Each arrow is plotted at the appropriate liquid composition point and points directly away from the composition of its equilibrium solid. There is no easy way except by inspection of determining how this set of short arrows should be joined up to make specific fractionation paths. However, the results of sketching in a set of curves which as far as possible remain parallel to the arrows at all points in the field is given in Figure 5.3b. With this set of curves it is possible to predict equilibrium crystallisation paths and these can in principle be checked experimentally so that a test of the accuracy of the fractionation paths does exist.

The fractionation paths can be used to predict the sequence of fractional crystallisation as illustrated in Figure 5.3b. A bulk composition A crystallises the solid a initially (the tangent to the fractionation curve at A cuts X–Y at a) and perfect fractionation leads to the evolution of the

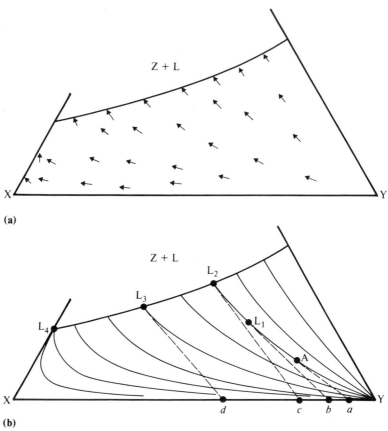

Figure 5.3 (a) Arrows showing directions of infinitesimally short segments of fractionation paths in the $XY_{ss} + L$ field of Figure 5.1. (b) The arrows of (a) are used to sketch a set of fractionation curves for the whole field.

liquid along the fractionation path until it reaches the boundary curve at L_2. The fractionating solids at any stage are found by projecting the tangent to the fractionation curves back to the XY_{ss} join. For example, when the liquid has composition L_1 the solid fractionating has composition b. When the liquid first reaches the boundary curve the fractionating solid is c. Subsequently the liquid evolves down the boundary curve fractionating both XY_{ss} and Z. The composition of fractionating XY_{ss} at any stage is found by projecting the tangent from the fractionation curve which reaches the boundary curve at the liquid composition point. For example, when the liquid has composition L_3 the fractionating solids are Z and d. The final liquid of fractionation, which is produced only in an infinitesimally small amount is L_4 while the final fractionating solids are Z and pure X.

Equilibrium crystallisation paths. During equilibrium crystallisation of composition in the $XY_{ss} + L$ field the liquid composition, the bulk composition, and the composition point of the solid remain co-linear. A curved liquid path results and is related to tie-lines as shown in Figure 5.4. Here the bulk composition A while above the liquidus temperature consists of liquid L_1. Crystallisation commences with the precipitation of solid a. As crystallisation proceeds the solid reacts with the liquid and is made over to the more X-rich compositions, b, c, and d. At all stages the solid co-exists with a liquid lying on the line projected through A, for example at a particular temperature the system consists of $L_2 + b$, at a lower temperature of $L_3 + c$, and at a still lower temperature of $L_4 + d$. During cooling the length of the tie-line L–A increases continuously while the tie-line from A to the solid composition becomes shorter. Measurement of these lengths shows in the present example that when the system consists of $L_4 + d$ it is 90% solid.

The above discussion illustrates the principle of the tie-line which pivots on the bulk composition so leading to a curved liquid path. Such a path can be determined experimentally for a specific composition by carrying out runs at a number of temperatures and analysing the co-existing phases (or calculating the liquid composition by the liquidus temperature method outlined earlier). However, the path can also be read from the phase diagram providing that sufficient information on fractionation curves is available.

The principle of reading the liquid path depends upon finding all the tangents to fractionation curves which project back through the bulk composition point. The path of the equilibrium liquid is then interpolated between the bulk composition point and the points where

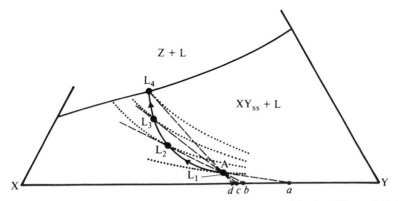

Figure 5.4 Illustration of equilibrium crystallisation path in the $XY_{ss} + L$ field of Figure 5.3. Dotted lines are fractionation curves. Dashed lines are tangents to fractionation curves at points labelled L.

appropriate tangents meet the fractionation curves. In this way the liquid path always fulfils the condition that liquid, bulk composition, and equilibrium solid are co-linear.

Equilibrium melting. Equilibrium melting liquid paths are of course the reverse of those of equilibrium crystallisation. In Figure 5.5 the bulk composition A consists below the solidus of a mixture of pure Z and an XY_{ss} of composition a. The first liquid on melting must therefore lie on the boundary curve between the $Z + L$ and $XY_{ss} + L$ fields at L_1 where the tangent to the fractionation curve which meets the boundary curve at L_1 projects across to a. Melting proceeds with the migration of the liquid up the boundary curve and the making over of a to more Y-rich compositions as Z and XY_{ss} melt together. When the liquid reaches L_2 the tangent to the fractionation curve which meets the boundary curve at this point back projects to b, passing through the bulk composition point A. This signifies that melting of Z is complete and the liquid now leaves the boundary curve and takes a curved path across the $XY_{ss} + L$ field to A as described in the previous section. The last solid to

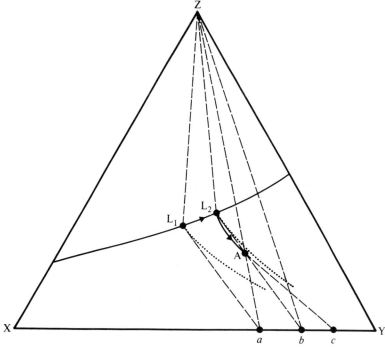

Figure 5.5   Illustration of liquid path during equilibrium melting for a bulk composition A lying in the $XY_{ss} + L$ field of the system illustrated in Figures 5.1–5.4. Symbols as in Figure 5.4.

melt is *c*, found by projecting the tangent to the fractionation curve which passes through A. In the figure this fractionation curve has been omitted for clarity but the tangent A–*c* is shown. Phase proportions can be determined by the lever rule and show for example that the system is 66% melted when the liquid leaves the boundary curve at L_2.

Fractional melting. Just as certain aspects of fractional melting in binary systems with solid solutions cannot readily be given quantitative treatment so the information available in a ternary of the type discussed above is inadequate for detailed prediction of behaviour during fractional melting. However, initial liquids of fractional melting are the same as those of equilibrium melting and are hence readily identified. In general, compositions lying in the Z + L field will give final liquids which have the composition of pure Z while other compositions will produce liquids evolving to pure Y. The latter compositions, those lying initially in the XY_{ss} + L field, will produce liquids which evolve along the boundary curve until Z has finished melting, and then cross the XY_{ss} + L field until the co-existing refractory XY_{ss} has been made over to pure Y and the liquid itself reaches Y. The former group produces liquids which evolve along the boundary curve until they reach the eutectic on the Z–Y binary. At this point melting of XY_{ss} can go to completion with the melting of the last crystals of pure Y composition and the liquid is then free to evolve along the binary melting the remaining Z until it reaches the composition of Z itself.

Boundary curve minima. In the system discussed above, the boundary curve between the primary phase fields of the two phases shows no maximum or minimum. This arises because the X/Y ratio of the liquid on the boundary curve is always greater than that of the co-existing solid assemblage. In a system where this ratio can change from greater than that of the co-existing solids to less than that of the co-existing solids via an intermediate stage where the ratio in solids and liquid is the same, a thermal maximum or minimum develops on the boundary curve.

Figure 5.6 illustrates a case where a minimum is present in a binary showing complete solid solution (analogy – the alkali feldspar system, $NaAlSi_3O_8$–$KAlSi_3O_8$) and this behaviour persists into the ternary when another phase, not showing solid solution, is present. Note that for a minimum to be present on the boundary curve in the ternary a similar behaviour in the binary is not actually a prerequisite. As we have seen in the previous chapter it is not necessarily safe to predict the behaviour of a ternary from known binary relationships, nor is the reverse procedure safe. However, in the best known example of a system showing a

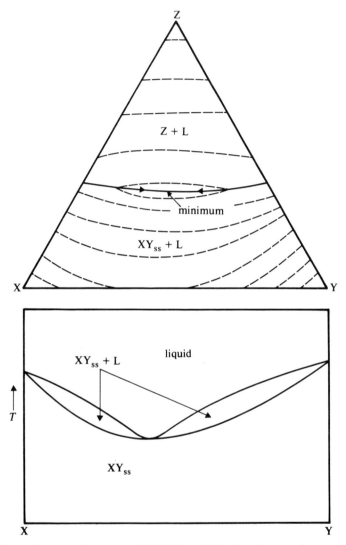

Figure 5.6 Hypothetical system X–Y–Z in which there is complete solid solution between X and Y and there is a minimum on the boundary curve between the Z + L and XY_{ss} + L fields. The corresponding binary for X–Y is shown below. The system forms an analogue for albite–orthoclase–silica.

boundary curve minimum (albite–orthoclase–silica) the minimum is also present in the binary.

Figure 5.7 shows isothermal sections for the system illustrated in Figure 5.6. Notice the two sets of three-phase triangles (L + XY_{ss} + Z) pointing towards each other indicating that X-rich liquids lying on the

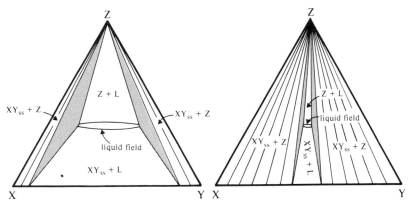

Figure 5.7 Isothermal sections for the system illustrated in Figure 5.6. High temperature (left), low temperature (right).

boundary curve equilibrate with XY_{ss} having a lower X/Y ratio than themselves, while Y-rich liquids on the boundary curve equilibrate with solids with higher X/Y ratios. At temperatures just above those of the minimum the three-phase triangles must clearly become very narrow and the minimum itself is generated at the point where the two sets of triangles close up and meet as a straight line. The unique liquid lying at the minimum on the boundary curve has a position such that it is co-linear with Z and the XY_{ss} with which it is in equilibrium.

Fractionation paths in the $XY_{ss} + L$ field must be qualitatively as shown in Figure 5.8, that is, consisting of two sets separated by a so-called 'unique fractionation curve' originating at the minimum on the X–Y join. Actually this curve is in a literal sense no more unique than any of the others. It is not necessarily straight, nor does it necessarily meet the boundary curve at the thermal minimum, the latter being located at the point where a specific fractionation curve meets the boundary curve while running in such a direction that the tangent at this point intersects Z (condition of co-linearity of Z, XY_{ss}, and equilibrium liquid). The reader is recommended papers by Osborn and Schairer (1941), Tuttle and Bowen (1958), and Hamilton and MacKenzie (1965) for further details of this topic.

Boundary curve maxima. Thermal maxima on boundary curves are generated in systems containing an intermediate compound which shows solid solution. They are analogous to maxima discussed in the previous chapter though their exact location depends upon tie-line information. Figure 5.9 shows a liquidus projection and isothermal section for such a ternary, and the binary for the join containing the intermediate compound. Note that although the thermal maximum *m* in

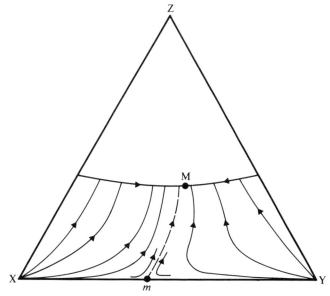

Figure 5.8   Sketch of form of fractionation curves for XY_{ss} + L field in system illustrated in Figures 5.6–5.7.

the binary coincides with the composition point of the pure intermediate compound AB, the maximum m' on the boundary curve P–Q does not necessarily lie exactly on the AB–C join. As in previous discussions its position is governed by the co-linearity of the liquid and the two co-existing solid phases.

Ternary solid solutions

Solid solutions discussed up to this point have been binary in that their compositions were expressible in terms of two end-members. Ranges of composition of such solutions are linear in ternary liquidus projections. However, many solid solution series are at least ternary (for example the feldspars) and hence solid solution ranges appear as areas in triangular composition projections. No new principles of diagram reading are necessary in the presence of ternary solid solutions, though in practice phase relations may be complex.

Figure 5.10 gives a diagrammatic example of an isothermal section of a system showing ternary solid solution in four phases. It is based on a figure by Biggar and Clarke (1971) which represents pyroxene phase relations in the quaternary system $CaO–MgO–FeO–SiO_2$ but nevertheless serves as an excellent illustration of the principles involved

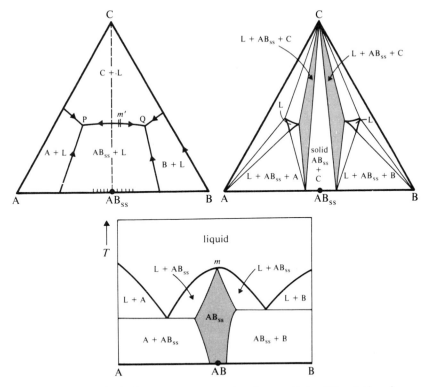

Figure 5.9 Hypothetical ternary system showing partial solid solution in an intermediate compound. The boundary curve between the C + L and AB_{ss} + L fields has a maximum m' which does not in general coincide with the join between C and pure AB.

in ternary solid solutions. Note the various fields representing the co-existence of three specific solid phases and the way in which the limits of solid solutions change discontinuously at each of these compositional points.

The system diopside–albite–anorthite

The equilibrium diagram for this system (Bowen, 1915) is given in Figure 5.11. Phase relations are analogous to those discussed at the beginning of this chapter (Figs 5.1–5.5).

The principal effect of adding diopside to the system albite–anorthite is to lower liquidus temperatures; but the effect of the additional component on the albite–anorthite ratios of co-existing liquids and plagioclase is not very large. Hence fractionation paths in the Pl_{ss} + L

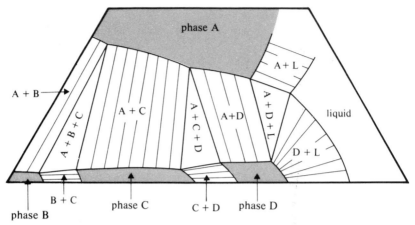

Figure 5.10   Isothermal section showing ternary solid solution fields (shaded). The diagram is based on Biggar and Clarke (1971), the original representing pyroxene phase relations projected into the pyroxene quadrilateral within the quaternary system $CaO–MgO–FeO–SiO_2$. The phases represented here as A–D are Ca-rich clinopyroxene, protoenstatite, orthopyroxene, and pigeonite, in the original.

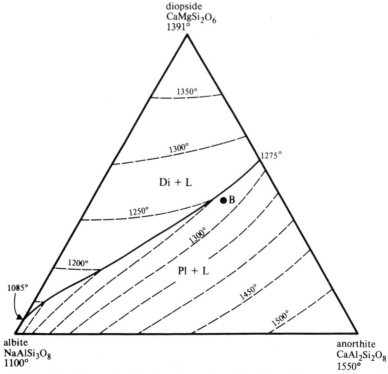

Figure 5.11   Liquidus projection for the system diopside–albite–anorthite at 1 atmosphere pressure (after Bowen, 1915).

field are qualitatively like those shown in Figure 5.3 and the boundary curve proceeds down-temperature all the way from the diopside–albite join. Liquids lying in the general vicinity of the point B of Figure 5.11 are simple analogues of natural basalts and fractionate to a trachyte-like end-point at the albite–diopside eutectic by the precipitation of diopside and a plagioclase with continually increasing albite content. The analogy between this behaviour and those of many natural volcanic series is clear.

The system nepheline–kalsilite–silica

The 1-atmosphere liquidus projection for this interesting system is given in Figure 5.12. Phases appearing on the liquidus are the silica polymorphs cristobalite and tridymite, their liquidus fields separated by an isothermal boundary at 1470 °C, the alkali feldspars, showing

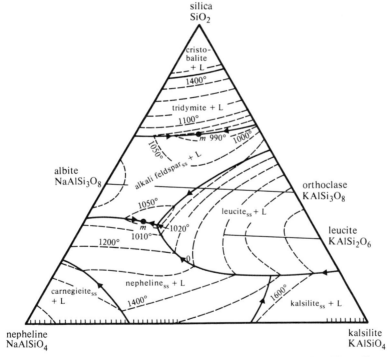

Figure 5.12 Liquidus projection for the system nepheline–kalsilite–silica at 1 atmosphere pressure (after Schairer 1957 and Fudali 1963). Ticks on base line represent approximate limits of nepheline and kalsilite solid solution. Solid solution is complete along the albite–orthoclase join and partial from leucite towards more sodic compositions (solid line). Boundary curve minima are labelled *m*.

complete solid solution between albite and orthoclase, leucite, showing extensive solid solution towards more sodic compositions, and nepheline and kalsilite showing extensive solid solution towards each other. The high-temperature polymorph of nepheline, carnegieite, also appears on the liquidus. Leucite has no sodium analogue in this system and the sodium mineral (jadeite) with the same formula, $NaAlSi_2O_6$ crystallises only under high pressures with a pyroxene structure. The zeolite analcite is compositionally similar to $NaAlSi_2O_6$ but requires OH groups to stabilise the structure and does not appear in the low pressure system in the absence of water. These facts explain why nepheline is the common sodic feldspathoid while its potassium analogue kalsilite is correspondingly rare. Only liquids more potassic than leucite crystallise kalsilite, and these are extremely rare in nature.

The system contains two minima, one on the alkali feldspar – tridymite boundary curve at 990 °C and the other on the nepheline – alkali feldspar boundary curve at 1010 °C. Compositionally these minima correspond to rhyolitic and phonolitic liquids respectively.

The alkali feldspar join, where it crosses the alkali-feldspar + L field constitutes a thermal divide and fractionation paths within the alkali feldspar field are shown diagrammatically in Figure 5.13. The alkali feldspar join can only be crossed by liquids which start on the silica-poor side of the join and then fractionate leucite. Such starting compositions are rare and in any event increasing pressure (see next section) severely reduces the size of the leucite field. Hence for most liquids the alkali feldspar join is a divide which clearly separates silica-undersaturated from silica-oversaturated liquids.

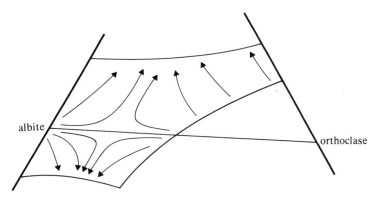

Figure 5.13　Sketch of fractionation curves for the alkali feldspar + L field of the system nepheline–kalsilite–silica (Fig. 5.12). (Partly after Hamilton and MacKenzie 1965.)

The albite–orthoclase join. Proper understanding of the system nepheline–kalsilite–silica is aided by a consideration of phase relations in the pseudobinary join albite–orthoclase and particularly the effects of water and confining pressure on these compositions. Reactions in the system are extremely sluggish in the absence of water and therefore most experimental work has been carried out with this component present. Additionally, the natural granitic and syenitic plutonic rocks for which the system forms a simple analogy clearly crystallise under a confining pressure in the presence of significant quantities of water. Experiments are most easily carried out with charges in sealed capsules containing enough water to saturate the phases present. After quenching, glasses are found to be vesicular indicating that a vapour phase was present during the run. Under these circumstances the vapour pressure is clearly the same as the total confining pressure, which is expressed as $P_{fluid} = P_{total}$ or $P_{H_2O} = P_{total}$ since the vapour phase is normally almost pure water. The solid phases in the absence of hydrous minerals such as amphibole and mica contain insignificant amounts of water in such experiments but the melt may contain substantial quantities of dissolved water. Hence water must be considered as an extra component and the system albite–orthoclase–H_2O is thus part of the quaternary system nepheline–kalsilite–silica–H_2O.

Phase relations for the pseudobinary at 1-atmosphere dry are given in Figure 5.14. Crystallisation of compositions in the leucite + L field such as P in Figure 5.14 is illustrated in Figure 5.15. As soon as leucite begins to crystallise the liquid leaves the binary albite–orthoclase join and takes a curving path across the leucite + L field until it meets the alkali feldspar – leucite boundary curve at Q. Alkali feldspar *b* now begins to crystallise and the equilibrium leucite *a* begins to resorb. Further cooling is accompanied by the making over of both alkali feldspar and leucite solid solutions to more sodic compositions. For example, at some lower temperature the liquid R co-exists with the solid phases *a'* and *b'*. During this stage the tie-line from feldspar to liquid approaches the bulk composition P continuously, indicating the continuing absorption of leucite. Eventually the liquid reaches the albite–orthoclase join at S and resorption of leucite is complete. The equilibrium feldspar at this stage has composition *c*. Further cooling involves only alkali feldspars and behaviour is thus truly binary from this stage on. Solidification takes place subsequently with the loss of liquid before it reaches the liquidus minimum at M.

Compositions more potassic than *c* in Figure 5.14 do not encounter the feldspar + L field during cooling. These solidify by the loss of the liquid and leucite simultaneously as the equilibrium alkali feldspar composition migrates along the alkali feldspar join and over-runs the bulk com-

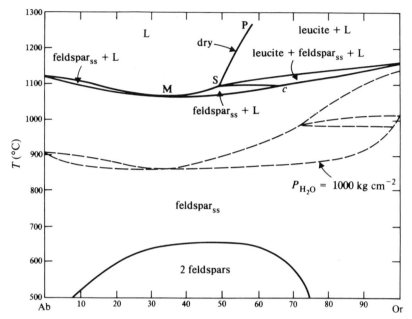

Figure 5.14   The system albite–orthoclase. Heavy lines refer to the dry system at 1 atmosphere, dashed lines to the water-bearing system at $P_{H_2O} = 1000$ kg cm^{-2}. All fields at the latter pressure include a vapour phase. (After Tuttle and Bowen 1958.)

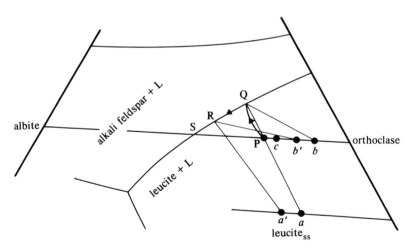

Figure 5.15   Illustration (not to scale) of the crystallisation of bulk compositions falling in the leucite + L field of the system nepheline–kalsilite–silica (Fig. 5.14).

position point, the last liquid still being on the feldspar–leucite boundary curve.

The effects of water pressure on the system are illustrated by Figures 5.14 and 5.16–5.18. Liquidus temperatures are dramatically lowered, the effect being due to the increasing solution of water in the melt with pressure. The leucite field is suppressed so that orthoclase melts congruently at 3 kbar. Meanwhile the effect on the solvus is only very slight since the solid phases take up very little of the extra component. At about 4 kbar solvus and solidus intersect. Below this pressure liquids solidify to a single feldspar which may exsolve to give a perthite or antiperthite at lower temperatures. Above this pressure liquids of appropriate composition may solidify directly to two feldspars, though these may themselves show additional exsolution at lower temperatures.

Liquidus projections for the 'granitic' part of the system nepheline–kalsilite–silica are given in Figure 5.17 (the system albite–orthoclase–silica–H_2O) and show the suppression of the leucite field as noted above. Additionally, the depression of liquidus temperature results in the disappearance of first cristobalite and then tridymite from the liquidus, and the appearance first of high quartz and then of low quartz. At the same time the liquidus field of the silica minerals slowly expands and the minimum melting composition moves to more sodic compositions and falls in temperature (see Fig. 5.18). As the

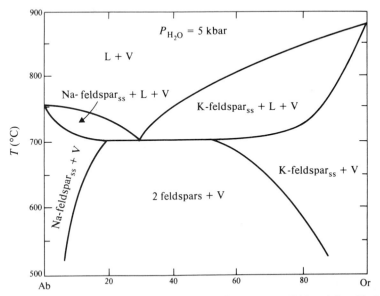

Figure 5.16   The system albite–orthoclase–H_2O at P_{H2O} = 5 kbar (after Yoder, Stewart and Smith 1957 and Morse 1970).

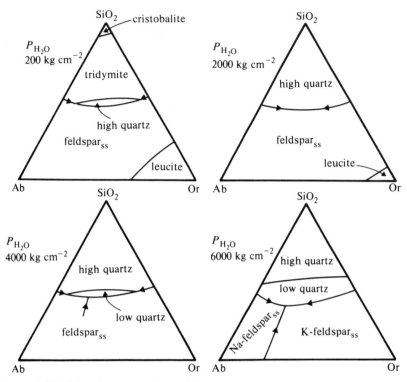

Figure 5.17   Liquidus projections of the system albite–orthoclase–silica–H_2O at different fluid pressures (after Tuttle and Bowen 1958).

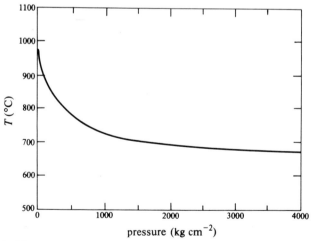

Figure 5.18   Temperature of liquidus minimum in the system albite–orthoclase–silica–H_2O as it varies with P_{H_2O} (after Tuttle and Bowen 1958).

liquidus minimum on the quartz–feldspar boundary curve drops below the temperature of the top of the solvus (at about 4 kbar) a boundary curve signifying a distinction between sodium-rich and potassium-rich alkali feldspar solid solutions appears. With increasing pressure this curve rapidly extends across the feldspar + liquid field so that at higher pressures the liquidus projection acquires the appearance of a normal eutectic diagram. The rather narrow pressure range round about 4 kbar where this boundary curve is incomplete shows an interesting crystallisation sequence for that very limited range of compositions which encounter the boundary, these being compositions lying on or very close to the unique fractionation curve leading from the minimum on the alkali feldspar join. At high temperatures these compositions crystallise a single feldspar with a composition lying somewhere near that of the minimum. When they reach the silica-poor end of the boundary curve the equilibrium feldspar must exsolve into two phases which differ from each other only slightly since the original feldspar has a composition rather close to that of the top of the solvus loop. Further cooling sees these feldspars become more disparate in composition as they move down opposite sides of the solvus while the liquid continues down the boundary curve to pick up quartz.

Some more detailed applications of the system albite–orthoclase–silica–H_2O to the study of granites will be given in Chapter 13. However, two important points deserve mention at this stage. Firstly it should be possible to make a distinction between those granites which contain one feldspar and those which contain two distinct feldspars and this distinction may be related to temperature of crystallisation and hence to pressure. Secondly because of the large effect of water on depressing liquidus temperatures, a rising water-saturated granite magma should show a strong tendency to crystallise as pressure drops allowing water to exsolve from the melt. This is the reason for the relative rarity of extrusive rhyolites. Acid magma approaching the surface will tend to devolatilise and solidify before eruption. In basic magmas, which are characteristically much drier, the effect is less important so that the extrusive habit is more characteristic.

Ternary feldspars

When the component anorthite is added to the system albite–orthoclase–H_2O complete solid solutions between albite and anorthite are created in addition to the already existing high-temperature solution of orthoclase and albite. Solid solution between anorthite and orthoclase is limited but nevertheless all solutions can be regarded as ternary rather

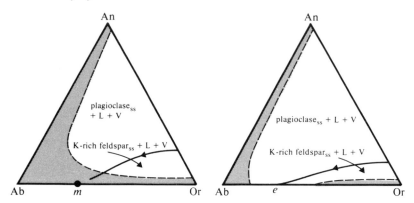

Figure 5.19   The system albite–anorthite–orthoclase–H_2O at ca. 2 kbar water pressure (left) and 5 kbar (right) based on Tuttle and Bowen (1958) and Yoder, Stewart and Smith (1957). A small leucite field is omitted in the first diagram. Shaded areas represent diagrammatically the limits of feldspar solid solutions co-existing with liquid. The point *m* is the minimum on the binary albite–orthoclase join, the pressure being low enough for solid solution still to be complete. The point *e* is the eutectic on this join generated at higher pressures when solidus and solvus intersect.

than binary. Liquidus diagrams showing approximate ranges of solid solutions are given in Figure 5.19 for fluid pressures of ca. 2 kbar and ca. 5 kbar. At the higher pressure a boundary curve separates the plagioclase + L and potassium-rich feldspar + L fields because liquidus temperatures are sufficiently depressed to keep them below the temperature of the top of the plagioclase–K-feldspar solvus. At the lower pressure, however, the boundary curve reaches a critical end point before it reaches the alkali feldspar join. This is analogous to the incomplete boundary curve between the two feldspar fields in the system albite–orthoclase–silica–H_2O at 4 kbar discussed in an earlier section (Fig. 5.17). However, in the present case high-temperature liquids co-exist with two discrete feldspars and low-temperature liquids, after the boundary curve has terminated, co-exist with only one, a reversal of the case previously discussed. A series of three-phase triangles linking co-existing feldspars and liquids is given diagrammatically in Figure 5.20 for the high-temperature (low-pressure) case. It illustrates two important points. Firstly, as temperature falls the co-existing feldspars approach each other in composition until at the point P they converge. The liquid co-existing with P lies at the termination of the boundary curve. At a lower temperature single feldspars such as *a* co-exist with liquids such as *b* which are no longer on a boundary curve. Secondly towards the low temperature side of the diagram the alkali feldspar solid

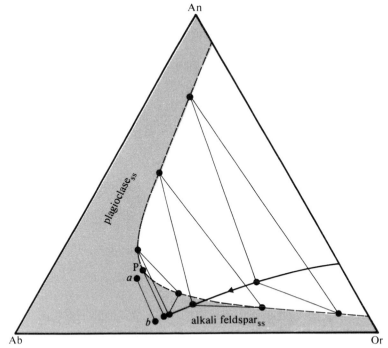

Figure 5.20 Illustration of crystallisation in the system albite–anorthite–orthoclase–H$_2$O (not to scale). Triangles represent liquid co-existing with two feldspars. P is the special case where the two feldspars converge in composition and co-exist with the liquid at the termination of the boundary curve.

solutions pass to the sodium-rich side of the boundary curve so that the equilibrium becomes resorptional down-temperature for plagioclase.

In Figure 5.21 the fractionation path of a liquid B analogous to relatively dry basalt is considered. Initially fractionation of calcic plagioclase (e.g. *a*) will give rise to a curving line of residual liquid evolution crossing the plagioclase + L field and meeting the boundary curve with the potassium-rich feldspar field at P. The fractionating plagioclase at this stage (*b*) is intermediate to somewhat sodic in composition while the liquid has become impoverished in the anorthite component and is trachytic. If the alkali feldspar which now begins to fractionate lies on the potassium-rich side of the boundary curve (for example at *e*) the latter is still co-precipitational and the liquid will fractionate down it precipitating two feldspars. Ultimately however the potassic feldspar (*d*) will cross the boundary curve to the sodic side and the equilibrium will become resorptional for plagioclase. The plagioclase at this stage is

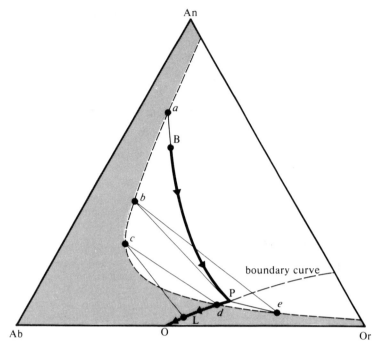

Figure 5.21   Illustration of fractional crystallisation of a 'basaltic' composition B in the system albite–anorthite–orthoclase–H₂O. The liquid line of descent is B–P (plagioclase fractionation), P–L (two feldspars fractionation), L–O (one alkali feldspar fractionation). The case illustrated refers to low water pressures (cf. Fig. 5.19).

Table 5.1   Run data at 1000 °C for the system A−B−C (see Exercise 1)

Composition	Phases present	Composition	Phases present
$A_{80}B_{15}C_5$	A^*	$A_{25}B_{60}C_{15}$	$B + L$
$A_{80}B_{10}C_{10}$	A	$A_{20}B_{40}C_{40}$	$C + L$
$A_{70}B_{25}C_5$	$A + B + L$	$A_{10}B_{85}C_5$	$A + B + L$
$A_{70}B_{15}C_{15}$	$A + C + L$	$A_{15}B_{60}C_{25}$	$B + C + L$
$A_{60}B_{20}C_{20}$	$A + C + L$	$A_{10}B_{45}C_{45}$	$B + C + L$
$A_{50}B_{35}C_{15}$	$A + L$	$A_5B_{40}C_{55}$	$B + C$
$A_{50}B_{25}C_{25}$	$A + L$	$A_5B_{30}C_{65}$	C
$A_{50}B_{15}C_{35}$	$A + C$	$A_{10}B_{15}C_{75}$	C
$A_{40}B_{30}C_{30}$	$A + L$	$A_{15}B_{20}C_{65}$	$A + C + L$
$A_{35}B_{45}C_{20}$	$A + B + L$	$A_{10}B_{35}C_{55}$	$C + L$
$A_{30}B_{35}C_{35}$	$C + L$	$A_{20}B_{25}C_{55}$	$C + L$

*A and C signify solid solutions of these phases, not pure phases.

likely to be a potassic oligoclase (*c*) while the other feldspar is an anorthoclase or sanidine (*d*). The fractionating liquid then leaves the boundary curve and takes a curved path fractionating one feldspar only until it
reaches the alkali feldspar join. The behaviour described is typical of
rather dry volcanic series evolving from basalt to trachyte, though many
other patterns are obviously possible, depending on the starting composition and the pressure at which fractionation takes place.

Exercises

1. Table 5.1 gives run data obtained at 1000 °C for a series of compositions in
   the ternary system A–B–C. Melting points of the pure phases A, B and C
   have been determined as 1360 °C, 1370 °C and 1490 °C, respectively. Experiments carried out in the three binary joins limiting the ternary show that A
   can dissolve up to 24% of B and 15% of C, B shows no detectable solid
   solution towards either A or C, and C can dissolve up to 37% of B and 15%
   of A. Binary eutectics are found at the following compositions and temperatures: $A_{45}B_{55}$ (1070 °C), $A_{52}C_{48}$ (1110 °C), and $B_{56}C_{44}$ (1125 °C).

   (a) Draw the isothermal section for 1000 °C.
   (b) Make a sketch of the liquidus projection showing thermal contours and
       probable maximum extent of ternary solid solution fields.

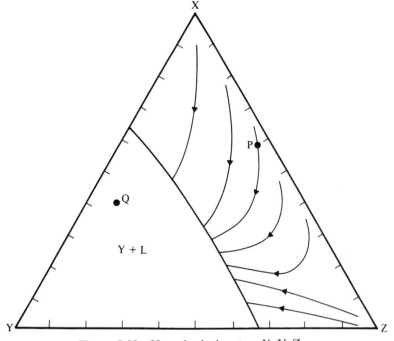

Figure 5.22 Hypothetical system X–Y–Z.

2. Figure 5.22 gives a liquidus diagram for a ternary system in which there is complete solid solution between X and Z while Y is a pure phase. Fractionation curves are given for the $XZ_{ss} + L$ field.

   (a) What phases does Q consist of when solid (give compositions)?
   (b) What is the composition of the first liquid formed on melting of Q?
   (c) If liquid P is allowed to undergo equilibrium crystallisation what is the composition of the residual liquid when the phase Y first begins to crystallise?
   (d) How much liquid is present at this stage?
   (e) Is there a liquidus maximum or a liquidus minimum on the boundary curve separating the $Z + L$ and $XZ_{ss} + L$ fields? What is the composition of the liquid at this point?

6

The interpretation of two-element variation diagrams

The preceding chapters covering simple experimental systems give an indication of the ways in which natural magmas *might* be expected to evolve by processes involving liquid–solid fractionation during melting and crystallisation. This knowledge is part of the essential background to hypothesis formation. It enables us to make sensible initial guesses as to possible ways in which particular magmatic series may have formed. We also require a method of examining the *compositional data* of natural rocks which will enable the formulation of precise hypotheses and thus refine and test the initial ideas. However, this is only part of the problem and we shall see in later chapters how a consideration of the *petrography* of rocks and of *melting experiments* carried out on natural samples both contribute further to the refinement of hypotheses.

The subject of phase diagrams is returned to in Chapter 8 and readers who prefer an uninterrupted diet might care to omit Chapters 6 and 7 at this stage. In the present chapter we consider variation diagrams in which one chemical parameter is plotted against another. The Harker diagram (oxide versus SiO_2) discussed in Chapter 2 is a specific example, but more generally it is useful to plot many of the possible pairs of elements against each other. The single most important property of these diagrams is that the lever rule, familiar in the reading of phase diagrams, is directly applicable to them. Thus, just as in a ternary liquidus projection it is possible to calculate graphically the way in which a liquid composition changes as a particular mineral is removed from it, so it is equally possible in a two-element variation diagram. More generally such calculations will be referred to as **mixing calculations**.

The mixing calculation provides a graphical solution to such problems as the mixing of two magmas, or changes in liquid composition during crystallisation, partial melting, and contamination. Most calculations in which a material of known composition is added to, mixed with or removed from, another material of known composition can be solved by this means. The graphical method has dimensional limitations, but the principles may be extended into computer methods for the solution of more complex problems. Thorough familiarity with the graphical method forms an ideal basis for the understanding of computer-based methods, which, however, are not dealt with in this book.

The mixing calculation demands that the diagram is capable of showing points representing the compositions of both liquids and minerals. With regard to the latter it is important to stress that *all* possible minerals must be capable of representation in the diagram if the method is to have maximum flexibility and scope. This is the main reason for choosing two-element variation diagrams as the basis for the analysis of variation covered in this chapter. A variation diagram based on the Fe/Mg ratio, for example, while useful for some purposes cannot be used here. On a plot of Fe/Mg against, say, SiO_2, minerals such as olivines, pyroxenes, biotites, and amphiboles can all be represented by composition points, but there is no such point for plagioclase, since it has an Fe/Mg ratio which is for practical purposes indeterminate. Clearly *any* variation diagram in which one of the plotting parameters is a ratio suffers from this defect.

Principles of the mixing calculation

In Figure 6.1 the two chemical parameters A and B represent weight percentages of oxides or parts per million of trace elements, or any other

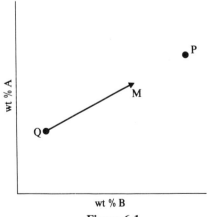

Figure 6.1

weight expression of analytical data. It is not necessary that both parameters are expressed in the same form. As a bulk composition P is added to composition Q the resultant mixture evolves along the straight line Q–P. For any specific mixture such as M the proportions of the two end-members P and Q are given by:

$$\frac{\text{Weight of P}}{\text{Weight of Q}} = \frac{\text{QM}}{\text{PM}}$$

and the percentage amounts of the end-members are:

$$\text{Weight per cent of P} = \frac{100 \text{ QM}}{\text{PQ}}$$
$$\text{Weight per cent of Q} = \frac{100 \text{ PM}}{\text{PQ}}$$

Note that this relationship only applies to two-element plots and does not hold true for diagrams where one of the parameters is a ratio. For example, apart from the disadvantages already noted, if a plot is made of, say, MgO against Al_2O_3/SiO_2 (the type of relationship used in the diagrams of Murata 1960) the locus of mixtures is in general curved, not linear (see also pp. 355–8).

Mixing calculations and the formulation of crystal fractionation hypotheses

Given a series of volcanic rocks showing a good chemical coherence, that is to say, comparatively little scatter, and clear trends in variation diagrams, a number of possible mechanisms of fractionation may come to mind. The petrological literature is, however, rich in descriptions of igneous rock sequences, particularly layered intrusions, where fractional crystallisation has clearly played an important, probably a dominating, role. Hence the chemical data from volcanic series may profitably be examined for evidence of crystal fractionation. If a hypothesis is erected on the basis of chemical data it must, however, be tested against petrographic or other evidence. As a simple example, if the chemical variation suggests that a particular series of lavas has evolved by the fractionation of olivine (e.g. by the progressive removal of olivine from a crystallising liquid), it is essential to demonstrate that the liquids concerned are actually capable of crystallising olivine. This test is usually a simple matter of petrographic examination (see Ch. 7) or may be subject to experimental study. Conversely, however, a hypothesis often emerges first from a consideration of petrography, and is then tested against chemical data. The interplay of these different lines of evidence in the hypothesis forming and hypothesis testing process cannot be over-emphasised. The complete hypothesis is internally consistent from all points of view. In the strict logical sense even such a hypothesis cannot be proved to be correct (a good hypothesis is one which has not been proved wrong *yet*) but many petrological hypotheses satisfactorily and simply explain the behaviour of a very large number of individual chemical elements and it is difficult actually to believe that there is some quite different explanation of the observed facts. Finally it must be stressed that, in view of the known importance of crystal–liquid fractionation processes, any series of normal igneous rocks should be examined first with such processes in mind. Only after crystal–liquid mechanisms have been rejected by the evidence, or found to be insufficient to explain all the evidence, is it reasonable to consider other processes such as pneumatolytic differentiation, diffusion etc. In practice, however, it seems very difficult to demonstrate that any substantial series of rocks has been generated by such processes because, even if they are operative, they must function against the heavy background 'noise' of normal crystal–liquid fractionation processes. The problem for example of detecting pneumatolytic transfer of alkali metals in a magma which is already undergoing crystal fractionation is obviously a difficult one.

In the discussion which follows it will be seen that crystal fractionation mechanisms can be very complex in natural systems, and we shall therefore approach the formulation of fractionation hypotheses from chemical evidence via a series of successive approximations. The reader may find it helpful to refer occasionally to Appendices 1 and 5 where representative rock and mineral analyses are given.

Figure 6.2 shows the predicted path of a liquid as it crystallises different numbers of mineral phases. Figure 6.2a is the straightforward case similar to that already discussed in the mixing calculation. The mineral plotting at P crystallises from the initial liquid B and the residual liquid therefore follows the path towards L. The distance of L from B will depend on the amount of crystallisation – in the case shown about 30% of the original liquid has crystallised as P. In Figure 6.2b two minerals P and Q simultaneously crystallise in such proportions that the composition of the total extract is at E. The liquid path in this case proceeds away from E towards L. The proportions of liquid to total solid at any stage are given by the ratio EB : LB while the proportions of the two crystalline phases are given by PE : QE.

In Figure 6.2c three minerals, P, Q and R crystallise in proportions such that the total solid has composition E (note that whatever the relative proportions of the three minerals, E must lie *inside* the triangle

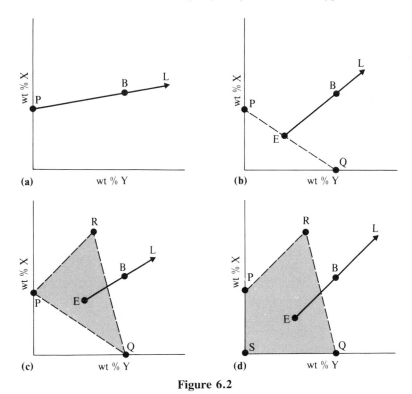

Figure 6.2

PQR). In Figure 6.2d four minerals, P, Q, R and S crystallise, in proportions given by E. Note again that E must lie inside the quadrilateral PQRS and as a consequence the liquid line of descent BL must always back-project through the same quadrilateral. In the general case any number of mineral composition points must enclose a polygon through which the liquid line must back-project. We term this the 'extract' polygon.

The term 'extract' has been used in inverted commas because, although the bulk composition of the total material crystallising is in fact extracted from the liquid in the models discussed, in other models this material may be *added*. For example in Figure 6.2a an initial liquid L could have crystals of P added to it to produce the magma B. The geometrical relations in the diagram are the same as when P is removed from B to give L. The term 'extract' will thus be used to mean the bulk composition of the solids throughout the following discussion, but it should be remembered that it does not imply a specific mechanism.

Let us now look at the problem from the other side. Given a known liquid path (which for the moment we assume is defined by the analyses of two rocks which we wish to relate) is it possible to determine the appropriate extract polygon? If it is, it constitutes the formulation of a crystal–liquid fractionation hypothesis to account for the liquid path.

Considering even the simple case shown in Figure 6.2a given that only the points B and L are known, it is immediately clear that no unique solution is to be expected, because any composition lying on the line BL projected in either direction is a possible solution. Petrographic assistance is required immediately, and the first test must clearly be to examine the plotted positions of likely minerals and mineral combinations relative to the line BL projected. However, even if, in a simple case such as that of Figure 6.2a, a likely mineral is seen to plot on the line (P in this case) it is by no means safe immediately to conclude that it is responsible for the fractionation. What has been demonstrated is only that the relationships of the *two* chemical parameters X and Y are consistent with this interpretation. All the chemical parameters must be similarly investigated before the mineral P may be taken as a near certainty. A convenient method of testing whether two rocks are related by the extraction of a single mineral is given in Figure 6.3 where a number of two-element plots are combined into one.

The complete testing of all element inter-relations is desirable in this type of study. In Figure 6.2d, for example, ambiguities are present in the single diagram of X against Y. Of the minerals shown, extract assemblages which are consistent with the orientation of BL are P + R + Q + S, P + R + Q, P + Q + S, R + S + Q, P + Q, R + Q and S + Q. For the solution of this type of problem the graphical technique of Figure 6.3 becomes cumbersome and it is better to keep the diagrams separate as in Figure 6.4. Here the case of two rocks B and L suspected of being related by an extract involving some or all of the minerals olivine, clinopyroxene and plagioclase is considered. In Figure 6.4a the possible extracts are seen to be Ol + Pl, Ol + Cpx or Ol + Cpx + Pl but not Pl + Cpx. Figure 6.4b excludes the possibility of Ol + Pl while Figure 6.4c excludes the possibility of Ol + Cpx + Pl. Therefore within the limits of the study (that is so far as the relations of CaO, MgO, Al_2O_3, SiO_2 and Na_2O are concerned) the hypothesis that the extract consists of Ol + Cpx is internally consistent. Notice also, an essential feature, that the proportions of olivine to clinopyroxene (approximately 50 : 50) are the same in each diagram, that is the line LB projected cuts the Ol–Cpx join at the same point each time.

As the number of phases participating in the extract increases, graphical methods become more difficult to apply. For example, even in the case of a three-phase extract the criterion that the mineral proportions implied are the same, becomes difficult to assess precisely. Semi-quantitatively, however, the method is still useful and may form a

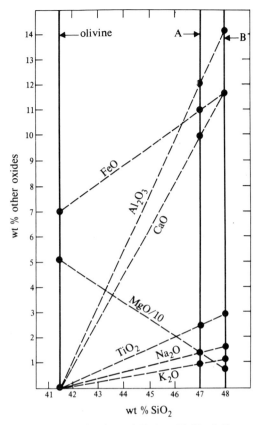

Figure 6.3 Oxides for two rocks A and B (see Table 6.1) are plotted against SiO$_2$ giving seven superimposed two-oxide plots. Back-projection to left for each line linking a particular oxide value in A and B shows that the two rocks are related by olivine control. Composition A can be converted to B by the removal of 15% of olivine.

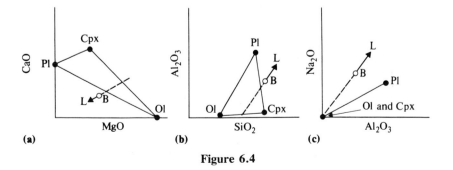

Figure 6.4

Table 6.1   Analyses of compositions used in extract calculation (Fig. 6.3)

	Olivine	Rock A	Rock B
SiO_2	41.5	47.0	48.0
TiO_2	–	2.5	3.0
Al_2O_3	–	12.0	14.2
FeO	7.0	11.0	11.7
MgO	51.0	15.0	8.5
CaO	–	10.0	11.8
Na_2O	–	1.5	1.8
K_2O	–	1.0	1.2

valuable part of initial studies. In Figure 6.5, for example, a case similar to that of Figure 6.4 is illustrated only here the extract consists of all three phases, olivine, clinopyroxene and plagioclase. The diagrams are clearly consistent with this hypothesis and show that the proportion of plagioclase in the extract must be comparatively high. This is a useful semi-quantitative conclusion. In fact, with the data given, a little consideration will show that the graphical procedure can be extended to calculate the mineral proportions more precisely (see later in this chapter) but in practice this is rarely worth doing.

Computer methods for solving complex mixing calculations have been devised by Bryan *et al.* (1969) and Wright and Doherty (1970). The essence of such calculations is the computation of a mixture of given minerals (the analyses of which form part of the input data) which will produce a best fit to the regression line calculated for the data points of the analysed rocks. Such methods are extremely useful, indeed essential, for many research studies, but when used for more and more complex phase assemblages become subject to the inherent limitations in the extract calculation concept discussed in the next section.

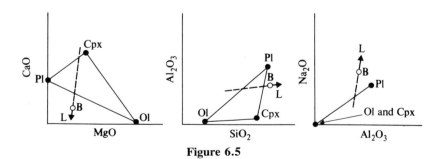

Figure 6.5

Inherent limitations. Certain factors place limits on the usefulness of even the most precise extract calculations. These are:

(a) In natural cases, as discussed previously under the heading of liquid line of descent in Chapter 2, two rocks may be *indirectly* related, that is to say they belong to similar but not identical liquid lines. An extract calculation based on them can therefore only give an approximate (though often useful) result, and efforts to refine the actual calculation may be meaningless.

(b) When the number of phases making up the extract is large (e.g. it is quite possible to find volcanic rocks with phenocryst assemblages such as olivine + clinopyroxene + biotite + plagioclase + alkali feldspar + magnetite + apatite + sphene) adjustments in the proportions of the phases can result in wide variations in the bulk composition of the possible extract. Computation will produce a best fit result but is also likely to produce a wide range of solutions which fit almost as well. In view of the limitation noted under (a) above there is not *necessarily* any reason to assume that the best fit answer represents the truth.

(c) It has been assumed in discussions so far that partial melting involves the addition of actual minerals to the liquid. While this is usually approximately true, complex solid solutions and reaction relationships mean that in some cases the increment of melt added in a given temperature interval during equilibrium melting may not have any exact mineralogical equivalent. Detailed examples of the solid solution effect have been discussed by Helz (1976) with reference to the melting of hornblende. Helz argues that partial melting trends can be distinguished from fractional crystallisation trends because in the former case calculated extracts do not coincide exactly with mixtures of observed or likely phenocrysts. In view of the statistical and other uncertainties involved in extract calculations this seems, however, to be taking an optimistic view of the situation. Nevertheless the general point is valid.

As an obvious example in which a reaction relationship introduces a potentially misleading element into an extract calculation consider the partial melting trend of a liquid in equilibrium with olivine and orthopyroxene at 1 atmosphere in a system such as $Fo-An-SiO_2$ (Fig. 4.20). Liquids evolving up-temperature on this boundary curve show the apparent progressive addition of $En + SiO_2$. Suppose such a series of liquids were available for study (each liquid would be produced at a different temperature from *different* volumes of the same general source) extract calculations would indicate possible fractionation of enstatite + quartz as opposed to the actual mechanism involving up-temperature removal of olivine and addition of enstatite.

(d) The input data may be imperfect as a result of analytical error or imperfect mineral separation.

(e) Processes other than crystal–liquid fractionation may have had a major or minor influence. These include secondary effects such as weathering. All these factors mean that a solution of the calculation, however precise, does not necessarily represent reality.

Application to series of volcanic rocks

The discussion above is strictly applicable to the solution of problems involving the relationship of two specific compositions. The reasoning is now extended to series of related compositions where three additional factors must be taken into account. These are firstly that the scatter of the data leads to some uncertainty as to the precise nature of the compositional trend, secondly that in a series of liquids evolving by crystal fractionation or partial melting, the compositions of the crystalline phases changes progressively as a result of solid solution, and thirdly that new phases may enter the crystallising assemblage or existing phases may cease to crystallise through the compositional range under study.

Statistical considerations. Taking first the question of scatter, Figure 6.6 gives TiO_2 versus MgO relations for a series of rocks from the Belingwe

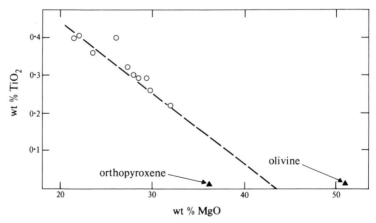

Figure 6.6 TiO_2 versus MgO for ultramafic lavas from the Belingwe Greenstone Belt, Rhodesia (Bickle *et al*. 1976). The least squares regression line is dashed. Triangles show mineral composition points.

greenstone belt of Rhodesia (Bickle *et al.* 1976) suspected of having been formed by the fractionation of olivine and/or orthopyroxene from an ultramafic basaltic liquid (or alternatively by the accumulation of these phases in a less ultramafic liquid). The question to be asked is: Does the back-projection of the trend indicate control by olivine, or by orthopyroxene, or by both? A linear regression line[1] indicates, by projecting to a point approximately mid-way between the composition points of olivine and orthopyroxene, that both phases are involved. However, this is no more than an indication, and the reader should speculate as to what might happen to the trend if, say, the number of analysed rocks were doubled. The problem may perhaps be solved by examining the relationship between other element pairs, or preferably by using more analytical data (see the later paper by Nisbet *et al.* 1977 for the results of a more detailed study).

The effects of solid solution. In Figure 6.7 a crystallising phase forms solid solutions between P and Q. An initial liquid L_1 precipitates a PQ solid solution having composition ss_1 and as the temperature falls and crystallisation proceeds the liquid changes composition to L_2 and the solid changes to ss_2. Note that under conditions of perfect fractional crystallisation the line linking L to ss is always tangential to the curve representing the locus of liquid compositions (cf. ternary diagrams in Ch. 5). If analytical data were available only for L_1 and L_2 an extract

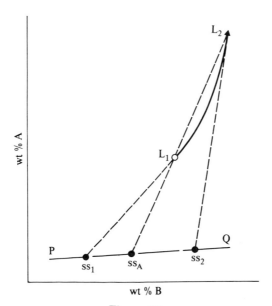

Figure 6.7

calculation would indicate that the material removed was ss_A. This is of course the *average* composition of all the PQ solid solution material removed over the temperature interval in question. The identification of ss_A would in practice be very useful in a petrogenetic investigation but such a conclusion might clearly be a half-truth. However, in most natural cases curvature of the liquid line of descent is not detectable because even if present it is not likely to be very pronounced and thus becomes masked by the scatter in the data.

Now consider a natural example in which analytical data are available for bulk compositions of lavas as well as the phenocryst compositions of a number of samples. Figure 6.8 shows plots of Al_2O_3 and CaO versus SiO_2 and CaO versus MgO for five lavas from Aden (Hill 1974) ranging in composition from basalt to trachyandesite. Each lava contains phenocrysts of olivine, clinopyroxene, and plagioclase, as well as minor amounts of an opaque phase which will be neglected for the moment. Note that on these diagrams the plagioclases and olivines show extensive ranges of solid solution and the clinopyroxenes show rather less. The evolution of the series can be seen in terms of a bulk compositional trend accompanied by a series of migrating three-phase triangles representing the phenocrysts. The analogy to ternary phase diagrams showing solid solutions is close.

Preliminary examination of the plots suggests that a series of extracts consisting of rather a large amount of plagioclase (50% or more of the total extract) with clinopyroxene and olivine is broadly consistent with the data (cf. Fig. 6.5). Parenthetically it may be noted that extracts consisting only of olivine + plagioclase at first sight also look feasible (the bulk compositional trend back projects across the olivine–plagioclase tie-lines in every case) but this must be excluded on the grounds of inconsistent proportions, that is to say, the very high proportion of plagioclase required in an olivine–plagioclase extract according to the CaO versus SiO_2 plot is not consistent with the much smaller amount required by the Al_2O_3 versus SiO_2 plot. Other possible extracts such as clinopyroxene + plagioclase or olivine + clinopyroxene are more readily rejected because these joins are not cut by the bulk compositional trend on one or more diagrams.

We now test in more detail the internal consistency of the olivine + clinopyroxene + plagioclase extract hypothesis. Consider first the three three-phase triangles which are available for a specific sample. We must determine whether these are internally consistent in terms of the mineral proportions of the extracts. This can be carried out graphically as shown in Figure 6.9. Here each triangle with its intersecting bulk compositional trend is transformed into an equilateral triangle, then the three equilateral triangles are superimposed (Fig. 6.10). If the triangles are

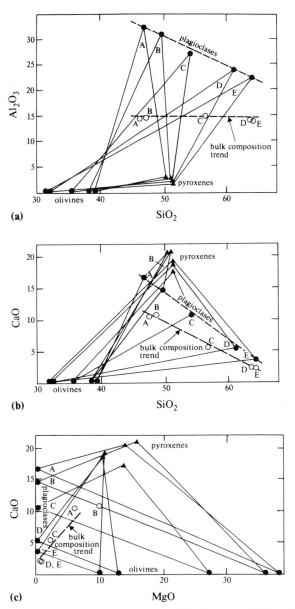

Figure 6.8 Variation diagrams for Al_2O_3 and CaO versus SiO_2 and CaO versus MgO for five Aden lavas (bulk compositions – open circles) and their phenocrysts (plagioclase and olivine – filled circles; clinopyroxene – triangles). Triangles A–E link phenocrysts in individual rocks A–E. Rocks are: A, B – basalts; C – basic trachyandesite; D, E – trachyandesites. Olivine compositions for A, C and E are estimates only; other mineral compositions are from microprobe analyses. Data from Hill (1974).

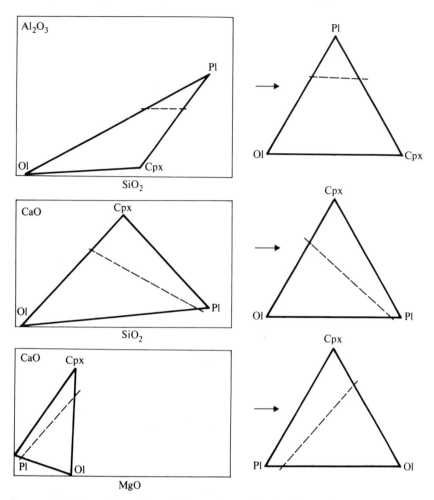

Figure 6.9 Transformation of three-phase triangles for rock E of Figure 6.8 into equilateral form. Dashed line represents bulk compositional trend which is also transformed, by maintaining the proportions in which it cuts the edges of the triangles. Full details of compositional representation in equilateral and non-equilateral triangles are given in Chapter 4.

internally consistent then the three superimposed bulk compositonal trends will intersect in a point which gives the mineral proportions which will satisfy all the data. In practice a triangle of error, or greater or lesser size, is usually obtained at this stage. Two such triangles of error are illustrated in Figure 6.10 for samples E and C of Figure 6.8. Best fit values of mineral proportions for the ratio plagioclase : clinopyroxene : olivine are then determined as approximately 55 : 26 : 19 and

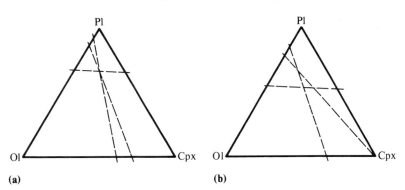

Figure 6.10 Superimposition of three equilateral three-phase triangles to determine best fit extracts, (a) for triangles E of Figures 6.8 and 6.9, (b) for triangles C of Figure 6.8.

$63:20:17$ respectively. Similar results, with varying sizes of triangles of error can be obtained for the other three-phase triangles of Figure 6.8.

The calculations confirm that the removal of phenocrysts actually present in the lavas is capable of driving compositions along the observed bulk compositional trend, at least as far as the relative behaviour of MgO, Al_2O_3, SiO_2 and CaO is concerned. Furthermore it suggests that the amount of plagioclase involved is large, and that this proportion may increase with fractionation (E is a more evolved rock than C). In terms of simple experimental data neither of these conclusions is unreasonable. Reference to the system diopside–albite–anorthite (Ch. 5) shows, for example, that all but the most basic liquids in equilibrium with diopside and plagioclase characteristically fractionate more plagioclase than diopside. Because of the possibilities of selective crystal settling (see a later section of this chapter) it is, however, unfortunately *not* possible to make the further check of finding out if mineral proportions suggested by the extract calculation coincide with phenocryst proportions.

An alternative way of interpreting the data is to suppose that the bulk compositional trend is created by the progressive melting of olivine gabbro. If a source rock consisting essentially of olivine, clinopyroxene, and plagioclase melts in a cotectic fashion (i.e. in the absence of a reaction relationship) as the system forsterite–diopside–anorthite (Fig. 4.21) suggests it might at low pressures, then progressive melting will consist of the addition of these three phases to the liquid. Extraction of the liquid at various stages, followed by a little cooling to produce the observed phenocrysts (the liquid being saturated with respect to all three phases) will produce a bulk compositional trend and phenocryst assemblages approximately, as observed. Without further information it is not easy to decide whether a crystal fractionation or partial melting model best fits the observed facts. What is certain, however, is that a crystal–liquid fractionation process of some sort, taking place at relatively low pressures (higher pressures would suppress the field of plagioclase stability), is capable of generating the observed features of the suite.

In the absence of precise compositional data for the phenocrysts the above analysis would clearly not be possible. But it should also be clear that a reasonable *guess* about mineral compositions is likely to give an approximate solution to the question of which phases are involved and their proportions. Reasonable guesswork in this case means that compositions of olivines and plagioclases may be estimated from optical data, or, if aphyric rocks are involved, a knowledge of the approximate composition of plagioclase or olivine which is normally found in lavas of similar composition may be employed.

Returning to the question of the minor opaque phase present as phenocrysts in the Aden lavas, the proportions in which this separates are low enough to have no substantial effect on the relationships of CaO, MgO, Al_2O_3 and SiO_2 discussed above. In other words with the diagrams so far employed this phenomenon is not detectable. However, Aden basalts contain approximately 3% TiO_2 which drops steadily to below 1% in the trachyandesites. Obviously a titanium-rich phase is also required in the extract (the clinopyroxenes mainly contain less than 1% TiO_2 which is clearly not enough) and this is represented by the opaque phenocrysts of ilmenite and titanomagnetite, which contain TiO_2 in the range 20–50%.

This illustrates the point that the hypothesis of olivine + clinopyroxene + plagioclase fractionation developed previously was, technically, wrong, because it did not explain all the known facts. It is, however, perhaps better to think of the original hypothesis as incomplete rather than incorrect.

Variation diagrams showing the entry of a new phase. The above discussion centred on a series of rocks in which the phase assemblage separating remained constant through a substantial range of fractionation. Next considered are variation diagrams in which a change of the fractionating assemblage takes place.

In Figure 6.11 no solid solution is involved, but here the initial liquid L_1 fractionates phase R, later joined by phase S. The liquid path is L_1–K–L_2, the inflection at K representing the incoming of S into the crystallising assemblage. For liquids between K and L_2 the proportions of R and S crystallising are represented by the point E. As in the previous case an extract calculation based only on analyses of L_1 and L_2 provides a conclusion which is a half-truth. The point E_A is what may be termed the **apparent extract** and although it represents accurately the material removed overall it does not coincide with a real extract at any stage.

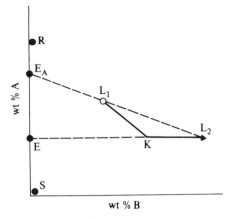

Figure 6.11

Variation diagrams for natural rock series which show inflections in apparent liquid line of descent provide some of the most powerful evidence for the operation of crystal–liquid processes during magmatic evolution. The sudden change of direction of the liquid path as illustrated in Figure 6.11 is *the* characteristic of liquids evolving by fractional crystallisation, or conversely by progressive partial fusion. It is not a characteristic to be expected from other processes such as volatile transfer and diffusion. Since large numbers of natural volcanic series show such inflections it is natural to conclude that their evolution has been strongly influenced by crystal–liquid fractionation processes. Plutonic cumulates also show general behaviour in accordance with the predictions of crystal–liquid fractionation theory and the phenomenon of **phase layering** is the plutonic analogue of inflections in the liquid line of descent.

Figures 6.12 and 6.13 show two good examples of natural volcanic series in which inflections may be detected in variation diagrams.

The precise formulation of extract calculations for series such as those discussed above depends first on the detection of inflections and then on the splitting of the analytical data into groups, each of which represents a different extracted mineral assemblage. An initial graphical study is almost indispensable, for using a computed regression line through a set of data in which an inflection in the liquid line of descent is concealed, can produce meaningless results. For research purposes computers with graph-plotting or graphic display facilities are finding increasing application in initial studies of this type.

Petrographic tests. Natural volcanic series containing one or more inflections in the apparent liquid line of descent are ideal for the petrographic testing of crystal–liquid fractionation hypotheses derived from analytical

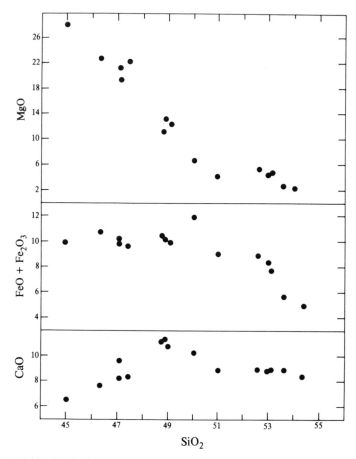

Figure 6.12 Variation diagram for basaltic and andesitic lavas of the New Georgia group, Solomon Islands (from data of Stanton and Bell 1969). Fractionation is dominated by olivine removal in rocks with $SiO_2 < 50\%$ and by the removal of olivine + clinopyroxene + magnetite in rocks with $SiO_2 > 50\%$.

data. If the liquid line of descent is divided into different parts at each point of inflection, each part may show a characteristic phenocryst assemblage consistent with the extract calculation for that part. For the series illustrated in Figure 6.13, for example, in those rocks which are porphyritic we should expect to find apatite phenocrysts only in those rocks which lie on the MgO-poor side of the P_2O_5 maximum (which takes place at about 4% MgO). Conversely apatite should be absent from the phenocryst assemblage of rocks lying on the MgO-rich side of the inflection.

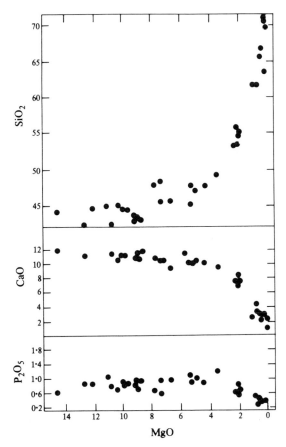

Figure 6.13 Variation diagram for lavas of Madeira and Porto Santo (after Schmincke and Weibel 1972).

So many examples are known where this sort of relationship does in fact hold that it is abundantly clear that the series concerned have evolved largely by crystal–liquid fractionation processes. At first sight this might be taken to mean that the specific process involved is fractional crystallisation, that is that the bulk compositions of the individual specimens represent successive residual liquids formed by the removal of crystallising solid phases. This is probably true in many cases, but the evidence that has been discussed so far is *also* compatible with an origin by progressive melting under some circumstances, as discussed briefly with regard to the Aden rocks earlier in this chapter. For example, referring again to the series represented in Figure 6.13, and considering the case of apatite, those rocks which contain apatite phenocrysts represent magmas which are saturated with respect to dissolved apatite. The MgO-poor liquids represented by analyses at the right-hand side of the diagram rather than being the products of advanced fractional crystallisation could in principle represent liquids produced by low degrees of partial melting. Melting proceeds with P_2O_5 enrichment in the liquid as apatite in the source rock contributes to the melt. Liquid drawn off and erupted at this stage is saturated with regard to apatite and will therefore produce apatite phenocrysts if it suffers even slight cooling. More advanced melts with MgO contents above 4% are no longer saturated with apatite (all the apatite in the source rock having been used up when the liquid reached the composition of the inflection point at 4% MgO) and hence will not produce apatite phenocrysts if subjected to slight degrees of subsequent cooling. Consideration of these possibilities shows therefore that the coincidence of chemical and petrographic evidence so far discussed is strictly consistent with an origin of the liquids by fractional crystallisation *or* by partial melting but does not in itself distinguish between the two possibilities with logical certainty.

In passing, however, it must be noted that the above ambiguity only exists between fractional crystallisation and partial melting processes which have taken place under approximately the same conditions, for substantial changes in phase relations may take place with changing pressure. Since petrographic and other criteria commonly indicate that the phenocrysts in lavas originate under conditions of comparatively low (near-surface) pressure, it follows that, if such series have originated largely by partial melting rather than by fractional crystallisation, the partial melting must also have taken place at a low pressure. However, this is not likely to happen commonly because the temperatures necessary to induce partial melting are not normally developed at high crustal levels – hence, in general, volcanic series which owe much, though by no means necessarily all, of their variation to fractional crystallisation are probably more the rule.

The detection of inflections. In the variation diagrams discussed (Figs 6.12 and 6.13) inflections are sometimes visible in one two-element plot while being undetectable in another plot for the same series, e.g. the entry of apatite into the crystallisation sequence is marked by an inflection in the $MgO–P_2O_5$ plot but is not detectable in $Na_2O–SiO_2$. However, an inflection in one diagram must theoretically be matched by inflections in the others, because whatever the chemical parameters employed the extract polygon must change its geometry with the entry of a new phase. The only theoretical exception to this rule is the case where the phase or phases already crystallising contain neither of the elements employed in the plotting parameters and therefore plot at the origin (cf. olivine and clinopyroxene in Figs. 6.4c and 6.5c). In this case, the entry of a third phase such as magnetite, which also plots at the origin, will clearly produce no inflection in the liquid path. However, this special case helps to explain why in many natural cases inflections are undetectable. It makes the point that the magnitude of the inflection in a particular diagram is entirely dependent on the relative positions in the diagram of the two successive extracts (one without the new phase, the other with it) and the composition of the liquid. Figure 6.14 illustrates one case where a liquid crystallising olivine proceeds to olivine + clinopyroxene crystallisation. The effect on the CaO–MgO plot is profound while the effect on Na_2O–MgO is slight. Slight inflections of the latter type are usually lost within the scatter in variation diagrams for natural series.

Another important factor concerns the number of phases crystallising. If this is high, as is the case with many intermediate rocks, often the addition of a new phase to the assemblage will usually have a relatively small effect on the bulk composition of the total extract. Hence as regards major elements it is characteristic for intermediate volcanic series to give inflection-free variation diagrams even though there may

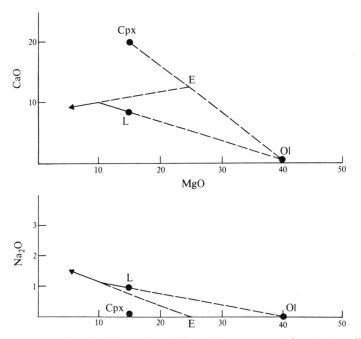

Figure 6.14   Starting liquid is L, the olivine–clinopyroxene mixture crystallising after the entry of clinopyroxene is E.

be some change in the assemblage crystallising. Conversely, basaltic series, in which the number of crystallising phases is characteristically low, frequently produce clearly inflected diagrams.

Porphyritic lavas. In the previous discussions the idea that the analyses of lavas can be taken to represent a liquid line of descent carries the implication that the analysed samples are either non-porphyritic or at least phenocryst-poor. In such cases the analyses clearly approximate to those of the liquid from which the samples crystallised (ignoring volatile constituents which may have been lost at this stage). However, most volcanic rocks are more or less porphyritic and to concentrate solely on non-porphyritic specimens during a petrogenetic study would be to ignore a large part of the available material.

Porphyritic lavas are frequently referred to as 'cumulus-enriched' – that is, they are inferred to be enriched in crystals which have originated elsewhere in the magmatic system. The bulk compositions of such samples do not therefore necessarily represent the compositions of liquids, though we shall see (p. 280) they may sometimes approximate to liquid compositions even though the crystal content is high. Variation diagrams which include the analyses of highly porphyritic rocks frequently show an increased scatter (e.g. see Fig. 6.15) but in many cases this is susceptible to rational interpretation and need not be discounted as a random and inexplicable phenomenon.

We note initially that when the analysis of a porphyritic lava departs notably from what appears to be the liquid line of descent (defined by the analyses of non-porphyritic and phenocryst-poor lavas) the departure is often consistent with the chemical composition and amounts of phenocrysts present. For example, the analyses in Figure 6.15 noted as being abnormally rich in plagioclase phenocrysts are also abnormally rich in Al_2O_3 and CaO, and poor in FeO relative to other rocks of the suite. These lavas may be thought of as being liquids from the normal liquid line (or lines) of descent to which plagioclase has been added. They are thus typical of the rocks termed 'cumulus-enriched'. However, further consideration of the mechanism by which such magmas might be formed shows that there are a number of other possibilities, and although the simple mechanical concept of adding plagioclase crystals to a liquid already present satisfies the compositional and petrographic data, it is by no means a unique solution. Some possibilities are listed below:

(a) A liquid from the normal liquid line of descent accidentally picks up plagioclase crystals formed as earlier precipitates from a different magma.

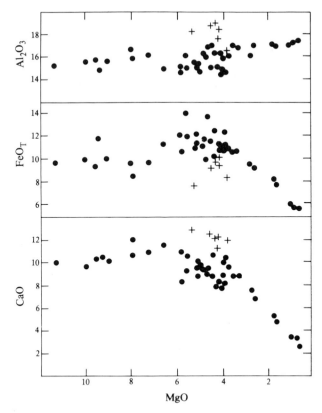

Figure 6.15   Variation diagram for lavas of Red Sea volcanic islands (after Gass *et al*. 1973). Rocks particularly rich in plagioclase phenocrysts (averaging 19%) shown by crosses. FeO$_T$ signifies total iron calculated as FeO.

(b) A liquid body crystallises plagioclase at the top of the chamber. The plagioclase sinks to deeper levels and this lower, plagioclase enriched magma is erupted.

(c) A liquid crystallises several phases such as pyroxene, olivine, and plagioclase. The olivine and pyroxene sink out of the liquid and the plagioclase (having a lower density) remains *in situ*. The upper part of the chamber is thus rich in plagioclase phenocrysts and depleted in ferromagnesian constituents. This is the process of **selective crystal fractionation** and the compositional effects are similar to (though not identical to) those of direct plagioclase addition.

(d) A variant of the above is possible in which the plagioclase floats upwards to produce crystal enrichment.

(e) A liquid rich in Al$_2$O$_3$ and CaO (not on the normal liquid line of descent) crystallises excess plagioclase and is erupted still containing the crystals. Removal of the crystals would lead to a liquid composition similar to that of the normal liquid lines of descent.

To distinguish with any degree of confidence between these possibilities (and others which may exist) may depend on the consideration of many lines of evidence apart from the compositional and petrographic data already discussed. Various arguments will be touched on in succeeding sections, the important point to note at present is that the compositional data alone do not necessarily lead to a unique hypothesis for the *mechanism* of formation of porphyritic rocks.

Porphyritic lavas whose compositions do not depart notably from the apparent liquid line of descent are, together with those types discussed in the previous section, also reasonably common. The simplest, though not necessarily the correct, explanation of such rocks is that they have undergone cumulus enrichment in *all* the phases crystallising, and that each phase has been added to the liquid in the same proportions as it crystallised in. A simple illustration is given in Figure 6.16 where two phases P and Q crystallise in such proportions that the bulk composition of the extract is E. The initial liquid L_1 undergoes fractionation of P and Q (for example by crystal settling) to produce a residual liquid L_2. Meanwhile the P and Q removed are added to a liquid the same as L_1 at a lower level in the chamber to produce a porphyritic magma of composition L_P which lies on the back-projection of the liquid line of descent and therefore in general appears to lie on it. In the limiting case when only one phase crystallises (e.g. in a picrite basalt magma crystallising only olivine) cumulus enrichment must inevitably lead to such a relationship, but in more complex cases there are a number of

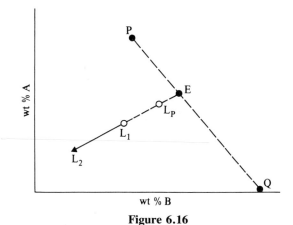

Figure 6.16

reasons why even simple crystal settling mechanisms will under certain circumstances not lead to cumulus enrichment in the correct proportions.

Selective crystal settling. Aspects of this phenomenon are considered in more detail in a later section. The present discussion is concerned with some of the effects as they influence the interpretation of variation diagrams. When a liquid crystallises the solid phases appear in perfectly definite proportions which are controlled by the bulk composition, the pressure, and the temperature. The rate at which crystalline phases are removed from contact with the liquid in which they originate (thus allowing the bulk composition of the original liquid to fractionate) is controlled, however, by mechanical factors which may affect various phases differently. We shall consider the case where crystals are removed by sinking, as a consequence of the density contrast between them and the liquid. Settling rates will be affected by crystal size and the viscosity of the liquid as well as the density contrast between the liquid and each mineral. Table 6.2 gives examples of approximate settling rates of different phases in a basaltic liquid (Wager and Brown 1968) calculated on the assumption that the crystals are spherical and that their settling velocity is adequately predicted by Stokes' Law:

$$V = \frac{2}{9} \, gr^2 \, \frac{(\rho_s - \rho_L)}{\eta}$$

where g is the acceleration due to gravity, r the radius of the sphere, ρ_s the density of the solid, ρ_L the density of the liquid and η the viscosity. Bottinga and Weill (1972) and Shaw (1972) have provided methods by which viscosities of melts can be calculated from compositional data, and Bottinga and Weill (1970) similarly calculate magma densities. The examples given in Table 6.2 illustrate the dramatic differences in settling rate which can occur in a particular instance.

Table 6.2 Settling velocities (metres per year) of crystals in basic magma (density: 2.58 gm cm^{-3}, viscosity: 3000 poises). Data from Wager and Brown (1968)

	Plagioclase		Olivine		Augite		Titaniferous magnetite	
ρ_s	2.68		3.70		3.28		4.92	
$\rho_s - \rho_L$	0.10		1.12		0.70		2.34	
radius (*mm*)	0.5	1.0	0.5	1.0	0.5	1.0	0.5	1.0
settling velocity (*m/year*)	5.7	23	64	256	40	160	134	535

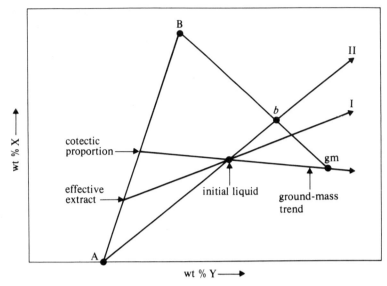

Figure 6.17   Compositional effects of selective crystal removal.

The principles of an approach to the solution of the compositional consequences of selective crystal settling are illustrated in Figure 6.17 where a simple two-phase extract is considered. Two minerals A and B crystallise in given proportions (the cotectic proportions) from an initial liquid. Making the simplifying assumption that no reaction is involved between crystals and liquid, the residual liquid trend is constructed and represents the groundmass compositional trend of porphyritic lavas as well as the bulk compositional trend of liquids showing complete crystal removal (i.e. perfect fractionation). If, however, an extreme case of selective fractionation is postulated in which crystals of A are entirely removed while B remains static then the bulk compositional trend II will be generated. This will represent porphyritic rocks with phenocrysts only of B, and as crystallisation proceeds the composition point of B, the bulk composition (*b*), and the groundmass (gm) will remain co-linear. If on the other hand both phases are removed but A is removed more efficiently than B, there will be an effective composition as shown, richer in A than the cotectic proportions, which is removed and will generate bulk compositional trend I. To investigate this trend and its associated rocks more fully, it is necessary to consider magma chamber models in more detail (see p. 272). However, we note that bulk compositional trend I must lie somewhere between trend II and the groundmass trend, and furthermore that the rocks are likely to contain phenocrysts of both A and B but B will be in excess relative to the proportions in which the two minerals crystallise.

Tie-line directions. One of the interesting consequences of selective crystal removal is that it leads to a discrepancy of direction between the bulk compositional trend of porphyritic rocks and the tie-lines which link bulk compositions to groundmasses. It is worth reflecting on this point for a moment, because simple, non-selective models of crystal fractionation inevitably imply a coincidence of these directions; that is to say, if the phenocrysts are removed from one composition the resultant is the composition of another, more evolved, rock in the same series. Non-parallelism therefore implies that the simple model is wrong. Such discrepancies can arise in a number of ways and are always important because they allow refinement of the simple model.

Figure 6.18 illustrates whole-rock-groundmass tie-lines for Aden lavas as they appear on the Al_2O_3 versus SiO_2 diagram. The four

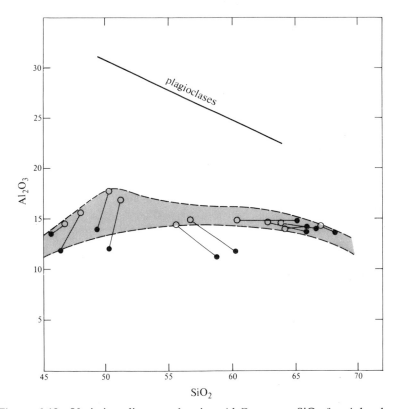

Figure 6.18   Variation diagram showing Al_2O_3 versus SiO_2 for Aden lavas. Shaded area is zone occupied by majority of analysed whole-rock samples. Open circles are bulk compositions of (l. to r.) four hawaiites, two basic trachyandesites, four trachyandesites and one trachyte, and are linked to groundmass compositions (small filled circles). Data from Hill (1974).

porphyritic hawaiites (46–51% SiO_2) show an extreme discrepancy between the tie-line direction and the bulk compositional trend. Two basic trachyandesites (55–57% SiO_2) are somewhat less discrepant while the remaining rocks (trachyandesites and one trachyte) show coincidence between the two directions. Comparison with Figure 6.8a and with other diagrams not shown here indicates that the phenocryst assemblages of the hawaiites contain a higher proportion of plagioclase than is required by any extract capable of producing the bulk compositional trend. The simple, though not the only, explanation of this is that plagioclase has fractionated less effectively than the other phases olivine, clinopyroxene and titanomagnetite. Whatever the reason, with increasing evolution the effect diminishes and is not detectable beyond the 60% SiO_2 level.

As a contrast to Aden, the calc-alkaline lavas of the Sidlaw Hills, Scotland (Gandy 1975) show a quite different pattern which gives no evidence of the operation of a selective process. Here (Fig. 6.19) the whole-rock-groundmass tie-lines for porphyritic basaltic andesites and andesites, containing phenocrysts of plagioclase and other phases including magnetite, olivine and clinopyroxene, show a scatter of directions which on average coincide approximately with the bulk compositional trend (increasing SiO_2 with constant Al_2O_3). For the shorter tie-lines it is probable that a substantial proportion of the directional scatter is caused by analytical error and approximations used in the calculation of groundmass compositions. Nevertheless both short and long tie-lines show the same pattern, that is SiO_2 enrichment in the groundmass coupled with variable behaviour of Al_2O_3. The three samples marked A, B, C were collected from different localities within

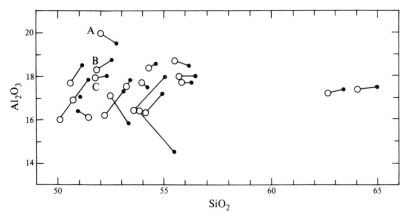

Figure 6.19 Variation diagram showing Al_2O_3 versus SiO_2 for lavas from Sidlaw Hills, Scotland. Symbols as in Figure 6.18. Data from Gandy (1975).

the same lava flow and suggest that variation in phenocryst proportions exists between one hand specimen and another. The average direction of these tie-lines is, however, approximately coincidental with the bulk compositional trend. Hence, apart from the analytical and other factors mentioned, the scatter of tie-line directions in general appears to be best interpreted in terms of a random variation of phenocryst proportions around the proportions in which the phases crystallised, on the scale of the samples analysed.

Before leaving the graphical treatment of selective crystal fractionation we note that its detection is subject to the same constraints as the detection of inflections already discussed. The effect on tie-line directions is entirely conditioned by the shape of the extract polygon and its position relative to the bulk compositional trend. Some arrangements, such as those discussed above, are very sensitive to the mineral proportions of the extract while others are not. In the Sidlaw lavas for example, tie-lines on Na_2O and K_2O versus SiO_2 diagrams show no substantial directional scatter.

Note

1. If the two oxides are designated x and y, the linear regression line has the form:

$$y = a_1 x + a_0$$

where, using the least squares method (n being the number of samples):

$$a_1 = \frac{\Sigma xy - \dfrac{\Sigma x \Sigma y}{n}}{\Sigma x^2 - \dfrac{(\Sigma x)^2}{n}}$$

and

$$a_0 = \bar{y} - a_1 \bar{x}$$

where

$$\bar{y} = \frac{\Sigma y}{n} \quad \text{and} \quad \bar{x} = \frac{\Sigma x}{n}$$

The coefficient of determination which expresses the closeness of fit to the line is

$$r^2 = \frac{\left[\Sigma xy - \dfrac{\Sigma x \Sigma y}{n} \right]^2}{\left[\Sigma x^2 - \dfrac{(\Sigma x)^2}{n} \right]\left[\Sigma y^2 - \dfrac{(\Sigma y)^2}{n} \right]}$$

Exercises

1. The analyses of three minerals and two rocks A and B are given in Table 6.3.
   (a) Calculate the bulk composition of a rock formed of 30% olivine, 20% orthopyroxene, and 50% plagioclase. N.B. this is best done by numerical methods, especially if a programmable calculator is available. The graphical solution is tedious.
   (b) Rock B is suspected of having formed from rock A by the accumulation of one or more of the listed minerals. Determine which mineral(s) fits the data best. What proportion of B consists of the accumulated mineral(s)? N.B. the graphical solution is ideal here unless simple inspection of the data immediately suggests a solution which can be tested numerically.

Table 6.3   Analytical data for Exercise 1

	Olivine	*Orthopyroxene*	*Plagioclase*	*Rock A*	*Rock B*
SiO_2	39.2	55.4	49.7	49.9	50.73
TiO_2	0.15	0.35	0.07	2.83	2.43
Al_2O_3	0.11	1.66	31.05	11.4	9.94
Fe_2O_3	0.7	0.64	0.86	3.0	2.65
FeO	19.8	6.39	–	9.74	9.24
MnO	0.27	0.13	–	0.16	0.16
MgO	39.4	32.9	0.05	10.4	13.78
CaO	0.27	1.50	14.60	9.72	8.49
Na_2O	–	0.14	3.12	1.97	1.70
K_2O	–	0.02	0.29	0.80	0.68

Table 6.4   Analytical data for Exercise 2

	A	*B*	*C*	*D*	*E*	*Ol rock B*	*Cpx rock C*	*Pl rock B*
SiO_2	50.2	51.0	51.8	52.5	52.9	39.3	50.9	46.3
TiO_2	0.90	0.96	1.00	0.63	0.41	–	–	–
Al_2O_3	13.0	13.8	14.8	13.6	13.0	0.7	2.9	33.4
FeO	8.9	8.0	7.2	7.1	7.1	19.8	11.1	–
MgO	10.8	9.2	7.1	6.6	6.6	37.7	15.6	–
CaO	9.0	9.9	10.6	10.3	10.0	–	17.3	17.3
Na_2O	2.1	2.3	2.4	2.5	2.6	–	–	0.05
K_2O	0.6	0.6	0.7	0.8	0.8	–	–	–
P_2O_3	0.38	0.42	0.46	0.52	0.40	–	–	–

2. Table 6.4 gives five whole-rock analyses of lavas from an individual volcano (analyses A–E). You may assume that they are representative of a much larger collection of data. Also given are mineral analyses which may be helpful, though it is not specified whether these are phenocryst or groundmass phases.
   (a) Plot variation diagrams from the analysed rocks using SiO_2 as abscissa. (Hint: make sure the scales of the diagrams are such that the mineral analyses can also be plotted.)
   (b) Using the above, give a brief outline of the fractionation history of the suite on the assumption that A is the parental magma and that evolution is by means of fractional crystallisation.
   (c) Assuming the analysed rocks contain sparse phenocrysts, predict which phenocryst phases should be present in each rock.

7

Petrographic aspects of volcanic rocks

Introduction

The normal preliminary petrographic study of rocks, that is to say the identification of the minerals and the examination of grain size and texture, enables rocks to be classified and named. The present chapter is not concerned with this routine exercise but with some of those other petrographic aspects of volcanic rocks capable of giving information of direct petrogenetic value. Amongst these features probably the most important is the phenocryst assemblage, that is to say, what it consists of and what the relations are between the various phases and the groundmass. The determination of the order of crystallisation is a related topic, and other features may give information about cooling rates, pyroclastic origins, and other diverse subjects of more or less petrogenetic interest. The field is, however, wide and ill-defined and we cannot hope to be entirely comprehensive in this treatment.

The phenocryst assemblage

The **porphyritic texture**, that is to say, large crystals (**phenocrysts**) set in a finer grained matrix (**groundmass**) is one of the most obvious characteristics of lavas and rapidly cooled hypabyssal rocks. This illustrates what may be termed the normal descriptive usage of the term phenocryst. Lavas with no phenocrysts appear to be very rare, though lavas with very sparsely distributed phenocrysts are fairly common. Most of the rocks described as non-porphyritic probably belong to this latter category, and it may be necessary to cut several thin sections before a phenocryst is encountered. Commonly, however, porphyritic rocks contain up to 30% or so of phenocrysts (expressed as a volumetric per cent of the whole rock) and lavas with 60% of phenocrysts or more are found occasionally. The fact that the great majority of rocks contain at least some phenocrysts has important petrogenetic consequences, for it demonstrates the rareness of superheated magmas (magmas above their liquidus temperature), at least in the near-surface environment. This places certain constraints on the amount of heat available for assimilation processes and also shows that the thermal conditions necessary for crystal–liquid fractionation are generally fulfilled.

The porphyritic texture is widely, and usually correctly, ascribed to the effects of a period of slow cooling during which the phenocrysts grew, followed by a period of rapid cooling (which is termed the quenching stage) during which the groundmass crystallised. It is this genetic interpretation of the texture which is of immediate interest because, in the case of lavas, it is frequently clear that the quenching stage coincides with the eruption of the lava onto the surface, while the period of slow cooling must have taken place underground. Crystals growing at this earlier stage are referred to as **intratelluric**, and since most processes of magmatic fractionation clearly take place in the intratelluric environment, a knowledge of what phases were present as solids within the magma prior to eruption is of great petrogenetic significance. If, for example, crystal–liquid fractionation is postulated, e.g. by a mechanism of simple gravitative crystal-removal, it is possible first to predict trends of liquid evolution, and secondly to predict types of cumulate rocks that may be formed. In either case ideas are obtained which help in the understanding of the origins of rocks found in association together. As discussed previously (Ch. 6) this problem can also be approached from chemical data and the two approaches are necessarily used in conjunction.

From the above the term **phenocryst assemblage** can ideally be defined to mean that group of solid phases which was in equilibrium with the liquid before the quenching stage. In this genetic sense the term carries the overtone that these are the phases which must be considered in the construction of simple crystal–liquid fractionation hypotheses. To determine the nature of the assemblage petrographically is at first sight simple, particularly when the groundmass is thoroughly quenched to a very fine-grained or glassy state. Figure 7.1 illustrates an acid pitchstone in which there is no difficulty in perceiving that the phenocryst assemblage consists of plagioclase, hedenbergite, fayalite, magnetite and quartz. Each of these is quite distinct in size from any of the tiny microlites in the groundmass. The thin section also incidentally illustrates a common feature of porphyritic rocks, that is the way the phenocrysts tend to clump together in what is called **glomeroporphyritic** texture.

Figure 7.2, a thin section of a phonolite, gives a clear illustration of one of the first difficulties encountered. The groundmass is more coarsely

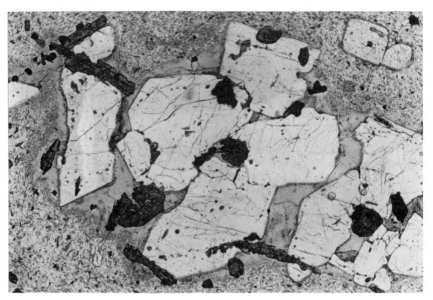

Figure 7.1 Glomeroporphyritic texture in a pitchstone. The phenocrysts are mainly plagioclase, with hedenbergite (elongated), fayalite, magnetite and quartz (upper right).

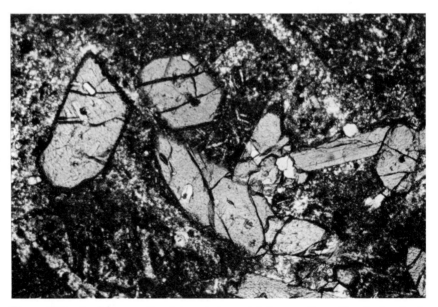

Figure 7.2 Aegirine–augite phenocrysts containing small apatites (colourless) in a phonolite. The texture is somewhat glomeroporphyritic and a few apatite crystals are included in the aggregates without being enclosed in pyroxene. Elongated pale crystals in groundmass are alkali feldspars.

crystalline than in the previous specimen and we note that the prominent clinopyroxene phenocrysts contain numerous small euhedra of a second phase, apatite. Clearly, unless a metasomatic process has been at work, if the clinopyroxene is of intratelluric origin then so is the apatite. The larger groundmass crystals visible are feldspars, and they can be assigned quite definitely to the groundmass because they are intricately intergrown with the fine clinopyroxene needles representing the quenching stage. Thus, intratelluric apatite crystals (phenocrysts in the genetic sense) are *smaller* than groundmass feldspars. There is thus some difficulty in reconciling the genetic and descriptive usages of the term phenocryst.

The problem is in fact quite general and not unimportant, since most minor phases, e.g. apatite, zircon, sphene and chromite, form small crystals even with slow cooling. However, discussions of the interpretation of chemical data already given (Ch. 6) demonstrate that minor phases may be of prime importance in the formulation of petrogenetic hypotheses because they tend to contain large quantities of elements, such as P, Zr, Cr and Ti, of which the fractionation trends can be documented just as clearly as those of major elements. Therefore the extent to which minor phases participate in phenocryst assemblages cannot be ignored. In passing we note that it is the inclusion of minor phenocrysts within larger crystals, together with the tendency for glomeroporphyritic aggregates to form, which is responsible for the efficient fractionation of many minor phases. Consideration of magma viscosities and crystal sizes suggests that left to themselves small apatite phenocrysts, small chromites etc. would have negligible settling rates and would in the absence of assistance from other much larger crystals separate from their parent liquids only with extreme reluctance.

With these considerations in mind it should be possible to make an accurate determination of the intratelluric phenocryst assemblage of a well-quenched rock. The assemblage will consist of the phases which form large crystals in clear distinction to the groundmass, and will include additionally minor phases enclosed within them. As the quenching stage, however, becomes less distinct it becomes progressively more difficult to be certain about the nature of the assemblage. Figure 7.3 for example shows an olivine-rich dyke rock in which it is still reasonably certain that the phenocryst assemblage consists of olivine + spinel, and the groundmass consists of clinopyroxene, plagioclase, and a late crystallising opaque phase, together with analcime (not visible in the photograph). This example still lies well within the range of correct petrographic interpretation but certain conditions of cooling lead to a texture known as **seriate** where, despite a large range of crystal sizes in any specific phase, there is no

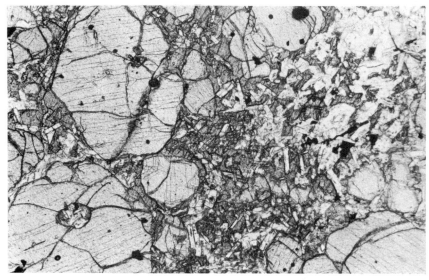

Figure 7.3 Picritic dolerite containing phenocrysts of olivine with spinel inclusions. The groundmass consists of plagioclase grown round by ophitic augite, and some magnetite. The large olivine (top left) is cut parallel to (001) (see text).

clear distinction into a bimodal size distribution. In such specimens uncertainty is likely to remain about the precise constitution of the phenocryst assemblage.

Seriate textures are not always readily identified in thin sections, and, in general, some care is necessary in the determination of crystal size. A random collection of equal-sized spheres would, for example, if sliced in a plane, produce a wide range of sizes of circular cross sections. The smallest circles are produced by the plane of section just grazing the surface of the sphere, so such sections may be identified by close examination of their gently shelving edges. Leucite-rich volcanics in which the phenocrysts are near-spherical commonly, however, show genuine seriate textures. In this case even very small leucite cross sections may be observed to have sharp (i.e. vertical rather than shelving) contacts against the groundmass. This shows that they are small crystals sliced through the centre rather than near-surface slices off much larger crystals.

Plagioclase feldspars of common tabular habit provide a different opportunity for the detection of seriate textures. Here the albite twinning on (010) is parallel to the plane of flattening of the crystals. Sections showing sharply defined albite twin lamellae therefore have rectangular outlines of variable length (measured parallel to the twin lamellae) depending on how the particular crystal has been sliced, but, if the crystals are all of the same size, of constant width. Thus seriate texture is readily detected if the elongate sections with sharply seen twin lamellae are of widely variable width.

Modal analysis of phenocrysts

By modal analysis is meant the determination of the amount of a mineral present in a rock, expressed usually as a volume per cent. Modal content of a mineral should not be confused with normative content (see p. 407). From the discussions in Chapter 6 it is evident that the interpretation of chemical analyses of volcanic rocks is strongly influenced by the nature and amounts of phenocrysts present. An analysis rich in Al_2O_3 may represent for example, an alumina-rich liquid (i.e. an aphyric lava) or it may represent a rock rich in accumulated plagioclase phenocrysts. Petrogenetically the distinction is important and it therefore follows that the analyses of volcanic rocks should be accompanied by petrographic descriptions, the minimum requirement of which is to give the amounts of the phenocrysts present.

Phenocryst proportions are best determined using a mechanical stage and a point-counter and measurement should exclude strongly zoned crystal margins formed by enlargement of the crystals during the quenching stage. Inhomogeneity of phenocryst distribution is encountered on many scales and for precise results it may be necessary to measure several thin sections from a single hand specimen. Studies of inhomogeneity on a larger scale within individual eruptive units are petrogenetically important and can give evidence of the existence of compositionally layered magma chambers and allow conclusions to be drawn concerning crystal distribution within them. Several ignimbrites described from western U.S.A. (see p. 273), for example, grade upwards from phenocryst-poor bases to phenocryst-rich tops suggesting eruption from chambers undergoing crystal settling.

Altered phenocrysts

In an imperfect world it is often necessary to examine rocks which are far from pristine, either because they are weathered or have undergone extensive late- or post-magmatic alteration or because they are slightly metamorphosed. Thus the petrographer needs skill in the identification of altered minerals, often completely pseudomorphed by low-temperature phases such as chlorite, serpentine, muscovite, and finely divided opaques. Pseudomorphs are recognised by a combination of shape and other morphological characteristics together with a knowledge of the common alteration products of each mineral.

Olivine, for example, shows characteristic shapes, of which the most readily identifiable is the section cut parallel to (100) (see Fig. 7.8). Another characteristic shape is shown in Figure 7.3, here a section cut parallel to (001). This crystal, an unaltered example, shows the typical curving internal cracks which are clearly visible picked out by opaque material in the altered examples of Figure 7.8. Olivine is also susceptible to marginal alteration (see Figs. 7.4 and 7.8). Of the common alteration products of olivine, apart from finely-divided opaques, the red-brown mineraloid iddingsite is a ubiquitous product of high-temperature (magmatic) alteration. Many basaltic rocks show marginal iddingsitisation of olivine phenocrysts accompanied by complete replacement of groundmass olivines. At somewhat lower temperatures complete pseudomorphing by low-relief high-birefringence green micaceous products is characteristic (the example in Fig. 7.4 is one of these). Serpentine minerals pseudomorphing olivine, though a characteristic feature of altered plutonic rocks, are not commonly observed in lavas.

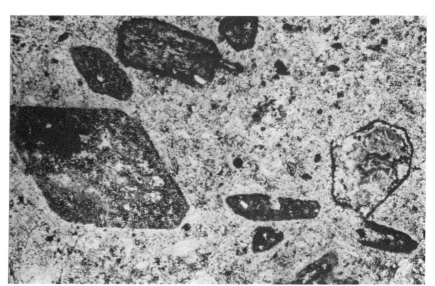

Figure 7.4 Pseudomorphs after various phenocrysts in an altered andesite. Hornblende (left) is represented by opaque granules. Note characteristic cross section shape. Biotite (top left) shows characteristic rectangular shape. Olivine (right) has dark rim of ore and centre of low relief micaceous material with a little ore.

Hornblende in lavas is very susceptible to alteration to a mass of fine-grained material full of finely divided opaque grains. In this case it is the characteristic shape, particularly that of the basal section, which is diagnostic (see Fig. 7.4). *Biotite*, also seen in this figure, shows a behaviour similar to that of hornblende.

Plagioclase feldspar is liable to alter to finely divided aggregates of white mica (sericite) and it is often notable that centres of crystals are more affected than edges. *Potassic feldspar* in contrast often becomes extremely turbid and charged with reddish opaque dust. Epidote is a common accessory mineral produced during the alteration of plagioclase feldspars.

Resorbed and mantled phenocrysts

It has been implicitly assumed up to this point that phenocrysts observed have been crystallising in the intratelluric environment from the liquids now represented by their groundmasses after eruption. The petrographic evidence of this is the euhedral form of the crystals. Many phenocrysts, however, show evidence of a reaction with the liquid which usually manifests itself in a corroded form or in the presence of a mantle (so-called reaction rim) of other phases armouring it and separating it from the groundmass. Even in lavas where the phenocrysts appear to be devoid of reaction features, a detailed study using large numbers of thin sections will usually reveal the presence of a few grains showing them.

Reaction has a variety of causes and it is frequently difficult to decide without extremely detailed studies what the cause may be in a specific case. We may, however, broadly identify two main types of reaction caused respectively by thermal or compositional disequilibrium, and by equilibrium resorption.

Thermal/compositional disequilibrium is presumably one of the most common causes of resorption. Many magma chambers probably exhibit vertical temperature zonation with the cooler liquids at the top where heat extraction is most effective. Crystals forming at high levels in the magma chamber will clearly not be stable at lower levels. Considering the case for example of a plagioclase crystal in a magma chamber saturated with regard to plagioclase at all levels, it is apparent that the more sodic crystals forming in profusion near the top will not be in equilibrium with the liquid if they sink to lower levels. Even so the development of resorptional features is by no means a foregone conclusion. More calcic plagioclase may mantle the crystal (reversed zoning) and prevent the internal sodic part from coming into contact with the liquid. The interior of the crystal thus effectively becomes a closed system of only binary type (system albite–anorthite) and as such exhibits much higher liquidus temperatures than the polycomponent system represented by the liquid. A sodic feldspar, if properly armoured, may thus show no signs of melting when immersed in a liquid much hotter than the one it originally crystallised from. Nevertheless in many cases there is evidently sufficient thermal disequilibrium (or else the armouring is not effective) for plagioclase crystals to begin to melt. Sometimes this is a marginal resorption leading to rounded crystals such as that illustrated in Figure 7.5 but perhaps rather more frequently the melting pervades the interior of the crystal and is apparently related to cleavage (Fig. 7.6). In most cases the subsequent quenching stage results in the overgrowing of the resorbed crystal by a mantle of fresh, non-resorbed, feldspar (Fig. 7.6).

Resorptional features of a similar nature are also caused by the change of *P–T* conditions as a magma moves towards the surface. In some cases this is due to particular phase boundaries being pressure-sensitive, as is the case with the quartz–feldspar boundary in the system SiO_2–Ab–Or discussed in Chapter 5. Analogous behaviour if extended to the natural system represented by granitic magma offers a ready explanation of the common, indeed almost invariable, resorbed nature of quartz phenocrysts in rhyolites, pitchstones and quartz-porphyries. Similarly, the sensitivity of the amphibole and biotite

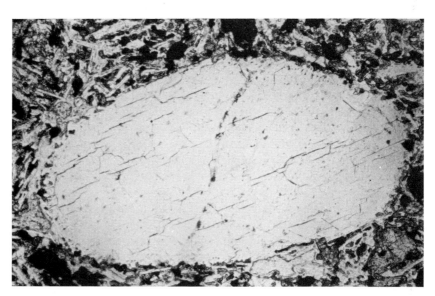

Figure 7.5   Basic plagioclase crystal showing marginal resorption in basalt.

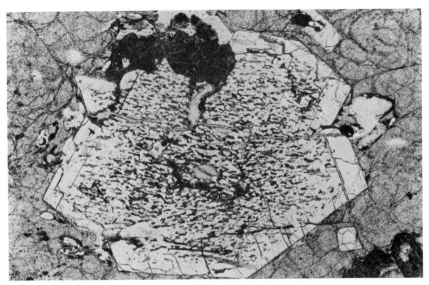

Figure 7.6 Intermediate plagioclase set in acid glass with perlitic cracks. The plagioclase shows extensive internal melting followed by overgrowth of equilibrium plagioclase. Dark crystals (top) may be identified by characteristic dark rim and curving crack patterns as semi-opaque pseudomorphs after olivine.

stability fields to the water content of the magma suggests that the very common presence of resorption features in these phases in volcanic rocks is connected with devolatilisation late in the eruptive history.

Resorption consequent upon a profound change in pressure is illustrated in Figure 7.7. Here a rounded mass of orthopyroxene crystals in a picrite basalt is jacketed by an overgrowth of clinopyroxene and olivine. Chemical and experimental evidence discussed in detail elsewhere (pp. 255–6) suggest that the orthopyroxene was in equilibrium with the liquid at a pressure in the region of 6–12 kbar. At near-surface pressures, however, the same liquid is in equilibrium with olivine and clinopyroxene. Quenching at low pressure has produced an overgrowth of clinopyroxene on the surface of the orthopyroxene and incorporated some olivine crystals as well. The minerals forming the rim are not essentially reaction products in this case and owe their presence mainly to the fact that clinopyroxene nucleated abundantly on orthopyroxene surfaces (the included olivine may be partly due to a concentration gradient in the liquid as in the case discussed below).

Equilibrium resorption of phenocrysts contrasts with the cases discussed above in terms of petrogenetic significance, though there are no certain criteria by which it can be recognised petrographically. One of the commonest situations in which equilibrium resorption is to be

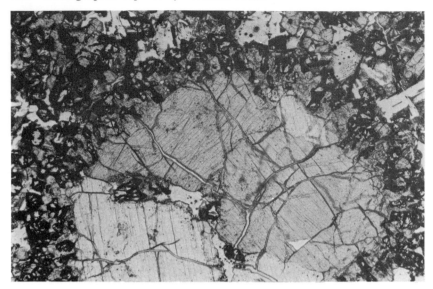

Figure 7.7 Rounded aggregate of orthopyroxene jacketed by overgrowth of clinopyroxene and olivine in picrite basalt.

expected is in tholeiitic liquids crystallising olivine. After initial crystallisation of excess olivine such liquids may resorb olivine down-temperature while crystallising orthopyroxene. The reaction may be written:

olivine + silica-rich liquid → orthopyroxene

Tholeiites showing rounded olivines of resorbed appearance are common, and these rocks characteristically contain more modal than normative olivine. Together these features suggest that the reaction referred to takes place quite commonly and is interrupted by quenching before it is complete. However, in the normal course of events the constituents required for the formation of the reaction product, orthopyroxene, appear to become disseminated in the liquid as the olivine dissolves. A concentration of orthopyroxene round the margins of the olivine is thus only seen if quenching takes place when there is still a concentration gradient in the liquid immediately adjacent to the olivine, that is, in general terms, an excess of MgO over the average concentration within the liquid (see Fig. 7.8). It must be emphasised that the case discussed may be interpreted as equilibrium resorption on theoretical and geochemical grounds and that the petrographic evidence does not bear directly on the reasons for resorption. It does, however, make the point, that evidence of resorption is not necessarily evidence of disequilibrium between the crystal concerned and its enclosing liquid.

Figure 7.8 Subhedral olivine crystals in picrite basalt showing extensive alteration to opaque granules along internal cracks and near margins. The groundmass shows a rim of pure orthopyroxene round the olivine but otherwise consists of plagioclase (colourless), clinopyroxene, minor orthopyroxene, and spicules of ore.

Xenocrysts

This term is used to describe crystals accidentally incorporated in the magma from a foreign source. Xenocrysts normally display resorptional or overgrowth features of the types discussed in the previous section and they may therefore only be identified with certainty when they consist of minerals entirely inappropriate to the magma composition, e.g. quartz grains derived from sandstone wall rocks in alkali basalt. Less exotic grains may occasionally be assigned a xenocrystic origin when their provenance is obvious. A basaltic lava flow from Little Aden (Cox *et al*. 1970) for example contains equilibrium phenocrysts of basic plagioclase (An_{80}) but additionally carries numerous crystals with margins of An_{80} and cores of An_{50}. Such strong reversed zoning is not a normal petrographic feature and in this case the crystals of An_{50} appear to be xenocrysts derived from a thick sequence of strongly feldsparphyric hawaiites which form the walls of the basalt vent.

Had the reverse been the case, that is hawaiite erupted through basalt, xenocrysts of plagioclase in the hawaiite but derived from the basalt would have been difficult to detect, for the subsequent outward zoning to An_{50} would not have appeared to be particularly unusual.

Furthermore, the basic cores of such crystals need not necessarily have been foreign but might have belonged to an earlier, higher temperature, part of the magmatic history. Indeed very often the detection of xenocrysts is hampered by the many possibilities, discussed in part in the previous section, by which magmatic crystallisation can give different assemblages at different pressures. It is often a matter of complex geochemical and petrological debate to decide whether the apparent xenocrysts in volcanic rocks are in fact such, or whether they are derived from higher pressure periods of crystallisation, or whether they represent refractory crystals from the source rocks. Such apparent xenocrysts, if large, are often simply termed **megacrysts** (to avoid genetic overtones) and are represented by olivine, orthopyroxene, clinopyroxene, amphibole, plagioclase, garnet, spinel and other minerals. Aggregates of grains are usually known as **nodules** and are frequently ultramafic in character. The term **cognate xenolith** is used for nodules which appear to have a genetic connection with the host liquid. They represent, for example, cumulate rocks disrupted and carried up by their own parental magma or its derivatives.

Order of crystallisation

The term 'order of crystallisation' refers to the sequence in which phases appear (and occasionally disappear) with falling temperature. It may refer to a particular magma represented by a single specimen or it may be applied to a fractionating series derived from a single parent magma. It is a topic closely related to those already discussed, and in the latter case is concerned simply with the determination of successive phenocryst assemblages in rocks which on chemical grounds can also be shown to belong to a single fractionation lineage. Assuming that the fractionation mechanism is reasonably well established by such studies, then the order of crystallisation determined can be applied directly to the prediction of the phase assemblages of associated cumulate sequences.

The crystallisation of a single rock presents a slightly different case because an equilibrium crystallisation path may be followed by the residual liquid. This is, however, an approximation in many natural cases since equilibrium between solids and liquids is rarely entirely maintained so that zoned phenocrysts are formed. During such crystallisation the bulk composition of the magma does not of course change, a feature which it shares with perfect equilibrium crystallisation. The usefulness of the determination of sequences of crystallisation in single rocks lies in the fact that the course of such crystallisation is frequently not substantially different from the course of fractional crystallisation. Thus the evidence from a *single rock* may be used to predict approximate likely fractionation trends and petrographic features of derivative magmas. The specimen illustrated in Figure 7.3 provides an appropriate example. Here the texture illustrates that olivine and spinel are the earliest phases to begin crystallising, and that subsequently plagioclase *begins* to crystallise before clinopyroxene (this is probably the best interpretation of the ophitic texture which looks superficially as though *all* the plagioclase crystallises before the clinopyroxene). Hence derivatives of the same magma produced by fractional crystallisation might be chemically characterised by the methods outlined in Chapter 6 and be expected to show phenocryst assemblages such as *olivine + plagioclase + spinel* but not *olivine + clinopyroxene + spinel*, though they may at lower temperatures show assemblages including *both* plagioclase and clinopyroxene. The only rigorous way, however, of determining the course of equilibrium crystallisation depends on finding several rocks all of the same bulk composition which have been naturally quenched at different temperatures. This is the basis of the quenching method in experimental petrology but is a condition difficult to meet in natural rocks.

Quenching textures

Very rapid crystallisation of magma during the quenching stage frequently produces distinctive textures and modes of crystallisation which are often readily recognisable. Crystals produced are usually small (but see below) and characterised by a great variety of more or less dendritic and skeletal forms. Crystallisation under such conditions takes place rapidly, albeit for a relatively short period, in response to considerable degrees of supersaturation. As a consequence the growth of crystal corners and edges is generally favoured over the growth of faces since, other considerations apart, these parts of the crystal are able to draw upon larger volumes of adjacent melt for supplies of the necessary constituents delivered by diffusion. Extremes of quenching of course prevent crystallisation and the melt becomes a glass. With somewhat lower cooling rates tiny and complicated branching dendritic crystals are formed (frequently encountered in experimental charges) while slower quenching will give rise to more solid crystals which nevertheless display some degree of skeletal growth. Cooling rate is, however, only one of the variables involved. Different phases respond in different ways and the composition of the magma is also important, particularly those features which have a marked effect on viscosity and hence on diffusion rates.

Of the minerals commonly encountered in volcanic rocks *olivine* is noteworthy for the frequency with which it forms skeletal crystals. It is in fact extremely difficult to quench melts rich in potential olivine to glasses, presumably because olivine has a relatively simple structure and thus crystallises readily from supersaturated melts. Quench olivine frequently forms crystals consisting of a number of parallel plates which may appear not to be connected to each other in the plane of a thin section. Quench crystals of this phase can vary from minute (dendrites in experimental charges) to comparatively large crystals. The most spectacular examples of quench olivine textures are found in the ultramafic lavas of Archaean terrains known as komatiites (Viljoen and Viljoen 1969a) where the texture, consisting of abundant skeletal olivine blades (see Fig. 7.9) is known as 'spinifex' (a type of grass, Nesbitt 1971). Excellent examples from the Abitibi greenstone belt of Canada are figured by Pyke *et al.* (1973).

Clinopyroxene also occasionally forms dendrites but more commonly rapid crystallisation gives rise to small prismatic crystals with hollow terminations. In longitudinal sections such crystals have a characteristic swallowtail appearance (Fig. 7.10). The prismatic, swallowtail habit is also characteristic of quench *plagioclase*, which also shows sheaf-like growths of acicular crystals (Fig. 7.11). Quench *apatite* is capable of

Figure 7.9 Spinifex texture (quench olivine) in a komatiite.

Figure 7.10 Quench crystals of augite showing characteristic swallowtail and hollow sections set in dark glass.

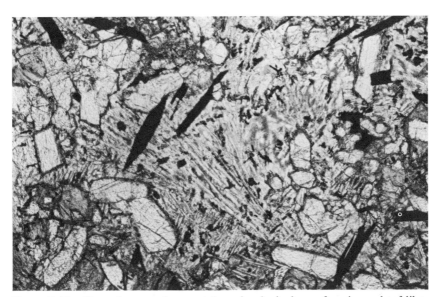

Figure 7.11 Curved quench crystals of plagioclase forming sheaf-like aggregate in basalt.

forming prisms with length to breadth ratios of 100 : 1 but which nevertheless retain an axial cavity for much of the length of the crystal. *Ilmenite* is another phase in which skeletal crystals are not uncommon, usually of a parallel-plate type.

From the petrological point of view it is useful to be able to identify quench textures because it provides additional evidence bearing on the question of which phases belong to the groundmass and which to the phenocryst assemblage. By no means all volcanic groundmasses, however, contain characteristic quench crystals and in such cases grain-size criteria alone must be used to identify the groundmass assemblage. Conversely, pre-existing phenocrysts occasionally act as a nucleus for skeletal growth but may be recognised by their large size and the confinement of skeletal features to their margins.

Occasionally, particularly in the case of olivine, large crystals which nevertheless display some internal skeletal features present a problem, as it is not easy to decide whether they belong genetically to the intratelluric or to the quenching stage.

Final products of solidification

A common feature of the groundmasses of lavas, and of some shallow-level intrusions, is the occurrence interstitially of glass (or its devitrification products), feldspathoidal minerals or zeolites. The formation of glass is commonly ascribed to extreme quenching of liquid caused by very rapid temperature fall. However, the composition and structure of the liquid are also important factors in determining whether or not a glass will form. Should the last liquid to crystallise be enriched in silica (and potassium) the consequent increase in its viscosity will impede diffusion to possible crystal nuclei and the formation of glass by supercooling will be favoured. On the other hand, sodium is known to reduce the viscosity of silica-bearing melts and its presence in some quantity will favour crystallisation, with the formation of sodium bearing feldspathoids or zeolites (e.g. analcite). Tholeiitic (i.e. SiO_2-saturated or oversaturated) basaltic lavas often exhibit **intersertal texture**, that is, the presence of wedge-shaped patches of usually brown or greenish glass fitted between the groundmass plagioclase and other minerals, whereas alkaline basalts contain feldspathoid or zeolite minerals in the same position. This is one of the ways in which basalts of these different types may sometimes be distinguished petrographically. In shallow intrusions products of late crystallisation are more readily recognised, especially in tholeiitic rocks, which are more likely to contain interstitial **micropegmatite** (a micrographic intergrowth of potash feldspar and quartz) than glass, presumably because of slower cooling, although an origin by devitrification cannot in some cases be excluded.

It should be noted, however, that not all cases of interstitial zeolite or silica minerals need be of primary (if late) crystallisation. Analcitisation of lavas and pyroclastics by post-consolidation gas action near volcanoes has frequently been reported. The petrographic test is to determine whether the mineral is solely interstitial or is also replacing groundmass phases such as plagioclase. Again, the silicification of lavas by fumarolic action is not uncommon, trachytes seeming to be particularly susceptible. Characteristic petrographic evidence of this secondary mineralisation is a very fine lacy texture of the silica mineral pervading the rock and tending to cross primary textures such as flow banding. Unawareness of this effect has led in some cases to the misidentification (using chemical data) of trachytes as primary rhyolites.

Certain volcanic rocks of the calc-alkaline suite, in particular the two-pyroxene andesites typical of mature island arc volcanoes, contain interstitial silica minerals which, though clearly of late crystallisation, are not amygdaloidal. Quartz occurs, but more noteworthy are the other polymorphs of silica, tridymite and cristobalite. The experimentally determined temperature of the inversion of cristobalite to tridymite is 1470 °C at 1 atmosphere, in other words well in excess of the *liquidus* temperature of the lavas concerned. Clearly the cristobalite in natural rocks must be of a lower-temperature, metastable, type and this is confirmed by the discovery by Peck *et al.* (1966) that in a Hawaiian lava lake it forms at about 800 °C.

Tridymite, too, is metastable at low temperatures and, given time, will normally invert to quartz. Whereas in some cases tridymite exhibits the long, curved form typical of quench crystals, in some shallow level granophyre intrusions in the Scottish Hebrides, it has adopted a stouter prismatic habit which is preserved after the inversion, i.e. a paramorph. In ordinary or plane-polarised light the crystals are clear and homogeneous but crossed nicols reveal a mosaic structure in which separate sectors of a crystal extinguish at different positions indicating that the inversion has been of the domain type (cf. Fig. 12.2).

Although no exact line can be drawn between volcanic and plutonic rocks as far as petrographic features are concerned, the slower cooling undergone by magmas in the plutonic environment favours the development of sub-solidus changes in many rock-forming minerals, and hence such changes are easier to detect. These features are considered in Chapter 12.

Other groundmass textures

Sizeable masses of volcanic glass frequently show a distinctive pattern of circular fractures on the millimetre or sub-millimetre scale known as **perlitic cracks.** These are presumably the effects of volume changes (Fig. 7.6) and may persist as ghost features after devitrification, itself usually signified by the occurrence of radially disposed acicular crystals forming **spherulites.** Either of these textures, or in rarer cases both, may thus help to identify the former presence of glass and hence, indirectly, of rapidly quenched liquid. It would, however, be unwise to assume that the devitrified material preserves the chemical composition of the original glass ($\equiv$ liquid) since glasses are noted for their water-absorbent capabilities, as witness the higher water content of pitchstones when compared with obsidians, and also for their susceptibility to chemical leaching.

Perlitic cracks and devitrification are characteristics of natural glass whether the glass formed as a lava flow or as a component of a pyroclast flow or fall deposit. Many volcanic rocks originally described as lavas have been reinterpreted as the deposits of glowing avalanches and a few cases of reinterpretation in the opposite direction are known (e.g. the rhyolites of M. Amiata, Italy). No one name for these deposits seems to satisfy everyone but **ignimbrite** is probably the best known and most euphonious, and this we shall use, remembering that its etymology implies a particular origin (from a 'glowing shower') which may describe only one feature of the eruption responsible.

Whereas lava flows tend to be megascopically homogeneous, ignimbrites characteristically display a zonation which is principally due to post-depositional processes. A typical pattern (but there are many variations) is one of unsorted and relatively unconsolidated pumice fragments forming the upper parts, grading down into a more compact zone with sporadic **shards** (small cuspate glass fragments), or **fiamme** (that is, streaky or sinuously curved pumice fragments with ragged ends) of dark glass. The rock here has an overall streaky texture known as **eutaxitic.** A zone of massive dark obsidian may then be present and the lowest parts of the ignimbrite may again consist of relatively loose pumice and ash. Phenocrysts, xenocrysts and xenoliths, cognate or accidental, are strewn through the whole mass.

A given ignimbrite, then, may show considerable variation throughout its mass and no single sample can be taken to represent the whole, either petrographically or chemically. A rock composed of sporadic, typically broken crystals, set in, and frequently draped by, glass shards which possess sinuous or cuspate outlines (see Fig. 7.12) may be readily recognised as an ignimbrite. However, difficulty in recognition

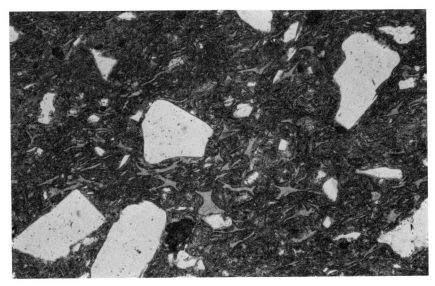

Figure 7.12 Ignimbrite with phenocrysts of quartz and feldspar (both colourless) and abundant shards forming clear areas of medium tone with indented (cuspate) curved outlines. The shards are deformed fragments of exploded vesicular glass.

increases in proportion to the degree of compaction and welding together of the glass shards (effected while they were still plastic) and the ultimate degree of welding into a massive obsidian produces a rock which megascopically shows little sign of its pyroclastic origin. Under the microscope, however, the obsidian may reveal curvilinear trains of opaque oxide granules which trace the intricate outlines of the original shards.

Other post-depositional processes which may obscure the primary features of an ignimbrite include devitrification of the glass and late-stage crystallisation of silica minerals from the vapour released as compaction of the vesicular fragments proceeded. Devitrification or slower cooling in unusually thick ignimbrites has in some cases resulted in the formation of granophyric intergrowth of quartz and feldspar.

Three particular aspects of ignimbrites signal the need for caution when these rocks are being considered as material for study by normal petrological and geochemical methods.

First, the evidently explosive nature of the eruption could cause dispersal of lighter fractions far from the main deposit ('aeolian differentiation'). Secondly, accidental xenoliths may form a sizeable component of ignimbrites and not all are large enough to be infallibly separated prior to chemical analysis. Thirdly, devitrification and recrystallisation could involve extensive transfer of certain chemical components within the ignimbrite.

It is quite possible that one or more of these factors may account for the compositional variations displayed by the glass fragments within a single ignimbrite which have otherwise been explained as the result of liquid immiscibility in the parent magma, and perhaps also differences between the composition of the glassy phase and the bulk composition of the ignimbrite.

In spite of these problems much important information has been gained from geochemical investigations. The commonest ignimbrite compositions are rhyolitic, rhyodacitic and andesitic, with rarer occurrences of alkaline types including high potash varieties among the Italian volcanics. Although basic pyroclast flows showing some welding have been described, these are very rare and do not contain extensive massive glass zones. Again it seems that composition of the original melt is an important factor in determining not only whether glass will form but also what the nature of an eruption will be.

Analyses of successively erupted ignimbrites have revealed systematic changes in chemical and isotopic composition, reflected for example in an upward decrease in normative orthoclase, that is, the reverse of what would occur in a normally differentiated magma chamber. The nature of ignimbrite eruption would seem *a priori* a more efficient mechanism than the eruption of lavas for emptying a magma chamber and thus an ignimbrite series could provide good material for studying a particular volcanic sequence (see Ch. 11).

8

Quaternary systems

Many of the fundamental principles of igneous petrology can be well understood after the combined study of ternary systems, petrography, and variation diagrams.

A consideration of quaternary and more complex systems is however useful in pursuing various aspects of petrology, particularly of basic, ultramafic, and granitic rocks, in more detail. In this and the succeeding chapters (9 and 10), attention is paid largely to basaltic systems. The study of granites is included in Chapters 12 and 13.

The reading of quaternary systems presents no more theoretical difficulty than that of ternaries, the principles involved being identical. However, the extra dimension introduces substantial difficulties in practice. Hence the treatment given here will not be comprehensive but will hope to serve as a foundation for readers who wish to study the subject in more detail.

Representation of composition

Composition in quaternary systems must be represented in three-dimensional space, the normal graphical device being the tetrahedron (Fig. 8.1). A bulk composition P, within the tetrahedron, has a composition in terms of the four components A, B, C and D as follows:

Per cent of A = 100 PQ/AQ
Per cent of B = 100 PT/BT
Per cent of C = 100 PR/CR
Per cent of D = 100 PS/DS

where the points, Q, R, S and T are found by projecting from each apex through P into the opposite face. This relationship is a direct extension into the extra dimension of the ternary principles given in Figures 4.1 and 4.2. The quaternary method given here is of course applicable to irregular as well as regular tetrahedra.

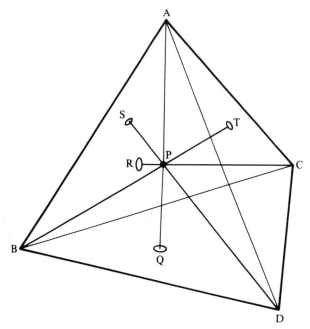

Figure 8.1   Representation of composition in a tetrahedron.

Sub-solidus joins

Quaternary systems in general contain more phases than ternaries and simply visualising the arrangement of two-, three- and four-end-member joins within the system presents substantial initial problems.

One of the best known quaternaries is $CaO-MgO-Al_2O_3-SiO_2$ (CMAS for short) illustrated in Figure 8.2 to show the positions of some of the geologically important phases which appear.

In the following discussion we shall assume that the diagram is plotted in molecular not weight proportions, though experimental work is normally reported in the latter. However, relationships are initially easier to visualise on the molecular basis.

Using a series of equations similar to those of chemical reactions, geometrical relationships can be investigated as in the following examples:

(a) What phase or composition lies at the mid-point of the join CS (wollastonite) – MS (enstatite)? Adding equal amounts of CS and MS together will clearly provide the solution, hence:

$$1 \text{ CS} + 1 \text{ MS} = CMS_2 \text{ (diopside)}$$

produces the solution.

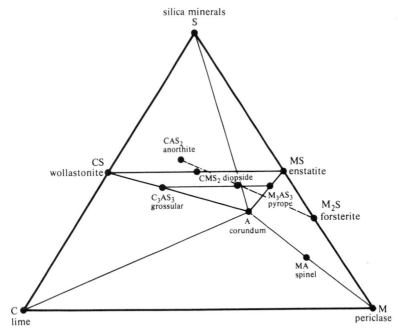

Figure 8.2 The system $CaO–MgO–Al_2O_3–SiO_2$ showing some important phases. Shorthand notation used is $C = CaO$, $M = MgO$, $A = Al_2O_3$, $S = SiO_2$. Hence, for example, MA (spinel) signifies $MgAl_2O_4$.

(b) Where does the join CAS_2 (anorthite) − M_2S (forsterite) cut the plane CS (wollastonite) − MS (enstatite) − A (alumina)? The composition of the point of intersection must be expressible in terms of $CAS_2 + M_2S$ in some proportion, and in terms of $CS + MS + A$. Hence, using p, q etc., to express the unknown coefficients we can write the equation:

$$pCAS_2 + qM_2S = rCS + sMS + tA$$

and in order that the equation shall balance for each component:

$$\text{for C} \quad p = r$$
$$\text{for M} \quad 2q = s$$
$$\text{for A} \quad p = t$$
$$\text{for S} \quad 2p + q = r + s$$

Since we are concerned only with proportions, let $p = 1$, then from the above, $r = 1$, $t = 1$, $q = 1$ and $s = 2$. Hence both sides of the equation sum to CM_2AS_3 which is the composition of the point of intersection. Further, since the garnet join (pyrope–grossular) runs from M_3AS_3 to C_3AS_3, this intersection point must lie on it, one third of the way along the join from pyrope to grossular for:

$$3CM_2AS_3 = 2M_3AS_3 \text{ (pyrope)} + C_3AS_3 \text{ (grossular)}$$

The above discussion is concerned with geometrical relations only and these two examples should suffice to illustrate the principles of the method. Further applications are considered later in the chapter.

Complexities introduced by solid solution have not been included in the above discussion. Clearly the complete sub-solidus diagram for a quaternary system consists of a number of sub-tetrahedra each representing the co-existence of four solid phases, and these are separated by three-end-member stable joins. In the absence of solid solutions the latter are planar triangles but if solid solution is present they develop into three-dimensional forms. Two-end-member joins may also occupy three-dimensional space if solid solutions are themselves ternary or quaternary.

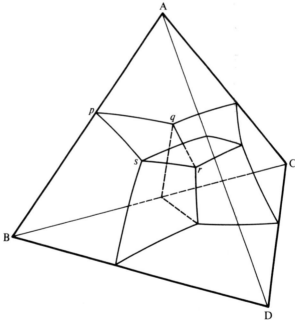

Figure 8.3   Perspective sketch of the simple quaternary system A–B–C–D. The upper part of the diagram is occupied by the A + L volume. This comes into contact with the B + L volume at the surface *p–q–r–s* which represents A + B + L. This surface meets the A + C + L surface in the boundary curve *q–r* which represents A + B + C + L. All four primary phase fields meet at the point *r* representing A + B + C + D + L.

Liquidus diagrams

In quaternary systems primary phase fields (i.e. the loci of liquids in equilibrium with one solid phase) become volumes which meet in surfaces representing liquid composition in equilibrium with two solid phases. Similarly any three primary phase volumes meet in a boundary curve representing liquids in equilibrium with three solids, while four primary phase fields must meet in a point. This point is a unique liquid equilibrating with four solids and represents a univariant (isobarically invariant) equilibrium.

A simple quaternary illustrates these features in Figure 8.3. Isotherms within primary phase fields are curved surfaces (Fig. 8.4) which can only properly be represented in three-dimensional models. Isothermal surfaces in adjacent primary phase fields meet at the boundary surface representing the locus of liquids in equilibrium with both solid phases and create thermal contours across it. The intersection of three such sets of isothermal surfaces obviously implies a specific temperature for each point on the boundary curve representing the locus of liquids in equilibrium with three solid phases. Isobaric invariant points where such boundary curves meet represent equilibria having specific and fixed temperatures.

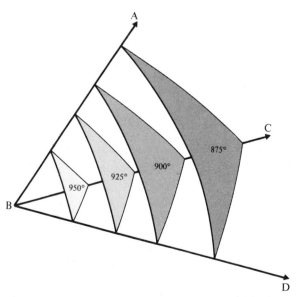

Figure 8.4   Perspective sketch of isothermal surfaces within the B + L volume of the system A–B–C–D (Fig. 8.3).

Crystallisation paths

Tetravariant equilibria. Equilibria between one solid phase and liquid have four degrees of freedom and are thus tetravariant or isobarically trivariant. Notice that the number of degrees of freedom in isobaric equilibria is the same as the number of dimensions needed to describe the locus of liquid composition, e.g. an isobaric trivariant equilibrium is equivalent to a primary phase *volume*, whereas liquid compositions in an isobaric divariant equilibrium (two solid phases + liquid) are represented by a *surface*. Liquid paths during crystallisation in tetravariant equilibria are simple projections away from the composition point of the crystallising phase. In the absence of solid solution these form a three-dimensional radiating set of rectilinear paths. Curvature appears if solid solution is present and requires tie-line information for its precise description.

Trivariant equilibria. Liquids crystallising on a boundary surface separating two primary phase volumes equilibrate with two solid phases and represent a trivariant equilibrium. Such equilibria may be co-precipitational or resorptional down-temperature for one solid phase (i.e. either of the form $A + B = L$ or $A = B + L$). The geometrical criteria for the distinction of these two types are an obvious extension of the Alkemade theorem. Using a system with an intermediate binary compound, Figure 8.5 illustrates various possibilities, including a case where the nature of the equilibrium changes with temperature.

Co-precipitational trivariant equilibria liquids will track across the surface following a linear path, in the absence of solid solutions, until they reach a divariant boundary curve limiting the surface. At this point the next phase will begin to crystallise. During equilibrium crystallisation in the resorptional case the liquid will behave similarly until resorption is complete, at which stage the liquid will leave the surface and enter the adjacent primary phase volume of the non-resorbed phase. Alternatively resorption may not be complete before the next down-temperature equilibrium is encountered. The geometrical conditions necessary for completion of resorption are illustrated in Figure 8.5. Under conditions of fractional crystallisation liquids will, of course, pass directly through the surface suffering only a change of direction as the fractionating solid phase changes.

Divariant equilibria. Divariant equilibria are represented by liquids lying on a boundary curve and in equilibrium with three solid phases. By analogy with earlier reasoning they may be classified as co-precipitational, mono-resorptional

($A + B = C + L$ type) or bi-resorptional ($A = B + C + L$ type).

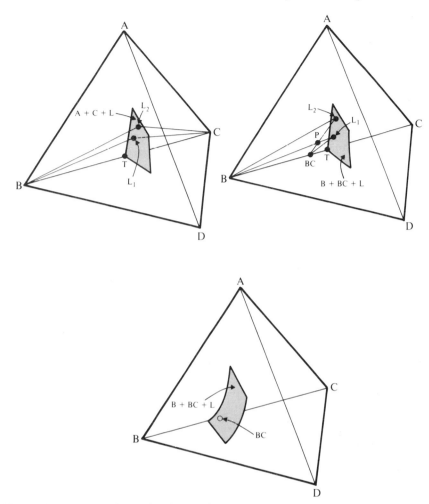

Figure 8.5 Perspective sketches of the system A–B–C–D to show crystallisation in trivariant equilibria. Top left: this is the co-precipitational case with two liquids shown on the surface B + C + L (shaded). The two triangles L_1–B–C and L_2–B–C are co-planar and can refer to any bulk composition lying in L_1–B–C. The back-tangent from L_2 through L_1 (the higher-temperature liquid) cuts the B–C join between B and C somewhere near T (the intersection will only be at T if the surface is planar). Top right: resorptional case. The shaded surface is B + BC + L where BC is an intermediate compound. L_1–B–BC and L_2–B–BC are again co-planar and refer to any bulk composition lying in their intersection P–B–BC. The back-tangent L_2–L_1 cuts B–BC *projected* near T. Bulk composition P has just reached the stage of complete resorption as the L–BC tie line sweeps upwards. The reaction is B + L = BC. Lower diagram: the surface B + BC + L is curved so that it is resorptional (upper part) and co-precipitational (lower part). BC composition point is behind the surface in this view.

Geometrically this depends on the relationship of the back-projection of the tangent of the liquid curve to the sub-solidus triangle representing the three solid phases as shown in the mono-resorptional example in Figure 8.6.

During equilibrium crystallisation in divariant equilibria the liquid will track down-temperature along the boundary curve if the equilibrium is co-precipitational or until resorption is complete in the other cases. If the equilibrium is of the mono-resportional type the liquid will leave the boundary curve along one of the adjacent trivariant surfaces on the completion of resorption, unless a univariant equilibrium is encountered first. In bi-resorptional cases the liquid will leave the curve and pass directly into an adjacent primary phase volume only if both phases complete resorption simultaneously. Normally it will behave as in the previous case. During fractional crystallisation liquids will pass across the curve onto an adjacent surface or directly into an adjacent volume, depending on the equilibrium type.

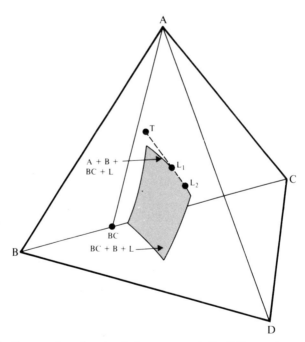

Figure 8.6 Perspective sketch of the system A–B–C–D to show criteria for resorption in divariant equilibria. L_1 and L_2 are successive liquids equilibrating with A + B + BC. The back tangent cuts the A–B–C plane outside the A–B–BC triangle signifying resorption of B. Had T been inside the triangle A–B–BC the equilibrium would have been co-precipitational.

Univariant equilibria. Liquids involved in univariant equilibria are the initial liquids of melting of four-phase solid assemblages or final liquids of equilibrium crystallisation. They are classified as co-precipitational (quaternary eutectics, $A + B + C + D = L$ type), and mono-, bi-, and tri-resorptional. Their nature depends upon the liquid position relative to the appropriate sub-solidus tetrahedron. If the liquid lies inside the tetrahedron the equilibrium is co-precipitational; if outside there are three cases in which the liquid 'sees' respectively one, two, and three faces of the tetrahedron (the mono-, bi-, and tri-resorptional cases respectively).

Heat extraction in univariant equilibria (N.B. the temperature is fixed) leads to the disappearance of the liquid in co-precipitational cases. In the other cases liquid will remain when the resorption of one phase is complete and will thus move on down-temperature along the appropriate boundary curve. For special bulk compositions which show the simultaneous resorption of two phases (bi- and tri-resorptional equilibria) liquid will pass directly onto an adjacent trivariant surface, as will all fractionating liquids in the bi-resorptional case. In the very special case in which three phases complete resorption simultaneously (tri-resorptional equilibria) the liquid will pass directly into the appropriate adjacent primary phase volume. All fractionating liquids encountering a tri-resorptional univariant equilibrium will of course also show this behaviour.

General relationships of univariant and divariant equilibria are shown in flow-sheet form in Figure 8.7.

Projection methods

Because of the practical difficulties inherent in the use of three-dimensional models and perspective sketches, quaternary studies are frequently presented as two-dimensional projections. Data points are normally projected from the composition point of a pure component or important phase into a convenient plane. Two such projections allow the precise location of a point in three-dimensional space. Thus in Figure 8.8 a point is shown projected from two different apices into the opposite face of the tetrahedron. The numerical equivalent of the graphical procedure is as follows: suppose the point is represented by $A_pB_qC_rD_s$ where the subscripts indicate the percentages of the four components A–D, then projection from A into the face B–C–D simply involves recalculating q, r, and s as percentages of the total $q + r + s$, and then plotting as in a ternary.

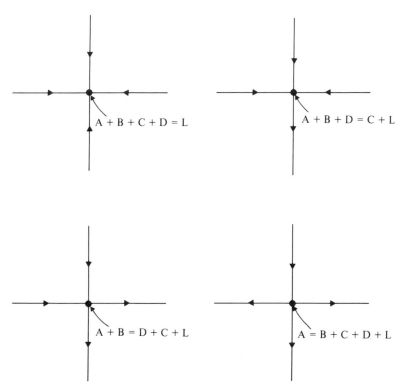

Figure 8.7 Flow sheet showing relationships of temperature arrows in divariant equilibria associated with four different types of univariant equilibrium. Each arrowed line represents an equilibrium of liquid and three solid phases but the nature (i.e. whether resorptional or not) of such equilibria is not related to the nature of associated univariant equilibria since the geometrical criteria are different and independent.

The original quaternary co-ordinates can be retrieved from two sets of ternary co-ordinates (e.g. $A_l B_m D_n$ and $A_o D_p C_q$) as follows: ratios of original amounts are preserved in the ternary co-ordinates in the form:

$$A/D = l/n = o/p$$
$$B/D = m/n \text{ and } C/D = q/p$$

Then if $D = 1$ the relative proportions of A, B, C, and D can be found. These are then recalculated as per cents to recover the original parameters.

Projection from the composition point of an important phase is frequently useful (see Ch. 9) and this is illustrated in Figure 8.9 where the surfaces bounding the primary phase field of the phase from which projection is made (A) are shown with thermal contours. The resultant

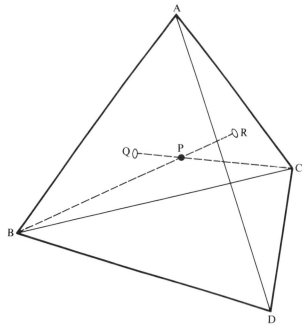

Figure 8.8 Perspective sketch of the system A–B–C–D to show projection of a quaternary composition P from B into A–D–C (point R) and from C into A–B–D (point Q).

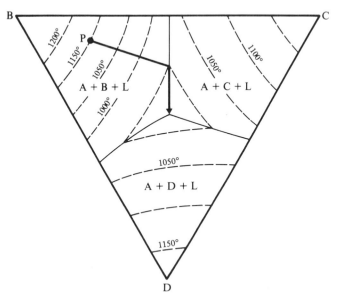

Figure 8.9 Projection from A of the system A–B–C–D into B–C–D (Fig. 8.3) to show surfaces bounding the A + L volume.

diagram looks superficially like a ternary liquidus projection, but there are several important differences.

Firstly, the crystallisation behaviour of a composition such as P can be read only if P lies in the primary phase field from which projection is made. The composition P′ in Figure 8.10 is coincident with P in projection but lies in a different primary phase field and therefore must show a significantly different crystallisation behaviour. In this case given the single projection (Fig. 8.9) and the information that P lies in the primary phase field of A the crystallisation behaviour can be read as follows:

(1) On cooling, P begins to crystallise A, but it is not possible to specify with the information given the temperature at which A appears. As A crystallises the residual liquid (in the absence of solid solution) will move directly away from A so that its position in projection *does not change*.

(2) When the temperature reaches 1150 °C (Fig. 8.9) the residual liquid will meet the surface representing the equilibrium A + B + L. Crystals of B will now appear and the residual liquid will move away from the point on the A–B join which represents

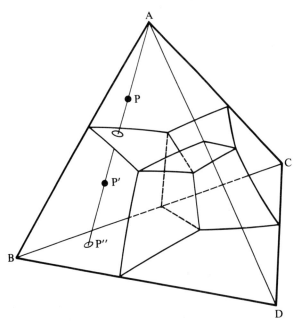

Figure 8.10 Projection of composition P from A (see Fig. 8.9). P and P′ coincide in their projected position P″.

the cotectic proportions. In projection this will appear as a liquid path moving directly away from B in the absence of solid solutions.

(3) At 1000 °C the liquid will reach the edge of the A + B + L surface, encountering the boundary curve A + B + C + L. Assuming this to be a co-precipitational equilibrium the liquid will now proceed down the boundary curve, crystallising A, B, and C until it reaches the univariant point where D begins to crystallise.

(4) In co-precipitational cases the liquid will now disappear and the composition will solidify. In more complex cases A will be lost by reaction and the liquid will disappear out of sight into a more remote part of the system, a different projection being needed to chart its progress.

A second important difference between this sort of projection and a ternary liquidus projection is that the lever rule cannot be applied directly to it since relative lengths of lines do not maintain their proportions when projected (see Fig. 8.11). When relative lengths of tie-lines are required, numerical methods provide the most convenient solutions (see Exercises).

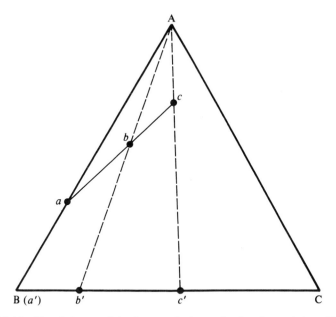

Figure 8.11 Breakdown of the lever rule in projection from A into B–C. The ratio $ab:bc$ is clearly not the same as $a'b':b'c'$. Knowledge of the original co-ordinates of a, b and c, however, provides a ready numerical solution (x parts of a + y parts of c = b).

The system forsterite–diopside–anorthite–silica

This system is part of the larger CMAS system (Fig. 8.2). The system will be used to illustrate some of the properties of quaternaries and also to act as an introduction to the crystallisation behaviour of basalts. The silica-poor part of Fo–Di–An–SiO$_2$ acts as an analogue for tholeiitic basalts and many ultramafic (peridotitic) rocks.

Four relevant ternary (or pseudo-ternary) systems are given in Figure 8.12 and on the assumption that the remaining bounding ternary Di–An–SiO$_2$ is a simple eutectic system this enables an approximate

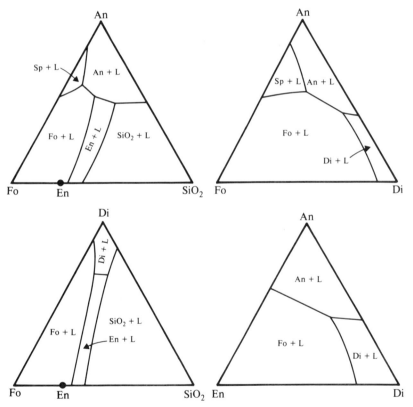

Figure 8.12 Four important joins in the system forsterite–diopside–anorthite–silica. For details of Fo–An–SiO$_2$ and Fo–An–Di see Figures 4.20 and 4.21. Data for Fo–Di–SiO$_2$ are after Bowen (1914) and Kushiro (1972b) but have been simplified with respect to the Ca-poor pyroxene field. Here 'En' includes the fields of both proto-enstatite and pigeonite. The two-liquid field has also been omitted. Data for En–An–Di are from Hytönen and Schairer (1961).

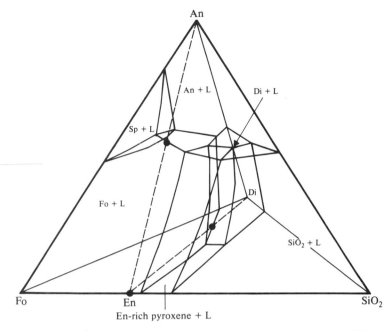

An

An + L

Di + L

Sp + L

Fo + L

Di

SiO₂ + L

Fo

En

SiO₂

En-rich pyroxene + L

Figure 8.13   Schematic perspective drawing of Fo–Di–An–SiO₂.

diagram of the quaternary to be constructed as in Figure 8.13. Phase relations have been somewhat simplified compared with published literature in respect of the pyroxene fields (e.g. in Fo–Di–SiO$_2$).

One of the most important features of the system is the pseudo-ternary nature of the En–Di–An join which contains a large field of Fo + L but no En-rich pyroxene field (the latter being the analogue of orthopyroxene or pigeonite in natural rocks). This implies that the entire En-rich pyroxene + L field lies on the SiO$_2$-rich side of the join and thus the reaction relationship Fo + L = En noted previously in ternary systems (e.g. Fig. 4.20) persists into the quaternary. As will be seen in Chapter 9 this relationship probably disappears with increasing pressure as the olivine field contracts. However, its presence at low pressures implies that solid mixtures of magnesian olivine, enstatite, diopside, and basic plagioclase (plagioclase peridotites or olivine–orthopyroxene gabbros) should give SiO$_2$-oversaturated liquids as their initial melting products. Similarly, because of the reaction relationships, basaltic liquids originating in the olivine field (as long as they do not lie on the SiO$_2$-poor side of Fo–Di–An) are capable of fractionating to SiO$_2$-bearing end products.

The projection from anorthite. As an illustration of the way in which the data from three-end-member joins can be used in the more precise delineation of a quaternary we consider the construction of the projection from anorthite. Inspection of Figure 8.13 suggests that the projection must look as shown in Figure 8.14 which can be thought of as the view of the inside of the tetrahedron an observer looking from the composition point of anorthite would have. The points ringed are those for which precise information can be obtained from the ternaries. Considering one of these (A), the point when the Fo + An +Di +L curve meets the Fo–Di–An face, reference to Figure 8.12 shows that its co-ordinates, when projected from An into Fo–Di can readily be calculated. This information can be used to find the precise position of A as shown in Figure 8.14. The other ringed points can similarly be transferred together with their anorthite contents derived by direct measurement in the ternaries. Complete information about the anorthite content is available for example for the lines A–B, C–D, and En–E which enables the sketching of the An contours on the An + Fo + L surface shown in this case. These should not be confused with thermal contours as used in Figure 8.9.

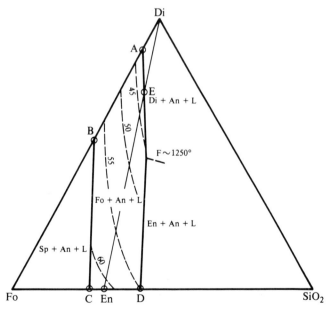

Figure 8.14 Projection from anorthite into Fo–Di–SiO$_2$ for the system Fo–Di–An–SiO$_2$. The Fo + An + L surface is shown contoured in An per cent.

The remaining piece of information which it would be highly desirable to have is the location of the univariant point F (Fo + Di + An + En-rich px + L). For this we turn to the isothermal sections provided by Hytönen and Schairer for En–An–Di. One of the critical sections is shown in Figure 8.15. Considering this section in a little detail and referring to Figure 8.13 for guidance, we recall that all starting compositions lie in the En–An–Di join and about half of them (those rich in En) lie initially in the forsterite field. Thus as cooling proceeds each of these compositions at some stage begins to crystallise forsterite so that residual liquids fan away from the Fo composition point and enter the SiO$_2$-oversaturated part of the system. These liquids divide themselves into those which will next encounter the surfaces of Fo + En + L, Fo + Di + L, or Fo + An + L respectively. At a somewhat lower temperature some of the liquids on the Fo + En + L and Fo + An + L surfaces will have reached their mutual boundary, the curve Fo + En + An + L, and similarly other liquids will have reached the Fo + En + Di + L and Fo + Di + An + L boundaries. However, a consideration of the Alkemade theorem (assuming that pyroxene solid

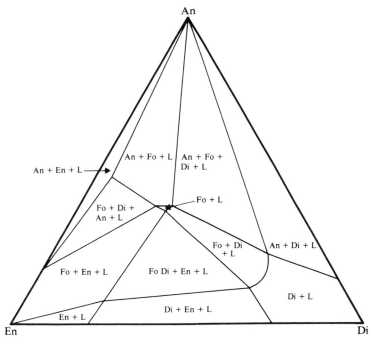

Figure 8.15 Isothermal section for En–An–Di at 1260 °C (Hytönen and Schairer 1961).

solutions do not extend right across the diagram) shows that the univariant point (F ≡ Fo + En + Di + An + L) must lie at a lower temperature than any temperature on the adjacent surfaces or boundary curves. Hence some liquids near this point may still be crystallising forsterite only while the other assemblages mentioned above exist in other bulk compositions. It is for this crucial temperature that the isothermal section of Fig. 8.15 has been drawn. In the centre is a small cuspate triangular field of Fo + L. Each corner of this field represents a liquid on a boundary curve, each edge represents the liquid on a boundary surface, while the field itself is that range of liquids still inside the primary phase volume of forsterite. All these liquids lie somewhere to the SiO$_2$-rich side of En–An–Di. Lines linking liquids to their appropriate equilibrium solids create a three-dimensional isothermal diagram consisting of a number of different contiguous three-dimensional volumes each representing a different assemblage. The isothermal section for En–An–Di is of course a specific two-dimensional slice across this figure.

Somewhere within the Fo + L field and thus obviously fairly closely located by the data given in the isothermal section is the specific liquid which will reach the invariant point F directly, without encountering any other equilibria *en route*. This liquid therefore gives a bearing on the position of F as illustrated in Figure 8.16. The information has been used to locate F in the plagioclase projection together with the assumption that the Fo + An + Di + L boundary can be assumed to project linearly through points A and E. The temperature of F is constrained by the data of the isothermal section to lie a little below 1260 °C.

Use of sets of projections in reading crystallisation sequences. Figure 8.17 shows the projections from plagioclase and olivine for the SiO$_2$-poor part of the system Fo–Di–An–SiO$_2$. Reference to Figure 8.13 will make it clear why the olivine projection has the form shown. A composition P (Fo$_{50}$Di$_{10}$An$_{20}$SiO$_{2\,20}$) is shown in both projections. In this case since the An + Fo + L and Fo + En + L surfaces are contoured it is clear that P lies in the primary phase field of forsterite (e.g. in the anorthite projection P, with only 20% An, must lie on the anorthite-poor side of the An + Fo + L surface, similarly with 50% Fo it lies on the forsterite-rich side of the Fo + En + L surface). Thus forsterite is the first phase to appear on cooling, and enstatite the second. Crystallisation of enstatite (accompanied in this case by forsterite resorption) must give rise to a residual liquid path running approximately away from the enstatite composition point in the forsterite projection as shown in Figure 8.17. Thus anorthite is the

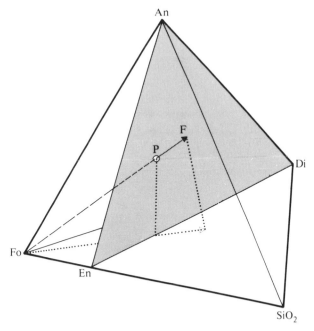

Figure 8.16 Projection of P (the small Fo + L field of Fig. 8.15) from forsterite gives a bearing on F, the univariant point. This is shown in the anorthite projection by the dotted lines.

third phase to crystallise as the residual liquid reaches the Fo + En + An + L boundary curve somewhere near Q. Q cannot, however, be located exactly without information on the composition of the enstatite solid solutions concerned. Since these are largely towards diopside, the projected position of the actual enstatite crystallising would lie displaced towards Di on the Di–SiO$_2$ join in the forsterite projection. Thus the actual projected liquid path would be qualitatively like that shown by the curve leading to Q′. In the case discussed, since P lies on the Fo-rich side of En–An–Di, resorption of Fo will not reach completion at any stage.

In contrast to the case discussed above where contours are available for important surfaces it is often useful to attempt to predict crystallisation sequences without this information. The preceding example also illustrates that it is only possible to predict accurately the identity of the second and third phases to crystallise in projections from the composition point of the primary phase. When this is not known, and contours are not available, an approximate prediction can be made as follows.

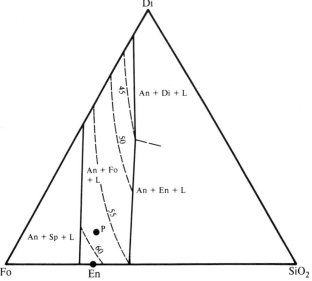

(a) Projection from anorthite

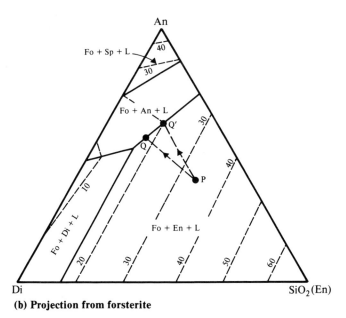

(b) Projection from forsterite

Figure 8.17 The system Fo–Di–An–SiO₂ showing projections (a) from anorthite, with contours in An per cent on An + Fo + L, and (b) from forsterite with Fo per cent contours on all surfaces.

Returning to P in Figure 8.17 the fact that it projects into the An + Fo + L field in the anorthite projection suggests at first sight that it lies either in Fo + L or An + L or on the surface An + Fo + L in between. Similarly, in the forsterite projection the position in the Fo + En + L field invites the speculation that P lies in Fo + L, or En + L, or on Fo + En + L (neglecting more silica-rich possibilities like tridymite + L). Of these two lists only Fo + L is common and therefore would appear to be the correct solution. With this decision made, the crystallisation path could be read from the forsterite projection as before. However, the initial conclusion will only be true if those surfaces out of sight below the surfaces shown in the projections lie parallel to the projection direction. In some cases this condition is approximately fulfilled but it cannot be relied on. Figure 8.18 illustrates the problem when the condition is conspicuously unfulfilled. Here a point P might appear to lie in A + L, B + L, or on A + B + L. However, because the B + D + L surface curves strongly, the point actually lies in D + L. Frequently, however, sufficient information is not available to be certain of the primary phase involved and it is only possible to make a guess on the lines of the above reasoning. The closer

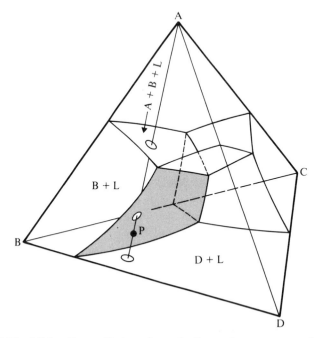

Figure 8.18 Misleading effects of projection when a concealed surface (shaded) is not parallel to the line of projection.

the point plots to the boundary of the field it projects into, the more likely such guesses are to be wrong.

Projection from a non-apical composition point. It is frequently extremely useful to project data from points other than the apices of the main tetrahedron. Firstly, as indicated previously, investigation of crystallisation is best carried out in projections from the composition point of the primary phase, and this does not always coincide with a pure component (e.g. enstatite in Fo–Di–An–SiO_2). Secondly it is useful to determine the relative positions of points and specific planes. For example, if the composition of liquid at a univariant point is known it is of obvious interest to know how its position relates to the appropriate sub-solidus tetrahedron, because this will enable the determination of the type of reaction involved. Equally well the position of a starting composition relative to various possible sub-solidus joins is important to know since this tells us the final products of equilibrium crystallisation. The solution to these latter problems consists of finding a projection point lying within the plane of immediate interest. The plane will then appear as an easily identifiable line in projection and projected points such as liquid compositions will be seen to lie either on one side of it or the other. A number of such projections will of course locate a point within a particular sub-tetrahedron.

The method is illustrated in Figure 8.19 where the sub-solidus joins for a system containing three binary compounds are illustrated. The problem posed is: given a known composition ($A_{48}B_{32}C_{10}D_{10}$ in weight per cent) somewhere within the system, which of the sub-tetrahedra does it lie in? It is likely from its co-ordinates to lie either in $A–CD_3–BC_3–AB$ or in $B–CD_3–BC_3–AB$ since these tetrahedra occupy most of the system. Projection from CD_3 into the plane A–B–C will solve this problem since the critical plane $CD_3–AB–BC_3$ will appear as an easily plotted line across it.

To project a composition from CD_3 it is only necessary to calculate the effects of removing a certain proportion of CD_3 from the bulk composition so that the resultant lies in the plane A–B–C. Thus where p, q, r etc. are the coefficient to be found:

$$p \times \text{bulk composition} - q\ CD_3 = r\ A + s\ B + t\ C$$

by analogy with the reasoning developed at the beginning of this chapter. Suppose the weight per cent compositions of CD_3, BC_3 and AB are respectively $C_{15}D_{85}$, $B_{20}C_{80}$, and $A_{52}B_{48}$ this becomes:

$$p\ A_{48}B_{32}C_{10}D_{10} - q\ C_{15}D_{85} = r\ A + s\ B + t\ C$$

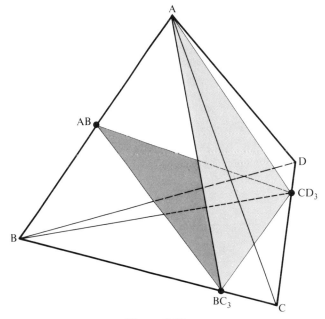

Figure 8.19

which gives

$$48p = r \qquad \text{for A}$$
$$32p = s \qquad \text{for B}$$
$$10p - 15q = t \qquad \text{for C}$$
$$10p - 85q = o \qquad \text{for D}$$

Letting $p = 1$ this gives the solutions $r = 48$, $s = 32$, and $t = 8.24$ which when recalculated to a sum of 100 to allow plotting in A–B–C gives approximately $A_{54}B_{37}C_9$. This is shown as point P in Figure 8.20 together with the various planes. It is clear that P must therefore lie in the sub-tetrahedron $A-CD_3-BC_3-AB$.

This sort of reasoning can be extended to any projection exercise in a quaternary system and in some cases will involve negative plotting parameters. For example in the present instance compositions lying on the D-rich side of the plane $A-B-CD_3$ can only project in A–B–C extended. Calculated projection parameters of such points will have negative values for C and may have A and B values >100. The application of this exercise to the determination of how a back-projected tangent to a liquid path when three solid phases are involved relates to the sub-solidus triangle is obvious. A negative

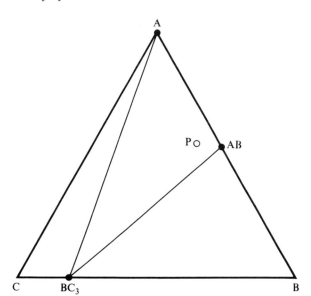

Figure 8.20   Projection of joins in the system illustrated in Figure 8.19 into A–B–C from CD_3.

parameter will indicate resorption for that phase, two negative parameters bi-resorption etc.

In conclusion it is worth remarking, however, that numerical solutions are best checked semi-quantitatively against a diagram or model, otherwise strange results can be obtained. For example in Figure 8.19 compositions more D-rich than CD_3 do not project into A–B–C at all (they lie between D and the plane $D = 85$, which is parallel to A–B–C). A numerical approach might however accidentally produce a solution for the other end of the projected line which does cut A–B–C extended.

Exercises

1. A projection from D into A–B–C for the true quaternary system A–B–C–D is given in Figure 8.21. A binary compound AC is present in the system. None of the phases shows solid solution. Describe the equilibrium crystallisation sequence of a composition P which lies in the primary phase field of D.
2. A quaternary system A–B–C–D contains a binary compound (BC) of composition $B_{55}C_{45}$ and a ternary compound (ABC) of composition $A_{45}B_{37}C_{18}$. Two liquids in equilibrium with BC, ABC, and pure B have compositions $A_{35}B_{23}C_{29}D_{13}$ and $A_{38}B_{19}C_{27}D_{16}$. By calculating the position

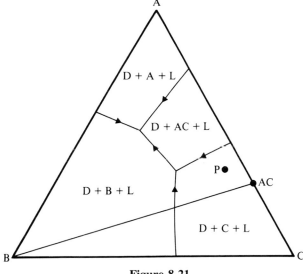

Figure 8.21

where the back-tangent through the two liquid compositions cuts A–B–C
determine the likely nature of the equilibrium B + BC + ABC + L.

3. Hytönen and Schairer (1961) in their study of the join En–An–Di in
   Fo–Di–An–SiO_2 found that certain enstatite-rich and diopside-rich starting
   compositions crystallised finally to assemblages of pyroxene (either
   enstatite-rich or diopside-rich) plus tridymite. What general conclusion can
   be drawn about the composition of these pyroxenes?

9
Experimental work on natural basaltic and allied rocks

Introduction

The experimental study of melting relationships in natural basaltic rocks dates from the late eighteenth century when Sir James Hall managed successfully to fuse samples of dolerite in support of the 'Plutonist' school of geological thought. However, systematic research has gained impetus only in the twentieth century, and particularly since 1962, the year which saw the appearance of the classic study by Yoder and Tilley– 'Origin of basalt magmas: an experimental study of natural and synthetic rock systems'.

The aims of experiments on natural materials are concerned, amongst other things, with the determination of crystallisation behaviour, which allows subsequent disciplined prediction of fractionation paths, and melting behaviour, particularly of ultramafic rocks, which may cast light on the deeper origins of magmas. Although at first sight experiments on natural rocks are obviously more relevant to petrology than are studies of simple systems, the diversity of potential starting materials leads to a different sort of uncertainty in the application of the results to general problems. One import of difficulty is to assess the role of volatile constituents such as H_2O and CO_2 and the oxidation state under which natural magnas evolved, since these can have an influence on phase relations. Thus experiments on natural rocks can only be relied on to produce directly relevant results if correct assumptions have been made about volatile constituents. Normally there is no exact basis upon which such assumptions can be made because volatile constituents are notoriously fugitive. Post-consolidation alteration may in addition affect further changes in volatile content and oxidation state. In addition, experimental charges often lose iron to platinum capsules. Despite these limitations, however, it is possible to derive a wealth of useful information from natural rock experiments. Many basaltic systems, for example, appear to have evolved under conditions approximating those of 'dry' experiments and direct testing of specific petrogenetic hypotheses evolved elsewhere is thus possible. The study of 'wet' and CO_2-bearing systems is both experimentally and conceptually more difficult and will be covered only in general terms in this work.

Atmospheric pressure experiments

Although it is obvious that most magmas must have had a long and possibly complex history of evolution under high confining pressures, the study of phase relations at 1 atmosphere is of prime interest because it confirms the importance of low-pressure crystallisation and fractionation histories. The almost ubiquitous presence of phenocrysts in lavas, the existence of layered intrusions, the evidence of effective crystal settling even after the extrusion of lavas, are all features pointing to the importance of low-pressure processes. The reason for the dominating effect of such processes is presumably the large difference in temperature between a magma and its surroundings when it reaches the surface or the near-surface environment. This is the stage of magmatic evolution during which fractional crystallisation is likely to be at its most effective. Because the low-pressure environment is now relatively well understood it might be thought lacking in interest, and it is certainly true that the investigation of high-pressure magmatic evolution is a current research preoccupation. However, high-pressure effects can best be detected if they appear as anomalies during low-pressure investigation. Hence low-pressure studies form an almost essential adjunct to other research.

Experimental methods. The quenching method is used as in the study of synthetic systems. Finely ground rock powder is placed in a metallic capsule, both iron and platinum having been widely used. In earlier experiments there was frequently no attempt to control f_{O_2}, the oxygen fugacity, and capsules were either open to the air or surrounded by a stream of argon. In short runs under these circumstances f_{O_2} is to some extent buffered by the presence of iron-bearing phases within the sample. Alternatively f_{O_2} may be buffered by an external buffer capsule containing mixtures such as nickel and nickel oxide which maintain f_{O_2} at a fixed value during the run since the internal platinum capsule is permeable to volatile constituents. Several such buffer mixtures are widely used; of these nickel–nickel oxide, quartz–fayalite–magnetite, and haematite–wüstite are the most appropriate to basaltic rocks. Gas mixtures flowing over the capsule are also widely used to produce f_{O_2}

control similar to that of buffer mixtures. Details of the general topic are given by Edgar (1973, Ch. 8). For present purposes we note that choice of f_{O_2} conditions for experiments will have a large influence on the temperature at which oxide phases such as magnetite crystallises in experiments but has a relatively smaller effect on the crystallisation of olivine, pyroxene, and plagioclases. Many petrogenetic conclusions reached are thus not seriously invalidated by uncertainty about the appropriate f_{O_2}.

The lavas of Kilauea, Hawaii. An excellent general account of variation in Hawaiian lavas is given by Macdonald and Katsura (1964). One of the volcanoes, Kilauea, is amongst the most intensively studied in the world and is characterised by the eruption of tholeiites, often rich in olivine phenocrysts. Compositional variation in the tholeiites is dominated by olivine control but some Mg-poor tholeiites show evidence of control by clinopyroxene and plagioclase fractionation. More evolved rocks are rare but include the oozes collected from drill holes in the Alae lava lake (Wright & Fiske 1971) illustrated in Figure 2.5. The volumetric abundance of olivine-rich rocks has suggested to most workers that olivine tholeiite magma has a parental status at Kilauea and that the more picritic magmas are formed by accumulation of olivine. Conversely, fractionation of olivine from the parent appears to lead to liquids in equilibrium with clinopyroxene and plagioclase, olivine being lost by reaction with the liquid. Thus fractionation of clinopyroxene and plagioclase is possible.

The above conclusions can be reached by a consideration of the petrography, compositional, time, and volumetric relations of the lavas. Experimental studies at 1 atmosphere (see Table 9.1) by Tilley *et al.* (1965) and Thompson and Tilley (1969) confirm that the model is

Table 9.1   Highest temperatures of crystallisation of major phases in Kilauean lavas (from Thompson & Tilley 1969)

Sample	Olivine (°C)	Clinopyroxene (°C)	Plagioclase (°C)
1959a	1321	1178	1159
1960–1	1273	1172	1159
1959b	1255	1172	1159
F14	1252	1171	1159
M1	1192	1176	1157
1960e	1166	1159	1159
S1$_\alpha$	1163	1163	1163
F6	1159	1159	1159
AL4	–	1083	1083

entirely plausible in terms of 1-atmosphere phase relations. Predicted 1-atmosphere phase relations from the above information would include the following:

(a) There should be a large group of rocks with olivine on the liquidus.
(b) If these lie compositionally on olivine-control, the second phase to enter during cooling should be the same phase in each case (either clinopyroxene or plagioclase) and should enter at an approximately constant temperature.
(c) Only a short temperature interval should separate the entry of the second and third phases otherwise a substantial stage of olivine + plagioclase or olivine + clinopyroxene fractionation should be detectable in the compositional variation. In fact some rocks related largely by clinopyroxene control have been identified by Murata and Richter (1966).
(d) More evolved rocks should show plagioclase and clinopyroxene on the liquidus without olivine.

Figure 9.1 shows the melting data obtained by Thompson and Tilley (1969) for a selection of Kilauean rocks, including as the most evolved (iron-enriched) sample a quenched ooze from the Alae lava lake (Peck *et al.*, 1966). In this diagram temperatures showing the first appearance of a phase during cooling are plotted against an Fe/Mg index which rises with fractionation. A plot of temperature against MgO content (falling with fractionation) has essentially the same form.

That the predictions made above are adequately fulfilled is clear from the diagram. In the more basic rocks a variable but substantial crystallisation interval of olivine is followed by the entry of clinopyroxene in each case at 1170–1180 °C, and shortly thereafter plagioclase at about 1160 °C (see Table 9.1). One specimen (1960e) shows the olivine entry only just in advance of clinopyroxene, while two (S1$_\alpha$ and F6) show the simultaneous entry of all three phases. The evolved specimen AL4 no longer has olivine on the liquidus but is saturated with regard to clinopyroxene and plagioclase.

The additional points KIpb and KIg are important because the latter is the separated groundmass of the former, a picrite basalt rich in olivine phenocrysts. KIg demonstrates that *liquids* within the olivine primary phase field at 1 atmosphere exist. Hence, from the above, the necessary conditions are fulfilled for olivine accumulation to take place to yield magmas more basic than KIg. Crystallisation of olivine from a liquid like KIg can give residual liquids with phase relations similar to those of 1960e, S1$_\alpha$ and F6, while fractionation involving olivine, clinopyroxene and plagioclase or clinopyroxene and plagioclase alone (it is not clear at

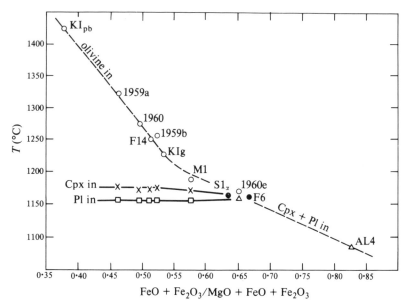

Figure 9.1   Crystallisation behaviour of Kilauea lavas at 1 atmosphere (after Thompson and Tilley 1969). Broken lines show approximate entry of each phase on cooling. Symbols show data of Table 9.1 for each sample: open circles – olivine liquidus, crosses – clinopyroxene, squares – plagioclase, filled circles – three phases enter together, triangles – plagioclase and clinopyroxene enter together. KIpb and KIg are referred to in the text.

what stage olivine disappears from the liquidus) can lead to liquids like AL4.

Basalts of Snake River and Skye. In contrast to the comparatively simple results obtained for Kilauea the basalts of the Snake River area of Idaho and of the Isle of Skye, Scotland, reveal in the melting behaviour the operation of more complex factors. These rocks have been studied by Tilley and Thompson (1970) and Thompson *et al.* (1972) and their experimental petrology has been reviewed by Thompson (1972a). Simplified diagrams after Thompson (1972a) are given in Figure 9.2 where the original data points have been omitted for clarity. The most magnesian rocks of both series show the order of crystallisation olivine–plagioclase–Cpx and both become cotectic with regard to olivine and plagioclase in somewhat more iron-rich compositions. Both then show plagioclase going onto the liquidus though this is more pronounced in the Skye series. Neither series shows clinopyroxene as a liquidus phase at any stage.

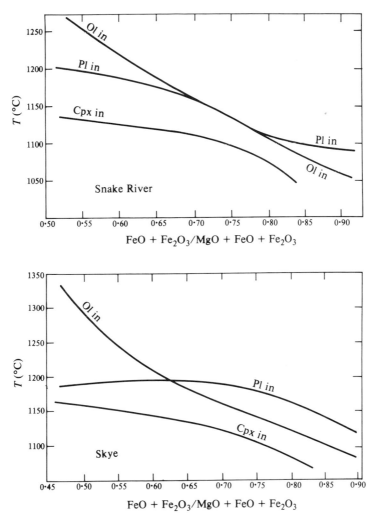

Figure 9.2 1-atmosphere melting relations of Snake River and Skye lavas (simplified after Thompson 1972a, original data points and magnetite entry curves omitted. Curves have been generalised slightly compared with the original).

Studies of compositional variation using the computer method of Bryan *et al*. (1969) led Thompson (1972b) to conclude that the removal of substantial quantities of all three phases, olivine, clinopyroxene and plagioclase is necessary to drive residual liquids along the Snake River and Skye chemical trends. None of these rocks, however, contains clinopyroxene phenocrysts, though olivine and plagioclase are common even when total phenocryst contents are very small. Hence two anomalies may be identified. Firstly, the phenocrysts present are not the assemblage required to model the chemical variation; secondly, olivine–plagioclase–phyric lavas nevertheless have plagioclase on the liquidus at a considerably higher temperature than that of olivine entry. The second effect may be due to the experimental runs having taken place at an anomalously high f_{O_2} since this tends to raise the fusion temperature of plagioclase and lower that of olivine. However, Thompson (1972a) argues that this effect is insufficient to explain the anomaly. Thus it appears that a simple explanation of these lava suites in terms of 1-atmosphere crystal fractionation is impossible, and recourse to higher-pressure phenomena is probably necessary. Thompson (1972a) reports that augite joins olivine and plagioclase on the liquidus in evolved phenocryst-poor Snake River basalt at 7–10 kbar anhydrous pressure and thus tentatively proposes that the compositional variation observed was generated at moderately high pressures by fractionation of these phases. High-pressure phenocrysts must subsequently have dissolved.

With regard to the plagioclase liquidus problem, this is a commonly observed feature of lava suites, even in *aphyric* rocks (cumulus enrichment in plagioclase *phenocrysts* does, of course, automatically lead to plagioclase on the liquidus) where it is readily explained by the postulation of modest P_{H_2O} during crystallisation, as indicated for example by the experiments of Yoder and Tilley (1962) and Nesbitt and Hamilton (1970). Hence the olivine and plagioclase phenocrysts observed in these suites, while presumably not originating at the high pressures at which the main fractionation took place are nevertheless presumed to be relics of a period of crystallisation under hydrous conditions at a pressure somewhat greater than 1 atmosphere.

A phase diagram for natural tholeiitic basalts. The examples from Kilauea, Snake River, and Skye discussed above illustrate the way in which 1-atmosphere experiments can be used in specific cases to test or formulate individual petrogenetic hypotheses. In this section we explore the problems involved in the construction and use of a phase diagram designed to predict the 1-atmosphere phase relations of tholeiites in general. This proves to be possible, though its use is at present subject to many limitations and uncertainties. It does, however, have many semi-precise applications and forms moreover a useful background to the study of high-pressure phase relations.

The basis of the method we shall use is the CIPW norm (see Appendix 3). All tholeiites by definition contain normative hypersthene, i.e. normative enstatite + ferrosilite (see Fig. 9.3), and the first task in adjusting an

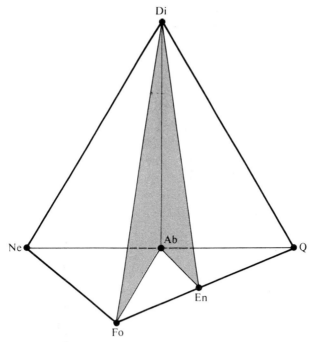

Figure 9.3 The normative basalt tetrahedron (Yoder and Tilley 1962). In addition to the normative minerals of the diagram most basalts also contain normative anorthite, orthoclase, ilmenite, magnetite and apatite. Alkali basalts occupy the sub-tetrahedron Ne–Fo–Di–Ab, olivine–tholeiites occupy Fo–Di–Ab–En, and quartz–tholeiites occupy Di–Ab–En–Q. The plane Fo–Di–Ab in this diagram is the plane of critical undersaturation and is equivalent to Fo–Di–An in $CaO–MgO–Al_2O_3–SiO_2$.

existing norm is to recalculate any hypersthene as an equivalent amount of olivine + quartz. Alternatively, if the norm is being calculated specifically for the purpose of this projection then hypersthene is not calculated in the first place, the magnesium and iron being allotted at this stage directly to olivine. The procedure for deriving four plotting parameters from an existing norm is as follows:

(a) Parameter 1 (plagioclase) is the sum of normative albite + anorthite.
(b) Parameter 2 (clinopyroxene) equals normative diopside.
(c) Parameter 3 (olivine) is any existing olivine in the norm plus the olivine derived from recalculation of hypersthene.
(d) Parameter 4 (quartz) is any existing quartz plus the quartz derived from recalculation of hypersthene.

Table 9.2 Calculation of plotting parameters for normative basalt system (example AL4, see Table 9.1)

Analysis		Standard CIPW norm			Recalculated norm	Plotting parameters as percentage
SiO_2	53.3		Q	5.64	10.14	12.7 – Q
TiO_2	3.36		Or	9.45	9.45	
Al_2O_3	12.37	Pl	{ Ab	27.77	{ 27.77	52.4 – Pl
Fe_2O_3	1.37		{ An	14.18	{ 14.18	
FeO	13.05		⎧ Wo	7.08	⎧ 7.08	
MnO	0.20	Di	{ En	2.30	{ 2.30	18.0 – Di
MgO	3.10		⎩ Fs	5.02	⎩ 5.02	
CaO	7.28	Hy	{ En	5.50	–	
Na_2O	3.26		{ Fs	12.52	–	
K_2O	1.57	Ol	{ Fo	–	{ 3.92	16.9 – Ol
P_2O_5	0.88		{ Fa	–	{ 9.59	
			Mt	2.09	2.09	
			Il	6.38	6.38	
			Ap	1.86	1.86	

(e) The parameters are summed and recalculated as percentages of the total to give the four plotting parameters. Other normative constituents such as orthoclase, magnetite, ilmenite and apatite, all normally present in tholeiite norms, are ignored. A worked example is given in Table 9.2.

The parameters calculated this way form an analogy to the synthetic system Fo–Di–An–SiO_2 discussed in the previous chapter, and to the Yoder and Tilley normative tetrahedron though Fo, Ab, and Di are now replaced respectively by Ol, Pl, and Cpx. The question is, can we use natural rocks of known 1-atmosphere phase relationships to delimit important equilibria within such a simplified natural system? By far the most useful projection is from SiO_2 into Ol–Cpx–Pl as long as consideration is restricted to compositions lying fairly close to the Ol–Cpx–Pl plane, for these are the basaltic compositions which we might expect to encounter olivine-, clinopyroxene- and plagioclase-bearing equilibria early in their crystallisation history, and conversely avoid equilibria involving silica minerals and hypersthene.

Figure 9.4 gives a projection of the relevant equilibria from SiO_2 into Fo–Di–An for the synthetic system Fo–Di–An–SiO_2. Data required for this projection are derived from Figures 8.12 and 8.17. An attempt to delimit the natural analogue of line A–B of Figure 9.4 (the locus of liquids in equilibrium with olivine, clinopyroxene and plagioclase) was made by Cox and Bell (1972) by taking the then available experimental data of Tilley, Yoder and Schairer (1963, 1964, 1965, 1967) for natural

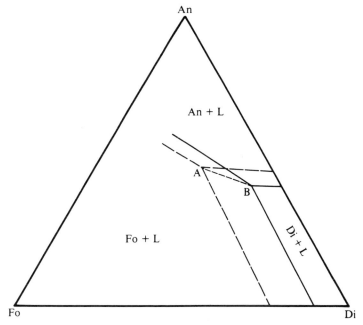

Figure 9.4 Silica-poor part of Fo–Di–An–SiO_2 projected from SiO_2. Heavy lines are phase boundaries in the plane Fo–An–Di, broken lines are the same boundaries in the plane $SiO_2 = 10\%$. A is the univariant point Fo + Di + An + En + L and A–B represents Fo + Di + An + L in projection.

basalts which showed the entry of all three major phases within 30 °C, thus approximately defining the equilibrium concerned. These rocks fall in the shaded area of Figure 9.5 to which lines analogous to the projection of the trivariant surfaces of Figure 9.4 have been added in order to show the expected projected position of rocks cotectic with regard to olivine + plagioclase, olivine + clinopyroxene and plagioclase + clinopyroxene respectively. The shaded zones on the diagram are thus zones of uncertainty in which melting relations cannot be predicted other than as 'approximately cotectic' with regard to the two or three phases concerned. Within limits to be discussed below, however, rocks which project outside the shaded areas can be assigned a primary phase with reasonable confidence. Providing also that crystallisation of the primary phase does not drive the residual liquid into the vicinity of the three solids + liquid area it should also be possible to predict the second phase successfully.

The uncertainty due to possible projection effects has been discussed above. The most important uncertainty however arises from variation in bulk composition, particularly the sodium/calcium ratio, the iron/magnesium ratio, and the potassium content. Consideration ofsimple

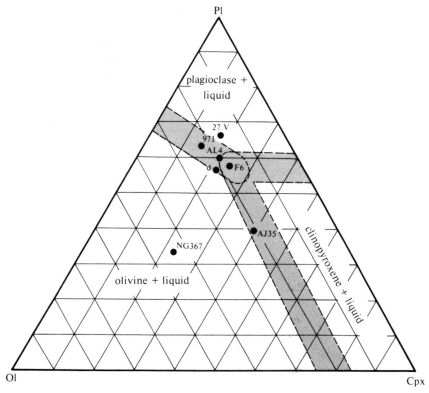

Figure 9.5 The silica-poor part of the simple normative basalt system Ol–Cpx–Pl–SiO$_2$ projected for SiO$_2$ into Ol–Cpx–Pl. The round area enclosed by a dotted line (round F6) is the area of rocks showing near-cotectic behaviour with regard to olivine, clinopyroxene and plagioclase of Cox and Bell (1972). Shaded zones radiating from this are expected projection areas of basalts equilibrating respectively with Ol + Pl (upper left), Ol + Cpx (lower) and Pl + Cpx (right). It is assumed that a large spinel field is absent from the Ol–Pl join in the natural system. Numbered points are rocks used in testing the diagram (see text).

systems such as forsterite–fayalite, diopside–albite–anorthite, and diopside–leucite–silica make it evident how strong the influence of these factors can be. Hence for practical purposes it is helpful to restrict projection to rocks which satisfy the following, somewhat arbitrary, criteria:

(a) Normative plagioclase at least as calcic as An$_{50}$.
(b) The ratio FeO + Fe$_2$O/MgO + FeO + Fe$_2$O$_3$ < 0.7.
(c) K$_2$O content less than 1%.
(d) In order to minimise the likelihood of encountering hypersthene- or silica-mineral-bearing equilibria early in the crystallisation sequence rocks with Q parameters > 10 should be avoided.

Restriction of projected rocks according to the above criteria is satisfactory for most purposes. However, some rocks which are far removed from cotectic composition may still give poor results. An extremely olivine-rich composition for example must crystallise a large amount of olivine before the second phase enters. It is the bulk composition of the groundmass at this stage which should satisfy the criteria.

To test the diagram the rocks in Table 9.3 (none of which was used in the compilation of the original diagram) have been projected into Figure 9.5. Crystallisation behaviour of each is predicted below by reading the diagram as a ternary, and then compared with experimentally determined phase relations.

NG367. Clearly in the olivine field the projected position suggests a long interval of olivine crystallisation followed by clinopyroxene. (Experimental result: Ol 1363 °C, Cpx 1205 °C, Pl 1155 °C.)

AJ35. Olivine and clinopyroxene should begin to crystallise at approximately the same temperature, and this might be expected to be a higher temperature than the Cpx entry at 1205 °C in the previous rock. Plagioclase entry should be at a lower temperature, approximately the same as that in the previous rock. (Experimental result: Ol and Cpx 1275 °C, Pl 1152 °C.)

971. Plagioclase and olivine should appear approximately together. Clinopyroxene should appear later at about 1150 °C. (Experimental result: Ol and Pl 1201 °C, Cpx 1138 °C.)

d. This rock is fairly close to cotectic with regard to all three major phases. However the sequence olivine, plagioclase, clinopyroxene might be expected. (Experimental result: Ol 1173 °C, Pl 1166 °C, Cpx 1113 °C.)

27V. Plagioclase should appear first, with olivine second. (Experimental result: Pl 1215 °C, Ol 1185 °C, Cpx 1175 °C.)

F6. This specimen should be approximately cotectic with regard to all three phases. (Experimental result: Ol, Cpx and Pl 1159 °C.)

Application of the tholeiite phase diagram. The most obvious application of the diagram is as a rather crude substitute for experimental work when facilities for the latter are not accessible. Some examples of the ways in which 1-atmosphere experiments are useful in petrogenetic reasoning have already been given and others will be considered in later sections. Before going on to discuss high-pressure phase relations in basaltic systems, an obvious extension of problems already presented, it is necessary to sound a cautionary note.

The perils of quaternary thinking. Analogies have been freely drawn above between the normative basalt system and the synthetic

Table 9.3   Rocks used to test normative basalt phase diagram

Analyses

	NG367	AJ35	971	d	27V	F6
SiO_2	47.12	47.33	46.94	44.35	50.51	50.74
TiO_2	0.43	2.26	1.37	3.40	0.83	3.35
Al_2O_3	9.10	10.52	17.15	12.61	17.52	13.57
Fe_2O_3	3.31	2.67	2.11	4.74	3.18	1.36
FeO	6.79	9.45	9.83	11.70	5.83	10.63
MnO	0.09	0.18	0.19	0.21	0.16	0.18
MgO	21.30	11.97	7.69	7.45	7.28	6.16
CaO	8.19	12.30	9.73	9.51	11.27	9.94
Na_2O	1.61	1.99	3.01	2.22	2.09	2.69
K_2O	0.85	0.75	0.43	0.26	0.43	0.67
P_2O_5	0.15	0.39	0.18	0.67	0.10	0.37

Parameters in $Ol-Di-Pl-Q$

	NG367	AJ35	971	d	27V	F6
Ol	43.1	21.7	22.4	20.6	14.6	16.0
Di	22.2	38.3	14.4	19.7	16.1	22.6
Pl	32.5	39.7	63.7	52.8	60.8	52.7
Q	2.2	0.4	−0.5*	6.9	8.5	8.6

Parameters in projection from Q

	NG367	AJ35	971	d	27V	F6
Ol	44.1	21.8	22.3	22.1	16.0	17.5
Di	22.7	38.4	14.3	21.2	17.6	24.8
Pl	33.2	39.8	63.4	56.7	66.4	57.7

	NG367	AJ35	971	d	27V	F6
$\dfrac{FeO + Fe_2O_3}{MgO + FeO + F_2O_3}$	0.32	0.53	0.61	0.69	0.55	0.66
normative $\dfrac{An}{Ab + An}$	0.52	0.51	0.55	0.56	0.68	0.51

NG367	Picrite basalt, New Georgia (Brown & Schairer 1968).
AJ35	Ankaramite, Anjouan, Comores Islands (Thompson & Flower 1971).
971	Basalt, Skye (Thompson *et al.* 1972).
*	N.B. slightly Ne-normative but this is not significantly different from, say, AJ35
d	Basalt, Snake River (Tilley & Thompson 1970).
27V	Olivine-bearing basaltic scoria, St Vincent (Brown & Schairer 1968).
F6	Basalt, Kilauea (Thompson & Tilley 1969).

Fo–Di–An–SiO_2, and it is thus tempting to think that the crystallisation of basalts can in general be dealt with satisfactorily in these terms. However, it is only the *initial* stages of crystallisation to which the phase diagram can be applied with any degree of certainty.

Specimen AL4 (Table 9.2) is an evolved Kilauean rock which does *not* satisfy the criteria for projection discussed previously. Its projected position in Figure 9.5 suggests that it might be cotectic with regard to olivine and plagioclase. Experimentally determined phase relations, however, show it to be on the clinopyroxene–plagioclase cotectic. This is an illustration of the way the important boundaries migrate within the simple basalt system with changing bulk composition. Comparison of Figures 9.4 and 9.5 will show that the composition cotectic with regard to all three major phases has migrated towards a composition richer in plagioclase and poorer in clinopyroxene in the basalt system relative to the synthetic system. The behaviour of AL4 indicates that this trend continues into rocks more evolved than basalt. The contraction of the plagioclase field is of course not surprising in view of the generally small liquidus field of albite compared with that of anorthite in many synthetic systems. Clearly, in order to comprehend the basalt system adequately it is necessary to think in terms of the behaviour of additional components, amongst which the build-up of iron and sodium in residual liquids must be important.

The nature of the **thermal divide** separating tholeiites and alkali basalts (Yoder and Tilley 1962) can profitably be thought of in these terms. Figure 9.3 illustrates that in quaternary terms the plane Ol–Cpx–Pl (equivalent to Fo–Di–Ab in the figure) separates silica-undersaturated basalts (nepheline-normative) from silica-saturated and oversaturated basalts (hypersthene-normative). No fractionation paths cross this plane and the locus of liquids equilibrating with Ol + Cpx + Pl has a thermal maximum on it. Strictly, the plane of the divide only approximates to the normative plane Ol–Cpx–Pl because the clinopyroxenes concerned, as a consequence of solid solution towards Ca-poor pyroxene and Ca-Tschermak's molecule, do not lie exactly at the Cpx composition point.

Consider now a liquid lying exactly in the plane of the divide and in equilibrium with all three major phases. In terms of the synthetic system Fo–Di–An (Fig. 9.4) this liquid lies at a ternary eutectic (B) and will solidify without change of composition as heat is extracted. Thought of in terms of AL4 discussed previously, in Fo–Di–An with Fe^{2+} and Na added the liquid will clearly migrate to more plagioclase-rich and clinopyroxene-poor compositions with cooling. Advanced crystal fractionation of such a liquid will clearly lead to an iron-rich feldspathic end product simulating a trachyte. If potassium is also considered as an additional component this effect will become even more extreme for the build-up of K_2O in the residual liquid must be even greater than that of Na_2O since it is barely affected by plagioclase removal.

Without elaborating further it should be clear that, far from leading to low-temperature liquids which are still basaltic, the original liquid must give rise to a thoroughly trachytic residue.

Consider now a slightly more SiO_2-rich version of the original liquid, one lying on the equilibrium Ol + Cpx + Pl + L but slightly displaced into the tholeiite field. At first sight, and thinking in quaternary terms, fractionation of such a liquid will give rise to residual liquids passing down-temperature along the Ol + Cpx + Pl + L curve and eventually reaching the equilibrium Ol + Cpx + Pl + Opx + L. Here olivine will be lost by reaction and fractionation of Cpx + Pl + Opx will lead ultimately to equilibria involving a silica-mineral. In a more complex system, however, the behaviour will be quite different. If the original liquid lies very close to the plane of the divide then the tie-line between liquid and the bulk composition of crystallising Ol + Cpx + Pl mixtures must be very short. Thus in order for the liquid to migrate significant distances down the boundary curve towards Ol + Cpx + Pl + Opx + L very large amounts of crystallisation must take place. While this is happening the build up of Fe/Mg ratio, Na_2O and K_2O must take place to a very substantial degree. Hence the liquid is likely to become trachytic long before it reaches Opx-bearing equilibria, and the basalt phase diagram rapidly becomes irrelevant. The Aden volcanics discussed in Chapter 6 form an excellent example of this sort of behaviour. The basic rocks are approximately silica-saturated. The differentiates crystallise olivine, clinopyroxene and plagioclase (+ an ore mineral) throughout and become quartz–trachytes. Olivine does not cease to crystallise until the extreme differentiate stage (many of the trachytes contain pure fayalite co-existing with quartz), and orthopyroxene never appears. Conversely those series of rocks which do show orthopyroxene phenocrysts are likely to be much more silica-rich even at the initial basaltic stage and they probably have not reached orthopyroxene-bearing equilibria as a consequence of an extensive period of Ol + Cpx + Pl crystallisation.

An alkali–silica diagram (Fig. 9.6a) is used to illustrate the fundamental properties of the thermal divide discussed above. Two basaltic liquids A and B represent the slightly over- and slightly under-saturated parents respectively. Both give rise to a trachytic residue (A' and B') at which stage alkali feldspar fractionation commences driving A' down towards the rhyolitic end-point R while B' migrates up towards P, the phonolite point. The line P–B'–A'–R is effectively a projection of the residua system (nepheline–kalsilite–silica, Fig. 5.12) into the alkali–silica diagram. A point between A' and B' represents the projection of the alkali feldspar divide into the alkali–silica diagram while P and R are the respective liquidus minima on the nepheline–alkali feldspar and quartz–alkali feldspar boundary curves.

If a range of basaltic parents straddling the divide were allowed to fractionate, the differentiates would be expected to split with two groups

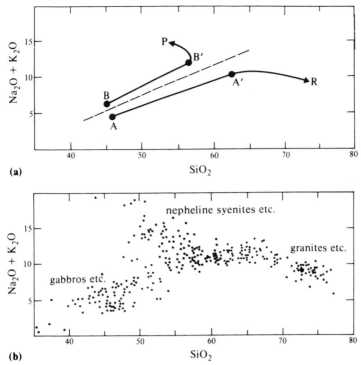

Figure 9.6 (a) Illustration of residual liquid paths in the alkali–silica diagram. A is a tholeiitic basalt, B an alkali basalt. A′ and B′ are the respective trachytic residual liquids when alkali feldspar first begins to crystallise. P and R are the phonolite and rhyolite end-points. The broken line is the projection of the plane of critical silica undersaturation (Ol–Cpx–Pl at basaltic end, Ab–Or at trachytic end, cf. Fig. A3.2 of Appendix 3). (b) Analysed rocks of the Gardar province, S.W. Greenland, (after Upton 1974), a series simulating the idealised behaviour shown in (a).

at the trachyte stage and give rise to both phonolitic and rhyolitic end products. A suite which appears to have evolved in this way, the Gardar province of S.W. Greenland, is illustrated in Figure 9.6b. The discussion by Upton (1974) of the Gardar province elaborates on the arguments concerning thermal divides given above.

To summarise, the Ol–Cpx–Pl divide (sometimes termed the olivine–gabbro divide) which separates alkali basalts and tholeiites at low pressures does not have the property of splitting basaltic fractionates into two distinct groups. Crystallisation does result in a small component of compositional movement of fractionates 'away' from the divide but the build-up of the non-quaternary components results in much faster evolution 'parallel' to the divide. The divide itself is only part of the divide Ol–Cpx–Pl–alkali feldspar in the larger system.

Crystallisation of alkali feldspars in a later, trachytic, stage of evolution is capable (in contrast to the basaltic stage) of producing two sharply divergent fractionation trends. Failure to appreciate the effects of the additional components has led several workers to suppose the latter also to be true at the basaltic stage. This is certainly not so, a thermal divide is literally a compositional 'plane' through which under specified conditions (in this case 1 atmosphere, dry) there are no fractionation paths. Under some circumstances fractionation paths may radiate rapidly away from the divide (trachyte case), under others may proceed almost parallel to it (basalt case).

Parenthetically, it should be added that when the 1-atmosphere, dry, conditions are not satisfied the divide can break down and be penetrated by fractioning liquids. The most common examples of this are series which are slightly silica-undersaturated at the basaltic stage but give rise to oversaturated residues as a result of fractionation of strongly silica-undersaturated phases such as hornblende or biotite.

Computer simulation of crystallisation paths

Nathan and Van Kirk (1978) offer a method of predicting the fractional crystallisation paths of a wide variety of silicate liquids and thus represents a very significant improvement on the limited graphical methods discussed previously. Nathan and Van Kirk use two sets of empirically derived equations to predict for a given liquid composition (a) the liquidus temperatures of a list of anhydrous minerals comprising magnetite, olivine, hypersthene, augite, quartz, plagioclase, orthoclase, leucite, and nepheline, and (b) the compositions of the minerals in terms of solid-solution end-members. The program selects the mineral with the highest liquidus temperature, determines its composition, and then repeatedly extracts a small amount of it to produce a residual liquid composition. Each time a new liquid is produced the process of identifying the phase with highest liquidus temperature is repeated. Ultimately the original phase is found no longer to have the highest liquidus temperature, signifying that the liquid has reached the edge of the primary phase field of the first phase and has crossed the boundary into another field. The program backtracks at this stage to locate the boundary precisely (two phases with the same liquidus temperature). Then the two phases are extracted together, again in repeated small increments, and phase proportions in the extract are adjusted to keep identity of liquidus temperature for the two phases. This process directs the residual liquid along the boundary curve concerned. Resorptional boundaries, alternatively, are readily detected because whatever the mixture of phases extracted the liquidus temperature of the resorbed phase persistently falls below that of the other phase or phases. It should be clear that given adequate equations in the first place, and sufficient computer time, such a computational method is capable of generating residual liquids (e.g. granitic) compositionally quite remote from the starting composition (e.g. basaltic). If refined in the future to include hydrous phases, and the results of high-pressure experiments, the method will eventually serve as a complete polycomponent phase diagram for igneous rocks. It will then be possible to model the crystallisation of any chosen composition under any desired conditions, and compare the results with natural rock data.

High-pressure experiments on dry materials

The experimental work discussed above has obvious and important applications to the study of the later stages of magmatic evolution which, in terms of the fractionation of compositions that can take place, is highly significant. It gives little or no information, however, about the deeper origins of magmas and their possible fractionation at any depth. High-pressure experiments are necessary in order to study these earlier stages of magmatism and their results can be applied to a wide range of petrological and geochemical problems.

Experimental apparatus for most high-pressure work consists either of an internally or externally heated pressure vessel in which pressure is applied via a fluid medium such as argon, water, or CO_2, or the piston–cylinder apparatus in which the sample is compressed by a hydraulic ram and is surrounded by a solid medium such as talc which will flow under pressure. The first type of apparatus is ideal for relatively low-temperature/pressure studies (e.g. temperatures up to 700 °C at 6 kbar or less, temperatures up to 950 °C at 0.5 kbar) and is stable during runs of long duration. The internally heated vessel is suitable for work up to about 10 kbar at magmatic temperatures, while the piston–cylinder apparatus can be used at very high temperatures and pressures (up to 1700 °C in the range 10–60 kbar). Most of the experiments relevant to the evolution of basalts at pressures simulating those of the Upper Mantle have been carried out in the latter type of apparatus. The disadvantage of this equipment relative to the fluid medium systems is that pressure cannot be measured with such accuracy and there is less stability. The reader is referred to Ernst (1976) for a more detailed description of high-pressure apparatus.

In the following pages we discuss some of the experimental results and their application to petrological problems. In so doing much simplification is involved and glossing-over of the many experimental problems. The presentation of summarised results given for example in Tables 9.5 and 9.7 disguises the complexity of the actual experimental results obtained in individual runs. The reader should consult original papers (e.g. Green & Ringwood 1967) where run data are given. It will be realised that not all runs on a particular sample are internally consistent, the identification of phases present is not always certain, nor is it always clear whether equilibrium was reached or whether the capsule leaked during the experiment, and so on. Wyllie (1971, p. 195) gives an interesting account of a discrepancy between two sets of data, where the issue concerned was thought to be the extent to which certain compositions lost iron to their containing capsules during experimental runs. A second example is discussed in Chapter 10. In short, a good deal of judgement and interpretation is involved in the production of the experimental data itself. The reader should try not to lose sight of this during the following pages.

Some typical results of experiments on dry basalts. The classic paper by Yoder and Tilley (1962) on the experimental petrology of basalts was followed by an equally important paper by Green and Ringwood (1967) in which the high-pressure phase relations of basalts were investigated in detail for the first time. The following discussion is based on their data.

Table 9.4 gives compositions and CIPW norms of two of the basaltic compositions investigated by Green and Ringwood while Table 9.5 gives a summary of their crystallisation behaviour at various pressures.

From inspection of Table 9.4, these are both fairly magnesian basalts, AL being slightly silica-undersaturated while OT is silica saturated and a generally somewhat more mafic composition. The 1-atmosphere experiments show them both to lie in the olivine + liquid field, OT having a somewhat higher liquidus temperature than AL. As the pressure is increased both compositions show very significant changes in crystallisation behaviour.

Considering first AL, a general increase in liquidus temperature is apparent, changing from an olivine liquidus at 1255 °C at 1 atmosphere to a garnet liquidus of 1430 °C at 27 kbar. At 9 kbar AL is still in the olivine field though the clinopyroxene field has evidently expanded somewhat relative to that of plagioclase. Several important changes take place between 9 kbar and 13.5 kbar; the olivine field must contract substantially while the pyroxene fields, especially orthopyroxene, undergo a rapid expansion. The disappearance of orthopyroxene down-temperature at 13.5 kbar suggests that the equilibrium Cpx + Opx + L is resorptional for Opx at this pressure. At 18 kbar and

Table 9.4 Chemical compositions and CIPW norms of two rocks investigated experimentally by Green and Ringwood (1967)

	Alkali olivine basalt (AL)	Olivine tholeiite (OT)
SiO_2	45.39	46.95
TiO_2	2.52	2.02
Al_2O_3	14.69	13.10
Fe_2O_3	1.87	1.02
FeO	12.42	10.07
MnO	0.18	0.15
MgO	10.37	14.55
CaO	9.14	10.16
Na_2O	2.62	1.73
K_2O	0.78	0.08
P_2O_5	0.02	0.21
Total	100.00	100.04

Norms		
Or	4.5	0.6
Ab	18.0	14.7
An	26.2	27.6
Ne	2.2	–
Di	15.7	17.0
Hy	–	12.3
Ol	25.8	21.9
Il	4.8	3.8
Mt	2.9	1.4
Ap	–	0.5

27 kbar orthopyroxene has disappeared from the liquidus and is replaced first by clinopyroxene and then by garnet.

The composition OT which is more magnesian and relatively more silica-rich shows a different sequence of changes which nevertheless have many important features in common with the behaviour of AL. The same general rise of liquidus temperature is observed and at the two highest pressures the crystallisation is as in AL dominated by garnet and clinopyroxene. As in AL a spinel field is encountered in the intermediate pressure range some way below the liquidus. However, in general in OT the early crystallisation of orthopyroxene is more important in the intermediate pressure range than it is in AL, which is of course not surprising in view of the more siliceous composition. However, the same expansion of the clinopyroxene field is still observable and clinopyroxene appears shortly below the liquidus at

Table 9.5   Melting relations of AL and OT (Table 9.4) at various pressures (data of Green & Ringwood 1967)

AL

1 atm.	Ol *in* 1255 °C, Pl *in* 1245 °C, Cpx *in* 1230 °C
9 kbar	Ol *in* above 1260 °C, Cpx *in* 1230 °C, Pl + Sp *in* 1200 °C
13.5 kbar	Opx + Cpx *in* 1290 °C, Opx *out* 1280 °C
18 kbar	Cpx *in* 1330 °C, Ga *in* about 1275 °C
27 kbar	Ga *in* 1430 °C, Cpx *in* 1415 °C

OT

1 atm.	Ol *in* about 1340 °C, Pl *in* 1270 °C
9 kbar	Ol *in* 1360 °C, Opx *in* 1300 °C, Cpx *in* 1260 °C
13.5 kbar	Opx *in* above 1400 °C, Cpx *in* about 1360 °C, Sp *in* 1300 °C
18 kbar	Opx *in* above 1425 °C, Cpx *in* 1410 °C, Opx *out* 1370 °C, Ga *in* 1360 ° C
22.5 kbar	Cpx *in* 1445 °C, Ga *in* 1420 °C
27 kbar	Ga *in* 1500 °C, Cpx *in* about 1475 °C

13.5 kbar and 18 kbar, reaching the liquidus by 22.5 kbar. The reaction relationship involving resorption of orthopyroxene is still apparently observable at 18 kbar.

The implications of the above observations for possible fractionation paths of basaltic liquids under pressure may be summarised as follows. Firstly, at pressures of ca. 25–30 kbar, corresponding to depths of 75–90 km, garnet and clinopyroxene would inevitably be important in any crystal fractionation process and this has substantial geochemical consequences particularly for minor elements (see Ch. 14). In somewhat lower pressures (ca. 12–15 kbar or 36–45 km depth) orthopyroxene has a large liquidus field and is capable of being the primary phase even in silica-undersaturated basaltic liquids. This implies that the thermal divide separating silica-saturated and silica-undersaturated liquids at low pressures (the olivine–clinopyroxene–plagioclase join) breaks down, thus allowing tholeiitic liquids to pass through the divide by fractionating orthopyroxene-bearing assemblages. Thirdly, the contraction of the olivine field with pressure carries the implication that initial melts of an olivine-bearing mantle must become more olivine-rich (picritic) with increasing pressure (all such melts must lie on the periphery of the olivine primary phase field and as this contracts melts must lie closer to olivine in compositions). A melt so formed, however, could proceed to the surface fractionating only olivine as the olivine field expanded with decompression.

Table 9.6 Analysis of peridotites

	1	*2*
SiO_2	43.86	45.62
TiO_2	0.39	0.20
Al_2O_3	1.96	2.55
Cr_2O_3	0.46	0.34
Fe_2O	0.69	1.88
FeO	6.21	5.89
MnO	0.14	0.13
NiO	0.25	0.25
MgO	39.82	41.41
CaO	1.68	1.90
Na_2O	0.14	0.19
K_2O	0.08	0.12
P_2O_5	0.03	0.03
H_2O+	4.06	–
H_2O-	0.06	–
Total	99.83	100.01

1. Garnet lherzolite nodule from kimberlite. KA64–16, Dutoitspan, Kimberley, South Africa. This is the sample experimented on by Ito and Kennedy (1967) and of which melting relations are given in Figure 9.7.
2. Average lherzolite nodule in kimberlite (Ito and Kennedy 1967). This is shown as AV in Figures 9.9 – 9.11 and Figures 9.13 and 9.14.

Experiment on a natural dry peridotite. Since basalt petrogenesis is concerned with melting processes as well as the crystal fractionation processes discussed above it is instructive to consider experiments on natural peridotites. These complement the studies of Green and Ringwood (1967) and give data for temperatures near the solidus. Table 9.6 gives the analysis of a natural peridotite investigated by Ito and Kennedy (1967). The phase relations determined at various pressures are illustrated in Figure 9.7.

Below the solidus this sample consists of olivine, orthopyroxene, clinopyroxene and spinel at low and moderate pressures. However, increasing pressure changes this to olivine, orthopyroxene, clinopyroxene and garnet (garnet lherzolite), this latter assemblage being the most common type of peridotite nodule obtained from kimberlite pipes.

The dramatic changes of liquidus phase seen in the basalts discussed previously do not take place in this composition which retains olivine as

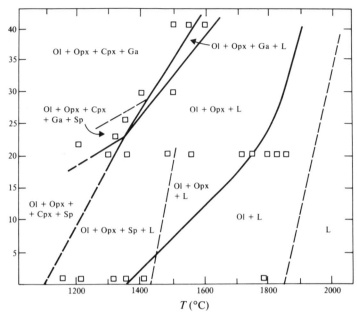

Figure 9.7   *P–T* diagram showing experimentally determined phase relations for peridotite KA64-16 (see Table 9.6). Small squares are data points (based on Ito and Kennedy 1967).

the primary phase even at 40 kbar. This is to be expected because of the very high normative olivine content of the peridotite relative to the basalts. However, below the liquidus, changes of phase relations in many ways analogous to those demonstrated by the basalts are still seen. The rapid rise in the orthopyroxene crystallisation temperature is prominent, and it is implied that the interval over which olivine crystallises alone is much reduced at high pressure. At 40 kbar clinopyroxene is the first phase to melt completely and is followed shortly by garnet. As discussed previously garnet and clinopyroxene fractionation might be expected to dominate the fractional crystallisation of *basaltic* liquids at high pressures. Here we see that the *melting* of garnet and clinopyroxene should be highly significant in the initial stages of melting of a peridotite source rock under pressure.

The polybaric phase diagram for dry natural basalts of O'Hara (1968). The experiments on basalts and peridotite discussed are only a few examples drawn from a substantial body of available data. Using these data O'Hara (1968) constructed a natural basalt phase diagram showing boundary curves for a variety of pressures up to 30 kbar. Like the diagram presented earlier in this chapter (Fig. 9.5) for basalts at 1 atmosphere the O'Hara diagrams are subject to limitations; indeed in the polybaric diagram the uncertainties are substantial since there are fewer experimental data and the solid solutions involving garnets and aluminous pyroxenes introduce further complexities. Nevertheless, if the diagrams are not interpreted too literally, they provide a superlative unifying framework in which to think about the issues and problems involved in the high-pressure phase relations of basaltic and ultramafic rocks.

The O'Hara diagrams are constructed by projecting natural rock analyses into a modification of the $CaO–MgO–Al_2O_3–SiO_2$ tetrahedron where plotting parameters are calculated as follows:

$C = $ (mol.prop. $CaO - 3^1/_3 P_2O_5 + 2Na_2O + 2K_2O$) × 56.08
$M = $ (mol.prop. $FeO + MnO + NiO + MgO - TiO_2$) × 40.31
$A = $ (mol.prop. $Al_2O_3 + Cr_2O_3 + Fe_2O_3 + Na_2O + K_2O + TiO_2$) × 101.96
$S = $ (mol.prop. $SiO_2 - 2Na_2O - 2K_2O$) × 60.09

In O'Hara's words (1968, p. 86),

> The objective of this mode of projection is to present results in a form directly comparable with weight percent plots and projections of phase equilibria in the system $CaO–MgO–Al_2O_3–SiO_2$. In this projection all albite and orthoclase molecules plot as the equivalent weight of anorthite (CAS_2) molecule, ulvospinel, hercynite, magnetite and chromite plot together with, and as the equivalent weight of, spinel (MA) molecule; all iron–nickel–magnesium olivines plot as the equivalent weight of forsterite (M_2S) molecule; all garnets plot at points along the grossular–pyrope join; and all (CaNa)(MgFe, Ni, MnAl, $CrFe^{3+}$)(Al, Fe, Cr, Si)$_2O_6$ clinopyroxenes plot at points along the diopside (CMS_2)–Ca-Tschermak's molecule (CAS) join, jadeite and aegirine plotting with CAS.

(The reader is advised to work carefully through Appendix 3 before attempting to confirm these statements.)

This projection system has one considerable advantage over the normative system used earlier in Figure 9.5 in that it deals with both silica-undersaturated and silica-oversaturated rocks. It is actually in many practical instances, however, disadvantageous to employ parameters so comprehensive that all the constituents of the analysis are used. The arguments advanced in Chapter 2 when discussing complex fractionation indices as used in variation diagrams apply here. If the effects of the different components cannot be individually distinguished no advantage is gained and some confusion may result. On the credit side all minerals can be represented and it is certainly convenient to

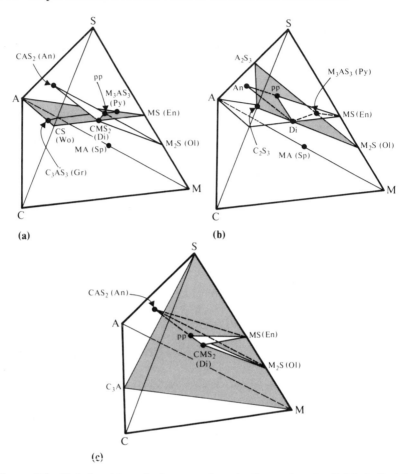

Figure 9.8 Relationships of planes and projection points in C–M–A–S. (a) Shaded plane is CS–MS–A. Piercing point of M_2S–CAS_2 join in this plane is marked 'pp'. The olivine–gabbro plane (Ol–Cpx–Pl) cuts CS–MS–A in the line Di–pp. Also shown is the garnet join pyrope (Py) – grossular (Gr). (b) Shaded plane is C_2S_3–A_2S_3–M_2S. Piercing point of MS–CAS_2 join marked 'pp'. The hypersthene gabbro (Opx–Cpx–Pl) and garnet–pyroxene (Opx–Cpx–Ga) planes are shown cutting C_2S_3–A_2S_3–M_2S. (c) Shaded plane is C_3A–M–S, shown here cut by the hypersthene gabbro and olivine gabbro planes.

have a rational series of lines and planes within the system on which all the pyroxene solid solutions, for example, can be plotted.

The principal projections used by O'Hara are:

(a) from *olivine* into the plane CS–MS–A
(b) from *enstatite* into the plane M_2S–A_2S_3–C_2S_3
(c) from *diopside* into the plane C_3A–M–S

These are shown relative to other important planes and joins in Figure 9.8. The method of calculation used in the projections is discussed in Appendix 4. For certainty in reading crystallisation sequences it is of course essential to project from the composition of the primary phase, hence the choice of important minerals as points to project from. The plane CS–MS–A used in the olivine projection is convenient because it contains all the pyroxene solid solutions as well as the garnet join. The other two planes are chosen to include important mineral composition points and to try to minimise projection distortion, though this is inevitably not always successful. Examination of Figure 9.8 for example shows that projection of many compositions (e.g. those lying near the anorthite–enstatite join) into $M_2S–A_2S_3–C_2S_3$ from enstatite is very oblique. A spherical array of basaltic composition points would thus project as an elliptical pattern into the plane. Such an ellipse would have a long axis pointing towards forsterite and thus give an entirely spurious impression of olivine fractionation.

The boundaries determined by O'Hara are given in Figures 9.9–9.11. Joins shown crossing these projections as lines may be located as planes in Figure 9.8. These are useful for identifying approximate normative character of projected points (see Appendix 3) though because of the complex projection parameters employed the sub-tetrahedra in which points lie in this version of the CMAS system do not always coincide exactly with their CIPW normative character. In the olivine projection into CS–MS–A nearly all tholeiites (defined normatively as containing hypersthene) plot on the enstatite-rich side of the line joining diopside to the olivine–plagioclase piercing point (this is the trace of the olivine–clinopyroxene–plagioclase plane as it cuts CS–MS–A), whereas nepheline-normative rocks (alkali basalts) plot on the enstatite-poor side. Inspection of Figure 9.8, however, shows that it is not possible to distinguish oversaturated tholeiites (Q-normative) from olivine tholeiites (Ol-normative) in this projection. These lie respectively in the sub-tetrahedra Fo–En–Di–An and En–Di–An–SiO_2. However, the diopside projection into $C_3A–M–S$ (Fig. 9.10) clearly distinguishes the two groups, as well as the alkali basalts. The Q-normative tholeiites mainly project on the SiO_2-rich side of the 'hypersthene gabbro divide' (the join En–An–Di) while the olivine tholeiites lie between it and the 'olivine gabbro divide' (Fo–Di–An). Alkali basalts lie on the silica-poor side of this divide in this projection.

An illustration of the way in which the phase boundaries are added to the diagrams is given in Figure 9.12 using the data given in Table 9.7. The data in the table refer to rocks all of which crystallise olivine as the primary phase at the pressures given. Hence it should be possible to make an olivine projection, showing as in the system illustrated

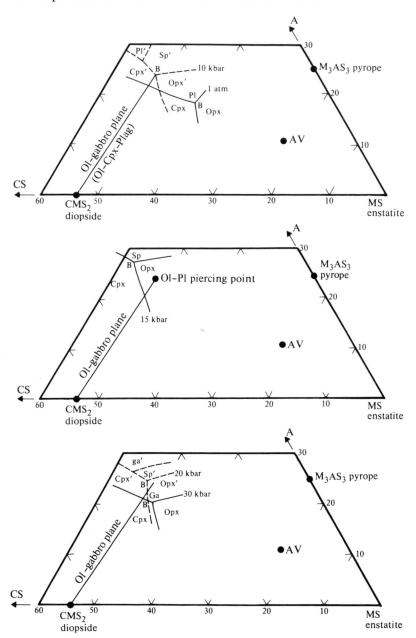

Figure 9.9 Phase boundaries at various pressures in the projection from *olivine* into CS–MS–A (O'Hara 1968). All fields contain olivine + liquid in addition to the phase indicated. The fields at 10 kbar and 20 kbar are labelled with primed symbols e.g. Opx'. AV is average peridotite (see Table 9.6). This diagram should be compared with Figure 9.8a.

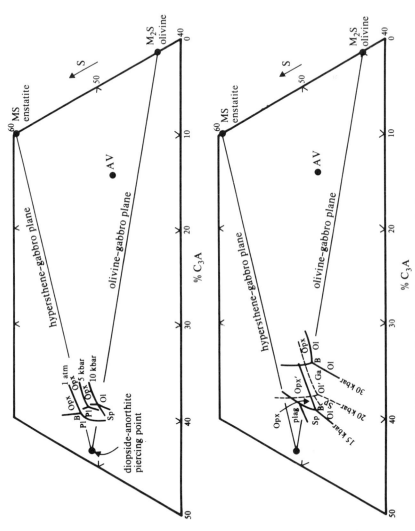

Figure 9.10 Phase boundaries at various pressures in the projection from *diopside* into C₃A–M–S (O'Hara 1968). All fields contain clinopyroxene + liquid in addition to phase indicated. Primed symbols used for 20 kbar fields (lower diagram). Comparison should be made with Figure 9.8c.

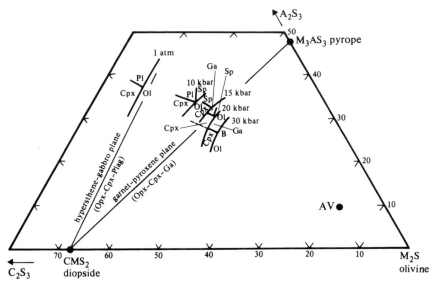

Figure 9.11   Phase boundaries at various pressures in the projection from *enstatite* into C_2S_3–A_2S_3–M_2S. All fields contain orthopyroxene + liquid in addition to phase indicated. Comparison should be made with Figure 9.8b.

previously in Figures 8.14 and 8.17 those fields which come into contact with the olivine primary phase field. For the 1-atmosphere data one knows that the three fields (olivine + clinopyroxene + liquid, olivine + plagioclase + liquid, and olivine + orthopyroxene + liquid) must fit together approximately as shown, and these boundaries are consistent with the third phase to enter. (For example, the residual liquid of KA must move approximately away from the En composition

Table 9.7   Melting relations of rocks at 1 atmosphere and 9 kbar

---

1 atmosphere

AL	Ol *in* 1255 °C, Pl *in* 1245 °C, Cpx *in* 1230 °C
OT	Ol *in* about 1340 °C, Pl *in* 1270 °C
KA	Ol *in* above 1385 °C, Opx *in* 1170 °C, Pl *in* 1155 °C, Cpx *in* 1140 °C

9 kbar

AL	Ol *in* above 1260 °C, Cpx *in* 1230 °C, Pl + Sp *in* 1200 °C
OB	Ol *in* above 1280 °C, Opx + Cpx *in* 1270 °C, Opx *out* 1250 °C
OT	Ol *in* 1360 °C, Opx *in* 1300 °C, Cpx *in* 1260 °C

AL	–	alkali olivine basalt, Green and Ringwood (1967), see Tables 9.4 and 9.5.
OT	–	olivine tholeiite, Green and Ringwood (1967), see Tables 9.4 and 9.5.
KA	–	hypersthene–olivine basalt, Kauai, Tilley *et al.* (1963).
PB	–	picrite basalt, Hawaii, Tilley *et al.* (1963).
OB	–	olivine basalt, Green and Ringwood (1967).

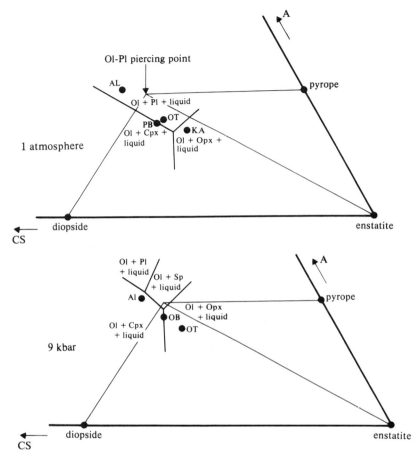

Figure 9.12   Projected positions of rocks given in Table 9.7 to illustrate method of positioning phase boundaries (olivine projection).

point after orthopyroxene has begun to crystallise so that it reaches the plagioclase field next, not the clinopyroxene field. This can, however, only be an approximate prediction in the absence of knowledge of the orthopyroxene composition.)

Similarly at 9 kbar the boundaries must be approximately those given. A consideration of the system in three dimensions (Fig. 9.8) indicates that the spinel field must lie between the plagioclase field and the alumina–enstatite join, while the phase relations of AL suggest that the olivine, clinopyroxene, plagioclase, and spinel fields must all come into contact. These examples are sufficient to illustrate in phase diagram form the expansion of the orthopyroxene field with pressure, already noted in the discussion of Green and Ringwood's experimental data.

The O'Hara diagrams are based on experimental data for nearly thirty rocks and show a degree of internal consistency which justifies their use as a generalised expression of phase relations, though when individual rocks are considered discrepancies can arise as a result of compositional factors too complex to find adequate expression in three dimensions.

General features illustrated by the diagrams. The contraction of the olivine field with increasing pressure is well shown in the projections from diopside and enstatite. As mentioned previously this implies that initial melts of peridotitic materials become more picritic with depth. The initial melt of peridotite consisting in the sub-solidus of olivine, orthopyroxene, clinopyroxene, and an aluminous phase (plagioclase at low pressure, spinel at intermediate pressures and garnet at high pressure) is shown by the point B on the diagrams.

With regard to the orthopyroxene field, in the range 5–10 kbar this has expanded so that point B lies between the hypersthene gabbro plane and the olivine gabbro divide (Fig. 9.10), implying that the reaction of olivine and liquid to give orthopyroxene, characteristic of low pressure tholeiitic liquids, has disappeared and that the hypersthene gabbro plane has become a thermal divide. At about 15 kbar however the orthopyroxene field has expanded sufficiently to penetrate the olivine gabbro plane. Hence this no longer forms a thermal divide and tholeiitic liquids can fractionate through it by precipitating ortho-pyroxene-bearing mineral assemblages. Above this pressure, however, the orthopyroxene field contracts again, pushed back by the expansion of the garnet field, so that B passes back to the silica-rich side of the olivine gabbro plane. Initial melts at 30 kbar thus have the low-pressure normative character of olivine tholeiites.

Advanced degrees of partial melting can be studied by reference to the point AV representing the average garnet lherzolite of Ito and Kennedy (1967), regarded as representing a generalised Upper Mantle composition. Disregarding solid solutions for the moment the melting path of AV can be read as shown in Figure 9.13. Here the initial melt is B, clinopyroxene is the first phase to melt completely and the liquid can then evolve along the curve B–C in equilibrium with garnet, olivine and orthopyroxene. At C the liquid, the bulk composition AV, olivine and orthopyroxene are co-planar signifying that the melting of garnet is complete. Liquid then evolves along the line C–AV across the olivine + orthopyroxene + liquid surface until one of the phases completes melting. Assuming this is orthopyroxene (this cannot be ascertained in this projection) the projected position of the liquid will reach AV at which point liquid, AV and olivine are co-linear. Melting is

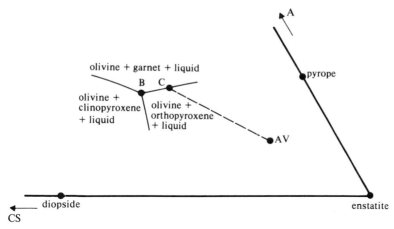

Figure 9.13 Reading of melting path for average peridotite at 30 kbar (olivine projection).

completed with the migration of the liquid towards the real position of AV at which stage all the olivine has melted. During this final stage the projected position of the liquid does not change.

O'Hara (1968) has used these diagrams to argue about the probable sequence of residual mantle compositions during progressive partial melting, with a view to explaining relative abundances of various nodule types in kimberlite. In the case discussed the residua are progressively

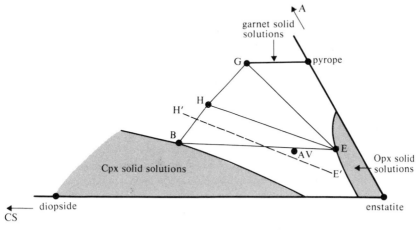

Figure 9.14 Solid solutions in clinopyroxene, orthopyroxene (shaded areas) and garnets seen in the olivine projection (after O'Hara 1968), and based on experiments in the pure CMAS system. H is the liquid co-existing with Ol, Cpx, Opx and Ga. It is argued that in natural rocks H–E is represented by H'–E'. AV is the average peridotite from Table 9.6.

garnet–Opx–olivine (garnet harzburgite) and then Opx–olivine (harzburgite) followed by dunite (olivine only). In practice the argument is complex and highly dependent on solid solution behaviour as O'Hara has shown. The issues involved are illustrated in Figure 9.14 where the extent of solid solutions in the synthetic (pure CMAS) system are shown. Bulk compositions projecting into the triangle H–G–E in the synthetic system will melt as described above to give garnet harzburgite residua, clinopyroxene being the first phase to disappear on melting. Compositions projecting into H–B–E on the other hand will lose garnet first and give rise to lherzolite residua (olivine–orthopyroxene–clinopyroxene). Compositions lying on the garnet-poor side of the line B–E will not contain garnet in the immediate sub-solidus and the initial liquid of melting will not be H but will lie somewhere on the olivine + orthopyroxene + clinopyroxene + liquid boundary (not shown in Fig. 9.14 but visible in Fig. 9.13). Viewed in terms of the solid solutions observed in the synthetic system the composition AV would not be a garnet lherzolite and would not melt as described. However, O'Hara argues that a line H′–E′ better describes the solid solutions and liquid composition in the natural system, in which case the previous discussion of the melting of AV is correct.

Returning to the normative character of partial melts, since AV is itself normatively an olivine tholeiite it follows that advanced partial melting at *any* pressure will at some stage give rise to tholeiitic liquids. Only in the approximate range 10–20 kbar will lower degrees of fusion give rise to silica-undersaturated liquids. Only at very low pressure will initial partial melts be silica-oversaturated (Q-normative tholeiites).

Applications of the O'Hara diagrams. As the previous discussion has attempted to show, the polybaric phase diagram provides an excellent basis for the consideration of possible melting behaviour of Upper Mantle materials and the ways in which magmas may evolve at high pressures. These possibilities have been examined in some detail in papers by O'Hara (1965), O'Hara and Yoder (1967) as well as by O'Hara (1968). Other papers (e.g. Strong 1972, Gass *et al.* 1973, McIver 1975) have attempted more specific interpretation of particular rock suites and for example have attempted to discover whether postulated parental liquids have the necessary compositions to represent initial melts of peridotite at reasonable depths. This type of application must, however, be highly speculative because the system is itself highly flexible. The variations in composition, pressure, and temperature (degree of partial melting) which are available generally preclude any sort of unique interpretation. When the role of water and other volatiles is also considered even more freedom is introduced. However, these limitations are inherent in most petrological research and a system such as the one discussed has at least the virtue of being a powerful tool for the construction of specific petrological models which can be tested in other ways.

A possible example of polybaric fractionation. Lava series which show obvious bulk compositional trends of high-pressure evolution are apparently rare. This is a consequence of the effectiveness of low-pressure processes operating in the near-surface environment, and the effect they have in obliterating earlier compositional features. Many trace element characters of lava suites are more resistant to change and often it is only this aspect of geochemistry which can be used to speculate about high-pressure stages of magmatic evolution (see Ch. 14).

The picritic lavas of the Nuanetsi area of Rhodesia, however, (Cox & Jamieson 1974) provide an unusual example of a suite which has apparently been erupted with abnormal rapidity and has hence avoided extensive low-pressure fractionation. These rocks were erupted carrying small olivine phenocrysts, often of a skeletal aspect, accompanied by very small clinopyroxenes. The matrix is glassy and carries abundant quench microlites of ore minerals, clinopyroxene, and, rarely, plagioclase. The only plausible low-pressure crystal fractionation trends would involve fractionation of olivine alone, or possibly olivine accompanied by clinopyroxene. However, the bulk compositional trend observed is incompatible with either of these, and is in fact far better described by olivine + orthopyroxene addition or removal. To reinforce this, a widespread though volumetrically insignificant group of orthopyroxene megacrysts is present in these lavas, each being overgrown by olivine and pyroxene (an example is illustrated in Fig. 7.7) and hence clearly armoured from, and not in equilibrium with, the groundmass. Thus a high-pressure stage of evolution involving olivine and orthopyroxene has been succeeded by a low pressure stage of olivine and clinopyroxene growth, the latter taking place shortly before eruption so that fractionation of the low-pressure phases has been minimal. Reference to the phase relations previously discussed illustrates the general plausibility of this hypothesis, since any liquid on the olivine + orthopyroxene boundary curve in the dry system will pass into the olivine field if the pressure is reduced. Many such liquids if then allowed to crystallise at low pressure will show the entry of clinopyroxene rather than orthopyroxene as the second phase.

Direct experiment can be used to try to determine the pressure at which these rocks crystallise olivine and orthopyroxene simultaneously, and hence to determine the depth at which the earlier fractionation took place. Preliminary experiments on almost anhydrous samples, however, show that the less magnesian rocks (MgO ca. 10%) lie on the olivine + orthopyroxene boundary at 6 kbar while the more basic samples (MgO contents up to 18%) do not equilibrate with both phases simultaneously until 12 kbar and above. The addition of water (see Ch. 10) would revise these pressure estimates upwards, but it does appear that the fractionation process is likely to have been polybaric and to have taken place over a substantial depth range. Whether the process involved was one of fractional crystallisation, partial melting, or simply equilibration of liquids of yet deeper provenance with wall rocks consisting of olivine and orthopyroxene, is not clear. However, a crystal fractionation process seems the least likely of these hypotheses because it involves a delicate coincidence between the amount of fractionation which has taken place and the depth at which fractionation has ceased for each particular lava (otherwise the correlation between depth of equilibration with olivine + orthopyroxene and MgO content would break down).

Exercises

1. Using the CIPW sequence (see Appendix 3) calculate weight and Eskola molecular norms for analyses A and B given in Table 9.8. Using Figure A3.2 estimate the degree of silica-saturation for each rock. Compare your results with the calculated values. (Hint: when calculating the Eskola norm the equations given under, for example, rules 8c and 8e in Appendix 3 need to be modified. For example, in calculating the distribution of silica between Ab and Ne, since the formulae are $NaAlSi_3O_8$ and $NaAlSiO_4$, the following relationship must apply:

$$x NaAlSi_3 + y NaAlSi = N.Na + N.Al + S.Si$$

where x = number of albite molecules, y = number of nepheline molecules, N = available Na and Al (these are equal), and S = available Si. Hence for Na we derive:

$$x + y = N$$

and for Si:

$$3x + y = S \quad hence \quad x = \frac{S - N}{2}$$

2. Using the norm, calculate the projection parameters of Rock C (Table 9.8) for the normative 1-atmosphere phase diagram (Fig. 9.5) and hence predict the 1-atmosphere crystallisation sequence.

Table 9.8 Analyses for use in exercises

	A	B	C	D
SiO_2	52.69	44.98	46.01	51.86
TiO_2	1.10	2.19	1.61	2.11
Al_2O_3	14.72	14.77	7.96	13.89
Fe_2O_3	1.50	3.70	2.54	3.00
FeO	8.71	7.69	9.09	8.44
MnO	0.17	0.39	0.18	0.14
MgO	6.63	5.35	20.19	6.99
CaO	10.96	15.40	10.49	10.77
Na_2O	2.00	2.51	1.26	2.40
K_2O	0.63	1.02	0.27	0.40
P_2O_5	0.13	0.51	0.17	0.22
Total	99.24	98.51	99.77	100.22

CIPW weight per cent norm of C

Or	1.67		
Ab	10.48		
An	15.29		
Di	Wo 14.96 En 10.90 Fs 2.64	28.50	
Hy	En 4.50 Fs 1.06	5.56	
Ol	Fo 24.57 Fa 6.63	31.20	
Mt	3.71		
Il	3.04		
Ap	0.31		

3. For Rock D of Table 9.8, calculate the O'Hara co-ordinates for the projection from olivine into CS–MS–A (Appendix 4) and predict the crystallistation sequence at 1-atmosphere using Figure 9.9. Supposing the rock represents the residual liquid of a more basic parent which has undergone only olivine fractionation, estimate the pressure at which the parental liquid could have been in equilibrium with garnet harzburgite (olivine + orthopyroxene + garnet).

10
Water-bearing basic rock systems

The effects of water on the phase relations of silicate systems are profound, as examples discussed in Chapter 5 have demonstrated. In this chapter the discussion of basaltic and peridotitic phase relations is amplified in the light of experiments on water-bearing systems. This topic is amongst those currently the subject of intense research activity which has substantially increased our understanding of possible upper mantle and deep crustal processes. However, it may have become clear in the previous chapter that even if we assume reasonable compositions for mantle materials, the P and T variables allow the postulation of many different paths of basaltic evolution and patterns of melting. Since water is a volatile constituent its magmatic abundance is not accurately recorded by the solid rocks available for collection, hence water content, like P and T is a variable about which we may speculate widely. Consequently the options open in the choice of hypotheses are broadened yet again, and the gap in our knowledge of what might happen in the Upper Mantle and lower crust and what does happen is very considerable. However, the experimental work is capable of suggesting specific hypotheses involving for example the role of hydrous minerals such as amphiboles and micas in magmatic processes. The geochemical effects of fractionation involving such phases are in principle distinctive so that it is in the detailed study of the geochemistry of natural rock series that much of the supporting evidence lies. At the present time, however, although several rather general consensuses of opinion exist (for example, that water is involved in the origin of andesites) there is still much debate about the details of most postulated processes.

Experiments on basaltic compositions

The characteristic features of basaltic phase relations in the presence of substantial quantities of water are firstly the depression of both solidus and liquidus temperatures relative to dry systems, and secondly the appearance of an amphibole stability field over a limited pressure range. Liquid compositions during low and moderate degrees of partial melting of basaltic samples are strikingly silica-enriched compared with their anhydrous counterparts and thus resemble rhyolites, dacites, and andesites. The production of silica-rich partial melts is not confined to amphibole-bearing compositions but as we shall see in a later section is also a characteristic of compositions involving solid phases familiar in the dry systems, that is olivine and orthopyroxene.

Fractional crystallisation of wet basaltic magmas is also clearly potentially distinct from that of dry systems since it may involve the precipitation of amphibole-bearing assemblages. Compared compositionally with their nearest dry analogues (the clinopyroxenes) amphiboles are mainly very silica-poor and contain relatively large concentrations of Fe, Na, K and Ti. Hence there is, even apart from minor element considerations discussed in Chapter 14, some real possibility of detecting amphibole fractionation in the evolution of natural suites. Many natural lavas are indeed erupted carrying amphibole phenocrysts, but falling pressure results in the resorption of this phase so that there may be a much larger group of lavas affected by amphibole fractionation in their earlier history but now showing no overt sign of it.

As an example from a detailed study of basaltic melting in the presence of excess water, the results obtained by Helz (1976) for the Picture Gorge tholeiite are given in Figure 10.1 where the modal mineralogy of the refractory residue is shown plotted against temperature. In the early stages of melting, feldspar is the main contributor to the melt and at a later stage amphibole. The increase in augite and olivine up-temperature indicates a reaction relationship between these phases, amphibole, and liquid.

Compositional features of successive melts in Table 10.1 illustrate the rhyolitic nature of the first melts and the evolution towards aluminous dacitic compositions with increasing temperature.

In a study of the phase relations of basanitic samples from Grenada, West Indies, at 5 kbar in the presence of excess water Cawthorn *et al.* (1973) obtained phase relations showing much in common with those determined by Helz in a distinctly different composition (see Table 10.2 for compositions of rocks used experimentally). The basanites show olivine and clinopyroxene together on the liquidus at about 1125 °C, with amphibole entering at about 1050 °C. Olivine ceases to crystallise at about 950 °C and plagioclase commences at 925 °C. When amphibole starts to crystallise the proportion of olivine and clinopyroxene in the charge is much reduced indicating, together with the disappearance of

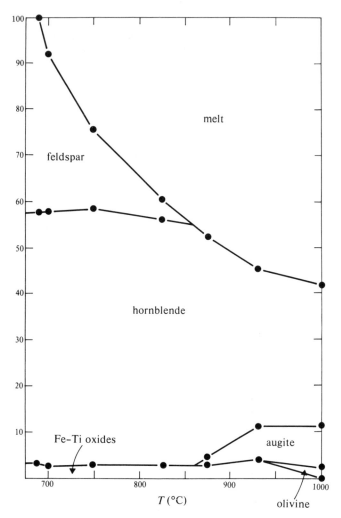

Figure 10.1   Modal composition (weight per cent) plotted against temperature for Picture Gorge tholeiite melted at P_{H_2O} = 5 kbar (after Helz 1976).

olivine below 1050 °C, the same reaction relationship between olivine, clinopyroxene and liquid noted by Helz. Cawthorn *et al.* comment that the existence of this reaction relationship implies that wet basaltic liquids undergoing fractional crystallisation at modest pressures are capable of fractionating amphibole alone. The analysed amphiboles in their study contain about 43% SiO_2 and up to 11% of normative nepheline. Fractionation of such an amphibole from a basanitic liquid containing more SiO_2 than this and less normative nepheline is obvi-

Table 10.1 Anhydrous melt compositions for Picture Gorge tholeiite (from data of Helz 1976)

	700 °C	875 °C	1000 °C
SiO_2	71.5	67.3	62.9
TiO_2	0.07	0.20	0.95
Al_2O_3	15.6	19.5	20.3
FeO	1.21	2.18	2.50
MgO	0.00	0.46	0.38
CaO	1.2	4.95	6.74
Na_2O	3.7	4.2	4.4
K_2O	3.7	1.1	1.1
P_2O_5	–	0.12	0.58

Table 10.2 Compositions and CIPW norms of experimentally melted rocks

	1	2
SiO_2	50.71	43.89
TiO_2	1.70	0.87
Al_2O_3	14.48	15.97
Fe_2O_3	4.89	9.96*
FeO	9.07	–
MgO	4.68	13.90
CaO	8.83	11.49
Na_2O	3.16	2.88
K_2O	0.77	0.67
P_2O_5	0.36	–
Q	0.73	–
or	4.55	4.00
ab	26.74	4.61
an	23.05	28.97
ne	–	10.84
Di	15.44	22.74
Hy	20.90	–
Ol	–	22.61
mt	3.13	4.56
il	3.23	1.67
ap	0.83	–

1. Picture Gorge tholeiite (Helz 1976).
2. Basanite (209) (Cawthorn *et al.* 1973).
* Total iron as Fe_2O_3. Norm calculated using ratio $FeO:Fe_2O_3 = 2$.

ously potentially capable of driving residual liquids through the olivine gabbro thermal divide to give rise to hypersthene- and quartz-normative products. The Grenada rocks do in fact show this compositional trend (basanites to quartz-normative basaltic andesites) and amphibole fractionation may thus be important in their evolution. A somewhat similar series of recent basanitic rocks, showing a marked decrease in silica-undersaturation in the less magnesian members, is found at Shuqra on the northern coast of the Gulf of Aden (Cox *et al.* 1977). The tectonic setting (rifted continental margin) could hardly be more different from that of Grenada (island arc) but the similarities are considerable. In the Shuqra rocks resorbed amphibole, spinel, clinopyroxene, and olivine megacrysts are widely distributed and it can hardly be a coincidence that two such silica-poor minerals as spinel and amphibole are associated with the unusual silica-enrichment trend of whole-rock compositions (see Fig. 10.2).

As an example of the postulated involvement of amphibole in partial melting Helz (1976, p. 177–80) develops the original suggestion of Gunn and Mooser (1971) that partial melting of an amphibole-bearing source has controlled the compositional evolution of the calc-alkaline rocks of the Valley of Mexico. Phenocrystic hornblende is absent from the andesites suggesting that, if amphibole fractionation is required to satisfy the compositional data, the fractionation should take place during partial melting rather than during crystallisation. However, as noted

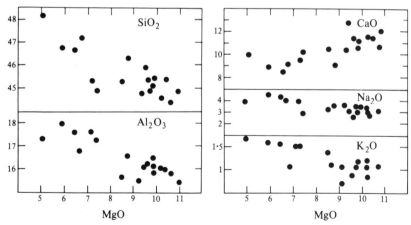

Figure 10.2 Variation diagram for basanitic lavas of Shuqra area, South Yemen (Cox, unpublished data). The rapid rise of SiO_2 with falling MgO requires a very silica-poor extract (fractional crystallisation models) or silica-poor refractory residue (partial melting models). Mixtures of amphibole, spinel, clinopyroxene, and olivine, all found as resorbed megacrysts, fit this and other compositional requirements.

earlier the absence of amphibole phenocrysts may be due to subsequent resorption and this type of argument is inevitably difficult to make watertight.

Experiments on synthetic peridotitic compositions

An extensive general review of the melting of wet peridotites based on experiments in a number of joins in the synthetic system $CaO-MgO-Al_2O_3-SiO_2-Na_2O-H_2O$ is given by Kushiro (1972a). In this and in other works by Kushiro particular attention is paid to the olivine + orthopyroxene + liquid equilibrium because of its relevance to partial melting of olivine- and orthopyroxene-rich mantle materials. In the following section Kushiro's work will be described first. In a later section, however, consideration is given to studies by Nicholls and Ringwood (1973), Green (1973) and Nicholls (1974) which cast doubt on some of Kushiro's conclusions. This later section is included as an example of the type of dispute which is frequent in the literature of experimental petrology and is as often as not a consequence of the experimental difficulties of the subject. In this case one of the important issues is that imperfect quenching of charges results in a misleading assessment of liquid (i.e. glass) compositions. In other conflicts, iron-loss to capsules, and non-attainment of equilibrium, involving, for example, failure to nucleate phases or failure to identify products as metastable, have all been bones of contention. It is unfortunate but inevitable that the non-specialist reader of research papers in experimental petrology is not always in a position to judge the validity of the results safely.

Figure 10.3 gives some of Kushiro's most important results showing unequivocally how in all three sub-systems ($Fo-Di-SiO_2$, $Fo-Ne-SiO_2$, and $Fo-Ca$-Tschermak's$-SiO_2$) the boundary between the olivine and orthopyroxene fields is shifted towards more silica-rich compositions in the water saturated systems (vapour-present systems) compared with the anhydrous. Note that for compositional reasons (absence of Fe, TiO_2, K_2O etc.) as well as temperature considerations, amphibole is not involved in these equilibria and the results seen are entirely due to the effect of water on the melt (see Fraser 1977 for general discussion).

Experiments carried out on compositions lying in the planes En–Di–Ab and En–Di–An show a crystallisation field of forsterite + liquid + vapour penetrating both up to at least 25 kbar. Thus all the experiments appear to indicate that some liquids in equilibrium with forsterite, enstatite and vapour should remain Q-normative up to at least this pressure. An analysis of a glass containing 12% normative quartz and in equilibrium with forsterite, enstatite, clinopyroxene and

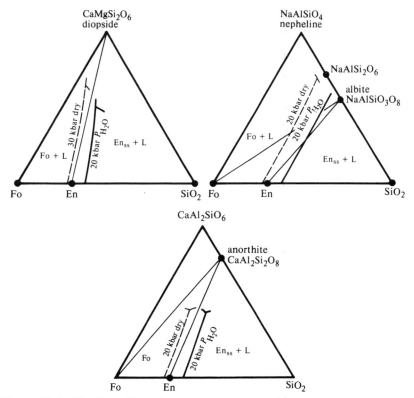

Figure 10.3   The Fo + En + L boundary at P_{H_2O} = 20 kbar compared with the dry system in three planes of the system CaO–MgO–Al$_2$O$_3$–Na$_2$O–SiO$_2$ (mainly after Kushiro 1972a). Note the consistent shift towards more silica-rich compositions when water is present.

vapour at 20 kbar apparently confirms that partial melting of the lherzo-lite assemblage gives a silica-oversaturated liquid under these con-ditions. The glass (Table 10.3) is deficient in too many constituents to be equated directly with any natural rock but it bears some similarity to an andesite in terms of its silica-content, though its low MgO suggests more dacitic affinities. As we shall see, it is the degree of silica-enrichment of there relatively high-pressure melts which is the main issue criticised by other workers.

The paper by Kushiro (1972a) gives an extended discussion of partial melting paths under vapour present, vapour absent (H$_2$O present but P_{H_2O} < total pressure) and open system conditions. He also discusses the interesting suggestion of Yoder (1971) that fractional fusion of hydrous peridotite can lead to the production of silica-rich liquid until H$_2$O is exhausted, followed by a jump to the production of basic liquid from the

Table 10.3 Glass representing liquid in equilibrium with Fo + Opx + Cpx + V at 1050 °C and 20 kbar (Kushiro 1972a)

Analysis		Norm	
SiO_2	54.6		
Al_2O_3	20.1	Q	12.0
MgO	0.92	ab	33.3
CaO	12.1	an	42.4
Na_2O	3.6	wo	9.8
Rest (H_2O)	~8.7	en	2.5

anhydrous residue. Another interesting point concerns the possible production of those rare lava compositions which are simultaneously rich in SiO_2 and MgO by advanced partial fusion of hydrous peridotite. Inspection of Figure 10.3a for example will indicate the enstatite-rich character of a melt which has migrated a substantial distance up-temperature along the Fo + En + L boundary. Before a more critical discussion of these results some details of experiments on peridotites are included.

Experiments on natural peridotites

Experiments on natural peridotites and related rocks in the presence of water and CO_2 have been carried out by Mysen and Boettcher (1975a,b and 1976) and others. The five compositions used by Mysen and Boettcher are given in Table 10.4 and it will be seen that they vary from a highly refractory garnet lherzolite, A, (high Mg/Fe, high total normative olivine + hypersthene, low normative feldspar) to a much less refractory garnet websterite, E, (a garnet–orthopyroxene–clinopyroxene rock). The vapour-saturated solidi of these compositions are compared with the liquidi of various lava types in Figure 10.4 and it is notable that the andesite + H_2O liquidus plots reasonably close to the solidus of the peridotites. This indicates the general plausibility of the hypothesis that andesites may be produced by the partial melting of hydrous peridotites. Also shown in Figure 10.4 is the postulated geotherm for shield areas (Clark & Ringwood 1964) which intersects the various solidi in the pressure range 20–35 kbar suggesting that the Low Velocity Zone of the upper mantle may be due to incipient melting of hydrous peridotite.

Mysen and Boettcher (1975a) also carried out experiments in which the fluid phase was a mixture of H_2O + CO_2 in various proportions. They discovered that for Rock B the solidus minimum rose from about 850° C when the fluid was pure water (see Fig. 10.4) to slightly over

Table 10.4   Compositions of rocks used by Mysen and Boettcher (1975a, b, 1976)

	A	B	C	D	E
SiO_2	45.7	43.7	45.10	44.82	45.58
TiO_2	0.05	0.20	0.13	0.52	0.80
Al_2O_3	1.6	4.0	3.92	8.21	13.69
Fe_2O_3	0.77	0.89	1.00	2.07	3.76
FeO	5.21	8.09	7.29	7.91	5.85
MgO	42.8	37.4	38.81	26.53	16.09
CaO	0.70	3.50	2.66	8.12	11.78
Na_2O	0.09	0.38	0.27	0.89	1.27
K_2O	0.04	0.01	0.02	0.03	0.02
		CIPW norms			
or	0.24	0.06	0.12	0.18	0.11
ab	0.76	3.22	2.28	7.53	10.48
an	3.42	9.18	9.43	18.32	31.69
Di	0.00	6.41	2.93	17.20	21.06
Hy	37.71	13.36	22.16	7.12	8.79
Ol	58.55	64.20	60.57	44.81	19.58
mt	1.12	1.29	1.45	3.00	5.57
il	0.09	0.38	0.25	0.99	1.52

A. Garnet lherzolite, Wesselton Mine, South Africa.
B. Spinel lherzolite, Hawaii.
C. Spinel lherzolite, Hawaii.
D. Garnet lherzolite, Hawaii.
E. Garnet websterite, Hawaii.

1000° C when the fluid had a mole fraction of H_2O of about 25%. They point out that variability in the composition of the fluid phase in the mantle could thus result in widely differing depths to the Low Velocity Zone.

Examples of determined phase relations are given in Figure 10.5. Both samples, the one peridotite and the other garnet websterite, show the limited stability field of amphibole characteristic of most basic and ultrabasic compositions (for another example see Lambert and Wyllie 1968). The role of amphibole as a phase involved in melting processes and basaltic fractionation is evidently restricted to a limited depth range within the lower crust and the upper part of the upper mantle.

The compositions of melts obtained by Mysen and Boettcher (1975a, b) are given in Table 10.5 and some considerable variety is present because of the temperature and $X_{H_2O}^V$ (mole fraction of H_2O in vapour) variables. Low temperature melts are generally andesitic in character when $X_{H_2O}^V$ is high (e.g. columns 1–3). In samples with $X_{H_2O}^V < 0.5$, however, low temperature melts are olivine- and, often,

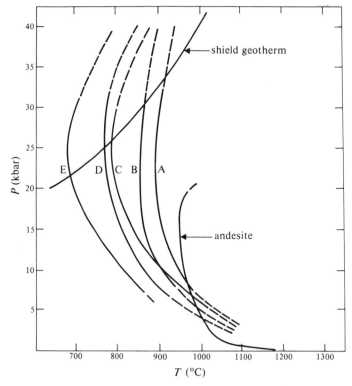

Figure 10.4 Solidi of water-saturated peridotites (A–D) and garnet websterite (E) (see Table 10.4) based on work of Mysen and Boettcher (1975a, 1976). Also shown are the shield geotherm and andesite liquidus (Mysen and Boettcher 1975a, Fig. 2).

nepheline-normative (e.g. column 4), though in general the silica saturation increases with temperature (e.g. columns 5 and 6). Reference to the anhydrous phase diagram discussed in Chapter 9 (Figures 9.9 – 9.11) will show that the effect of reducing $X^V_{H_2O}$ in the above experiments is to produce melt compositions qualitatively approaching those of the anhydrous system, for it is at pressures of ca. 15 kbar that anhydrous initial melts are at their most silica-undersaturated. Higher temperatures, implying higher degrees of melting, will of course reduce this effect since the bulk compositions of the starting samples are hypersthene-normative.

Phlogopite stability

Apart from amphibole discussed above the mica phlogopite is likely to be a principal repository for water in unmelted upper mantle rocks.

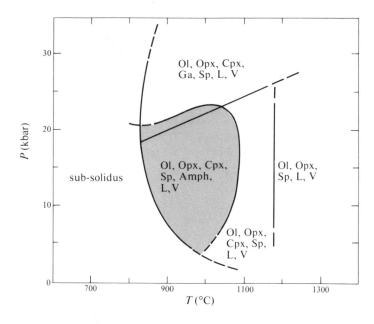

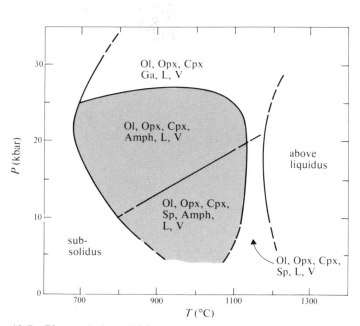

Figure 10.5 Phase relations of (a) peridotite B, and (b) garnet websterite (Table 10.4) from Mysen and Boettcher (1975a and b, 1976). Amphibole bearing assemblages above the solidus are shaded.

Table 10.5 Compositions of partial melts formed from rocks B and E of Table 10.4 at various pressures, temperatures and values of $X_{H_2O}^V$ (from data of Mysen & Boettcher 1975b, 1976)

	1	2	3	4	5	6
Sample	B	B	B	B*	E	E
$X_{H_2O}^V$	1	1	0.75	0.20	1	1
T (°C)	1050	1150	1150	1150	1130	1010
P (kb)	15	15	15	15	15	15
SiO_2	59.4	61.4	58.0	43.5	50.5	58.9
TiO_2	0.5	0.6	0.4	0.5	0.9	0.4
Al_2O_3	24.4	16.2	16.1	10.5	15.5	24.0
FeO^*	2.4	5.9	5.2	13.0	2.9	0.07
MgO	0.8	3.7	4.7	10.7	15.2	0.3
CaO	9.4	9.4	12.0	13.8	12.8	11.8
Na_2O	2.9	2.5	3.8	7.6	1.6	3.7
K_2O	0.2	0.1	0.5	0.4	0.1	0.1

CIPW norms						
Q	23.34	19.26	5.39	–	–	12.9
or	1.18	0.59	2.95	–	0.6	0.6
ab	24.54	21.16	32.16	–	13.4	31.3
an	42.06	32.69	25.40	–	34.8	48.6
ne	–	–	–	28.05	–	–
Di	6.31	11.57	26.38	32.19	22.6	3.8
Hy	2.11	13.49	6.96	–	12.7	–
Ol	–	–	–	25.54	14.1	–
il	0.19	1.14	0.76	0.95	1.7	0.8
wo	–	–	–	–	–	2.3
C	2.76	–	–	–	–	–

*Norm also includes 9.10% larnite and 1.85% leucite.

Phlogopite is the only common primary hydrous mineral observed in ultramafic nodules derived from kimberlite, although it is far from abundant. Consideration of Upper Mantle melting processes is incomplete without reference to this phase because despite its low concentration, it is, in the absence of amphibole, the most likely repository not only of H_2O but also of K, Rb and several other trace elements. The contribution of phlogopite to the liquid during low degrees of partial melting may be of great significance to the trace element geochemistry of basaltic liquids (see Ch. 14).

Experimental evidence bearing on phlogopite stability is sparse but Mysen and Boettcher (1975a) demonstrated by spiking their rock C with synthetic phlogopite and H_2O, that the assemblage amphibole + phlogopite co-exists with liquid over a wide temperature range in the pressure interval 10–20 kbar, and also that phlogopite is stable above the upper temperature stability limit of amphibole.

Experiments in the synthetic system $CaO–K_2O–MgO–Al_2O_3–SiO_2–H_2O$ by Modreski and Boettcher (1973) and Bravo and O'Hara (1975) demonstrate the stability of phlogopite up to 35 kbar in compositions analogous to those of the Upper Mantle.

The experiments of Nicholls and Ringwood (1973)

The previous sections have illustrated some of the main features of the phase relations of water-bearing basic and ultramafic systems under pressure. A conclusion beyond dispute is that water has an important effect on expanding the field of olivine crystallisation into silica-rich compositions relative to those in the dry system. A considerable debate has developed, however, concerning the exact extent of such silica-enrichment, and how it is affected by pressure. Any direct partial melt of the mantle must be in equilibrium with magnesian olivine. Under water-saturated conditions maximum silica-enrichment of the liquid is therefore to be expected, but this effect will be offset by pressure. Hence with increasing pressure there must come a point where any given silica-saturated or oversaturated composition will no longer be capable of equilibrating with olivine. This pressure represents the maximum depth at which such a composition could be produced by partial melting of an olivine-rich mantle. The pressure will be lower for andesites than for example silica-saturated tholeiites. The problem of the nature of partial melts can thus be approached from several different angles.

The direct approach, as illustrated by the work of Kushiro and many others, consists of analysing glasses with the electron microprobe. Kushiro *et al.* (1972), for example, carried out melting experiments on a natural spinel lherzolite and obtained a silica-rich glass at 26 kbar and 1190 °C under water-saturated conditions. This glass when recalculated anhydrous contained 68% SiO_2, 10.2% CaO, 0.6% MgO, and 1.1% FeO. It was thus very similar to the glass noted in Table 10.3 from a synthetic peridotite melt, and like it was apparently formed in the presence of residual olivine, orthopyroxene and clinopyroxene.

There are two alternative ways of checking this conclusion, which at first sight appears to support the early results conclusively. Nicholls (in

Nicholls and Ringwood 1973) made up an identical glass and studied its phase relations at the same pressure as Kushiro *et al.* His experiments showed that the liquidus temperature was 940° C, that is, 250° C lower than previously reported, and that the liquidus phase was clino-pyroxene. He reconciled the two sets of results by suggesting that the melt produced by Kushiro *et al.* had failed to quench properly, and that the glass analysed had been formed only after the crystallisation of silica-poor quench crystals. Experiments on similar compositions by Green (1973) indicated that the quench crystals were likely to have been metastable olivine and amphibole. The experiments suggest that melts produced by partial melting of the mantle may not be as silica-rich as had been supposed.

As an alternative to the check on liquidus temperature, Nicholls and Ringwood (1973) determined the nature of the liquidus phase with increasing pressure under water-saturated conditions in a number of compositions. They found that olivine disappeared from the liquidus in an almost exactly silica-saturated tholeiite at about 20 kbar and was replaced by garnet and pyroxenes. Preliminary results for a synthetic and a natural basaltic andesite indicated that olivine remained on the liquidus only to 5 kbar and 7 kbar respectively. These pressures give the maximum depths (about 60 km and 18 km) at which magmas of these compositions can be generated directly from olivine-rich mantle, and hence place considerable constraints on models of magma generation in island-arc environments. The reader is referred to Nicholls (1974) who reports the results of numerous additional experiments and reviews the important issues.

It is hoped that the above discussion illustrates one of the fundamental aspects of the methodology of experimental petrology, that is the necessity for major hypotheses to be confirmed by numerous experiments designed to approach the problem from different directions.

11

Compositionally zoned magma bodies and their bearing on crystal settling

After the substantial excursion in the last three chapters into experimental petrology we return to studies of individual rock sequences, working towards the study of plutonic rocks and layered intrusions given in Chapters 12 and 13. The theme of crystal fractionation at low pressures dominates this section of the book, and it must be emphasised again that the investigation of the high-pressure origins of magmas cannot take place in isolation. The possibility of substantial low-pressure modification of magmas is ever-present.

Evidence that magma chambers may be compositionally zoned is derived from volcanic eruptions, usually rather voluminous, which take place rapidly and deposit their products in an identifiable time sequence. If the duration of the eruption is short compared with the time scale of differentiation of the magma chamber then the study of the volcanic products can give an effectively instantaneous view of the interior of the chamber, at least as regards any vertical zonation it may show. The supposition is that the earliest parts of the erupted sequence are derived from the upper part of the chamber and that successive products come from deeper levels. Many such sequences have been studied and are characterised by the early eruption of the most differentiated magmas followed by progressively less evolved material. It is usual also to find that the phenocryst content of the eruption increases with time, indicating the concentration of crystals towards lower parts of the chamber. Eruptions of this type do much to reinforce the ideas of gravitative crystal settling which are derived from the study of layered intrusions (see Chs. 12 and 13).

The time element is crucial in the study of the eruptive products and it is thus not surprising that most of the evidence so far collected comes from pyroclastic rocks. Basal layers of tuffs clearly represent earlier parts of the eruption than stratigraphically higher layers so that vertical sampling of sections reveals the change in the nature of the erupted magma with time. The study of lavas is obviously more difficult because of the complex flow patterns they may show, though in some instances sequences of quite distinct superimposed flows can be demonstrated to belong to a single larger eruptive event.

It is also essential, in order that the evidence shall be preserved, that the eruption is not accompanied by excessive mixing of magma from different parts of the chamber. That zoned magma chamber sequences exist at all implies also a lack of convection in the chamber.

Examples of zoned eruptive sequences have been described from many parts of the world but particularly from acid ash-flow deposits in the western United States (Williams 1942; Ratté & Steven 1964; Quinlivan & Lipman 1965; Smith & Bailey 1966; Lipman *et al.* 1966; Giles & Cruft 1968; Byers *et al.* 1968). They are also known from Japan (Katsui 1963; Lipman 1967), New Zealand (Martin 1965; Ewart 1965), Ethiopia (Gibson 1970), and the Canary Islands (Schmincke 1969). Most of these localities are related to young orogenic areas though there is probably no particular reason for this other than the relative abundance of those acid magmas which characteristically give rise to air-fall and pyroclastic flow deposits. Zonation of eruption in the well-dated Icelandic acid to intermediate tephra eruptions of the Hekla volcano (Thorarinsson 1950; Thorarinsson & Sigvaldason 1972) is also well documented. The Aden volcano (Hill, 1974) provides a good example of zonation in a lava sequence several hundred metres thick.

An ash-flow sheet from Aso caldera, south-western Japan

As an example illustrating some of the features of a zoned ash-flow deposit the sheet described by Lipman (1967) from the Aso caldera is ideal. This deposit shows most of the typical features and has been investigated in sufficient detail to reveal some interesting additional complexities.

The Aso caldera was the site of voluminous late-Pleistocene ash-flow eruptions which are distributed over a roughly circular area some 120 km in diameter. At least three major ash-flow deposits are believed to have been erupted from the caldera and it is the youngest of these, designated Aso III which is the subject of Lipman's study. This unit is divisible locally into partially separate cooling units, the boundaries of which are marked by local concentrations of pumice blocks and xenoliths and are traceable over large areas thus giving some degree of stratigraphic control. Lithologically the deposit is partly welded and consists of pumice blocks making up about 25% of the tuff set in a matrix of crystals and glass shards. Small xenoliths of andesitic and sedimentary material are widely distributed. In gross composition the deposit grades upwards from phenocryst-poor rhyodacitic dacite to phenocryst-rich trachyandesite.

Phenocryst contents and groundmass glass compositions of pumice blocks from various stratigraphic levels were investigated by Lipman, the results for phenocrysts being given in Table 11.1. There are several notable features here, firstly the typical rising phenocryst content in the upper levels. However, it is interesting to note that hornblende is missing from the pumice blocks of the base of the deposit, and olivine and augite are only present towards the top. Moreover the composition of the plagioclase changes to a more calcic variety upwards in the sheet. Thus a simple model involving only the downward accumulation of crystals in a uniform liquid matrix is clearly inadequate to explain the zonation of the erupted products. The glass analyses confirm this (Table 11.2), that is the liquid matrix in which the phenocrysts were suspended evidently became more basic downwards in the chamber. A considerable part of the change in bulk composition of the

Table 11.1   Details of phenocrysts in pumice blocks of Aso III ash-flow sheet (after Lipman 1967)

Stratigraphic unit		Phases	Total phenocryst (%)	Average plagioclase
Upper unit	Upper part	hypersthene hornblende augite olivine plagioclase	20–25	An_{50}
	Lower part	hypersthene hornblende plagioclase	7–10	An_{45}
Lower unit	Upper part	hypersthene hornblende plagioclase	6–7	An_{45}
	Lower part	hypersthene plagioclase	5–6	An_{40}

Table 11.2 Analyses of glasses (recalculated free of H_2O and CO_2) from Aso III pumice blocks (after Lipman 1967)

	1	*2*	*3*
SiO_2	69.85	68.26	62.48
TiO_2	0.39	0.45	0.45
Al_2O_3	15.73	16.45	19.26
Fe_2O_3	0.81	0.69	1.08
FeO	1.26	1.79	2.04
MnO	0.10	0.11	0.07
MgO	0.54	0.63	1.82
CaO	1.53	2.44	4.90
Na_2O	5.24	5.01	4.51
K_2O	4.43	3.96	3.08
P_2O_5	0.11	0.21	0.30

Stratigraphic positions: 1. Lower part of upper unit. 2. & 3. Upper part of upper unit. An analysed glass from the lower unit is similar to 1.

pumice blocks is accounted for by the change of the matrix composition, and only some is attributable to variation in phenocryst content. These data, while suggesting strongly that the Aso III tuff was erupted from a vertically zoned magma chamber in which crystal settling was taking place, leave several unanswered questions with regard to the precise mechanism of differentiation, for example whether processes such as volatile transfer might have been operative as well as fractional crystallisation. It does seem certain, however, particularly from the evidence of the plagioclase compositions, that a temperature zonation was also present in the chamber. Some of these problems will be investigated further in a later section of this chapter. Before leaving the Aso ash-flow sheet some of the additional complexities noted by Lipman should, however, be mentioned to illustrate limitations inherent in the study of pyroclastic deposits. Firstly Lipman notes that the crystal content of the matrix is usually higher than that of the associated pumice blocks and there is thus evidently some mechanical sorting involved during the emplacement of the flow. Furthermore, at the base of the deposit the matrix does contain hornblende crystals, although these are absent from the pumice blocks at this horizon. Lastly, although at any particular horizon pumice blocks of a particular type predominate, there are usually some with characteristics of those belonging to other parts of the flow. These features all suggest that some degree of mixing and sorting of the deposit has taken place during the eruption, thus making the task of precise petrological interpretation substantially more difficult.

The Shamsan caldera sequence of Aden

The Aden volcano is the easternmost of a line of Plio-Pleistocene centres lying on the southern margin of the Arabian peninsula (Cox *et al.* 1970). These volcanoes characteristically exhibit early cone-building activity by trachybasalt, trachyte, and rhyolite flows, followed by the eruption of rhyolites, trachytes, and trachyandesites into calderas. Limited data suggest that the caldera eruptions generally show the same reversal of differentiation sequence with time that is characteristic of the ash-flow deposits discussed, that is early eruptions of rhyolite are succeeded by progressively more basic compositions. Caldera sequences are apparently erupted rapidly, each flow passing through the previously consolidated one and spreading over it. Confinement by the caldera wall is evidently the reason for the preservation of this exact stratigraphy which serves to document the progressive expulsion of the magma from the chamber beneath. The Shamsan caldera sequence in Aden has been investigated in detail by Hill (1974). A cross section of the volcano is given in Figure 11.1 and illustrates the distinction between the caldera flows, and those now preserved on the flanks of the volcano which overflowed from the caldera at a later stage of the eruption. The caldera flows are horizontal and can thus be arranged in precise stratigraphic sequence but the overflow lavas can only be assigned rather vaguely to a late stage of the eruption. Because of the emplacement of the younger rocks of the Tawela caldera the lowest flows of the Shamsan sequence are unfortunately not preserved.

An illustration of the gradual change in composition of the flows with time is given in Figure 11.2 where chemical data are plotted against topographic height. The lowest flows seen have SiO_2 contents

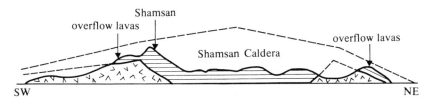

Figure 11.1 Sketch section across the Aden volcano to show the Shamsan Caldera rocks (based on Cox *et al.* 1969). Section is 3 km long, Shamsan stands 500 m above sea level (base of section). Horizontal line ornament – flows within the caldera. Inclined line ornament – overflows. Vee ornament – older cone rocks. Broken line indicates postulated former extent of Shamsan series. The younger Tawela caldera sequence which obliterates the central part of the Shamsan structure has been omitted for clarity.

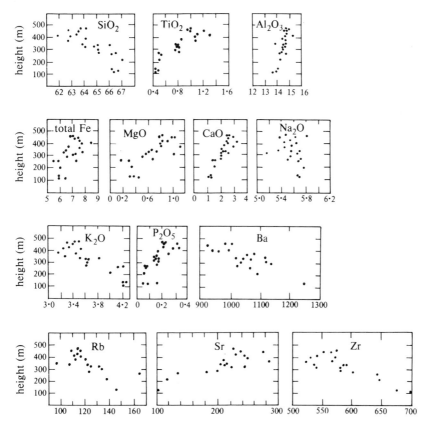

Figure 11.2 Variation in whole-rock composition plotted against height above sea level for Shamsan caldera rocks, Aden (after Hill 1974). Trace element values are in ppm.

approaching 67% but this drops to as low as 62% in the uppermost rocks. Samples from the overflow lavas have silica contents in the range 62.4–60%, showing that the trend towards reduced silica continued into the later stages of the eruption. As another example, CaO which reaches 3% in the highest caldera rocks lies in the range 3.4–4.2% in overflow samples. The whole sequence in fact shows a very typical fractionation trend with the elements K, Ba, Rb, Zr rising with SiO_2 and many other elements, e.g. Ca, Mg, Fe, Ti, P and Sr, falling.

Phenocryst in the Shamsan lavas consist of olivine clinopyroxene, plagioclase, calcic anorthoclase (subsidiary to plagioclase), opaque ores, and apatite. There is a regular increase in phenocryst content upwards through the sequence (Fig. 11.3) which lends credence to the crystal settling hypothesis for the underlying magma chamber. As in the Aso III

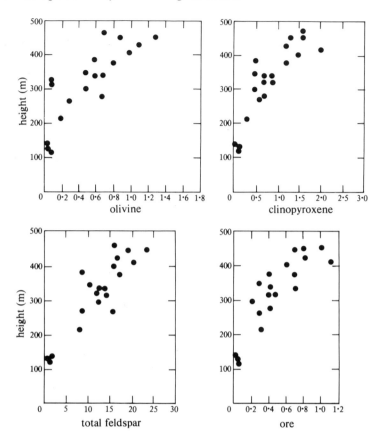

Figure 11.3   Variation in phenocryst content plotted against height above sea level for Shamsan caldera rocks, Aden (after Hill 1974). Values are in volume per cent. Because of the low concentration of some phases modal analyses were carried out on up to six standard-size thin sections per sample.

Table 11.3   Compositions of phenocrysts in Shamsan caldera sequence lavas (from data of Hill 1974)

Height above sea level	Fe/Mg ratio clinopyroxenes	Fe/Mg ratio olivines	An % in feldspar
Overflows	0.37	0.50	39
400–500 m	0.43	0.69	25
300–400 m	0.48	0.70	27
200–300 m	0.63	no data	13

Clinopyroxene and olivine data are averages. Because of strong zoning, the feldspar figure given is the most An-rich value for the height range.

ash-flow sheet the increase in phenocryst content is accompanied by the increasingly basic character of groundmass compositions and by changes in the composition of the phases (see Table 11.3) which indicate increasing temperatures of equilibration downwards in the magma chamber.

A model for crystal settling in a non-convecting magma chamber

Although as noted earlier the simplest type of crystal settling model (uniform liquid throughout) is not capable of explaining sequences like the Aso III sheet or Shamsan lavas, minor modifications to the hypothesis produce a plausible explanation of the salient geochemical and mineralogical features.

In essence, suppose that a magma chamber is filled with liquid and is then cooled with heat losses being greater towards the top of the chamber. As a result, more crystallisation takes place at the top than elsewhere so that the residual liquid composition must take on a vertical zonation, the more evolved liquid lying at the top of the chamber and the less evolved below. At the same time crystal settling produces a downward increase of crystal (phenocryst) content. These crystals may be made over to more basic compositions as they sink into hotter liquid, or, alternatively, depending on the cooling rate of the whole chamber, by the time they sink to a given level the liquid here may have cooled to a temperature similar to that at which the crystals originally formed and hence reaction will not take place. Still greater cooling rates will result in sinking crystals being made over to more evolved compositions or else zoning normally outwards. Irrespective of these possible variations the net result of the existence of a vertical temperature gradient in the chamber is to produce higher temperature (more basic) crystals at lower levels than exist at the same time at higher levels. Thus at any given time the chamber would consist of magma becoming progressively more basic downwards as a result of two factors, firstly the greater crystal content, and secondly the more basic liquid composition. Qualitatively this sort of model provides an adequate explanation for the main features of the sequences described, though even a relatively simple model like this is extremely difficult to translate into precise quantitative form. One petrographic feature not mentioned for example is that the Aden lavas characteristically show a glomeroporphyritic texture so that the quantitative modelling of settling rates, even based on measured crystal sizes, calculated viscosities, and experimentally determined temperatures, is highly uncertain. To this must be added factors such as nucleation rates, rates of reaction (making-over) between crystals and several others.

Nevertheless, the relationships between overall chemical change, phenocryst content, and height in the sequence, are so well established for the Shamsan rocks that it is difficult to believe that crystal settling has not been a highly important mechanism in their evolution, although the operation of other fractionation mechanisms in addition cannot be excluded. However, a puzzling question mark hangs over this series as a result of the strontium-isotope studies of Dickinson *et al.* (1969) and Carter and Norry (1976). The Aden volcanics in general do not produce a Rb/Sr isochron which is appropriate to their known eruptive age of 5–10 Ma (based on K/Ar dating) and appear to have been produced by the reworking of already fractionated older material. The Shamsan sequence itself produces a range of $^{87}Sr/^{86}Sr$ ratios which is not consistent with derivation from a uniform parent magma. How the evidence of classical petrology and isotope geochemistry can be reconciled has yet to be seen.

Compensated crystal settling

The compensated crystal settling hypothesis (Cox & Bell 1972; Krishnamurthy & Cox 1977) was introduced to explain the formation of highly porphyritic lava sequences which appeared to have evolved without any notable fractionation of the phenocrysts observed. The postulated mechanism is a variant of that described above with reference to sequences of the Aso III and Shamsan types.

The problem is identifiable in series of volcanics which show phenocryst assemblages which are not consistent with the fractionation of bulk composition observed. Two sequences of picrite basalt lavas, one from the New Georgia Group of the Solomon Islands, the other from the Deccan Traps of India (see references above) show the same feature, that is they have bulk compositions lying on olivine control lines while the rocks themselves contain abundant large phenocrysts of both olivine *and* clinopyroxene. Whatever the history of the olivine phenocrysts, it is clear that clinopyroxene cannot have fractionated in these liquids, either by removal or accumulation, because otherwise the bulk compositional trend would have been destroyed. Experiments on the Deccan samples show that clinopyroxene is the second phase to enter on cooling, and geochemical data show that there is a comparative uniformity of crystal and groundmass compositions. These rocks therefore appear to differ fundamentally from those described above in that the variation in composition is very largely accounted for by the variable total phenocryst content, not by the groundmass composition.

It follows that all these rocks were erupted inside a rather small temperature interval, somewhat below the entry of clinopyroxene, the more basic rocks (18% MgO) carrying very abundant olivine phenocrysts accompanied by clinopyroxene, the less basic ones (12% MgO) carrying much the same amount of clinopyroxene but substantially less olivine. Such a sequence therefore gives every appearance of having been derived from an original series of olivine-controlled liquids which have then undergone varying degrees of *equilibrium* crystallisation, and eventually all been erupted within the same narrow temperature range. However, such a postulate would appear to imply the temporary suspension of the laws of gravity, for so many lines of geological evidence indicate that crystal settling is extremely difficult to avoid.

The compensated crystal settling model is designed to allow equilibrium crystallisation and crystal settling to take place simultaneously. Compared with the previous model here it is postulated that vertical temperature gradation in the magma chamber is negligible, which implies that cooling is predominantly through the sides of the chamber rather than through the top. A dyke-like conduit would have the necessary geometrical properties for this to apply plausibly.

Imagine the chamber to be filled with magma at a certain time. Heat is extracted through the walls and crystallisation commences throughout. After a certain time interval suppose that an average crystal has settled through a distance S which is much smaller than the height of the chamber under consideration, H. Then from the top of the chamber down to depth S the magma will have suffered a net loss of crystals, while between depths S and H the crystal content will be uniform, and will equal the amount of crystallisation that has taken place. The net loss of crystals above S is balanced by crystals passing out of the part of the chamber under consideration through the arbitrary base at depth H. If the chamber is floored these may collect as cumulates or they may otherwise dissolve in hotter liquids below. Between depths S and H the bulk composition of the magma has not changed during this process. This is the zone of compensated crystal settling in which something approaching equilibrium crystallisation is possible. The term compensation refers to the fact that each crystal which is lost from a certain level is replaced by a similar one from above. If the chamber were suddenly evacuated the eruptive sequence would consist first of evolved phenocryst-poor liquid, as in the earlier model discussed, grading rapidly into the voluminous eruption of phenocryst-rich magma of uniform composition. As long as the vertical zone throughout which cooling takes place is substantially larger than the distance through which crystals have settled during the cooling period, the amount of porphyritic magma which still has the bulk composition of the original liquid will be large, and the amount of early evolved magma will be small.

Highly porphyritic volcanic sequences are common, and dyke-like feeders are particularly abundant in some basaltic provinces. To that extent the compensated crystal settling hypothesis, or side-cooling magma chamber model, is plausible; but it is not to date based on the wealth of descriptive data which supports the top-cooling model discussed with regard to the ash-flow sheets and Aden volcanics. However, it may prove to have some importance in the study of ultramafic magmas, for it suggests that not all highly porphyritic picrite basalts are formed by the accumulation of mafic phenocrysts in more normal basaltic liquids. Some of these may represent genuine primitive ultramafic liquids which have suffered advanced, quasi-equilibrium crystallisation in the near-surface environment (Cox 1978).

Exercises

1. A dyke-like magma chamber with parallel vertical sides is filled with liquid of MgO content 16%. Cooling takes place uniformly and after a certain period of time the liquid at all levels has crystallised 10% of olivine of average MgO content 50%. During this period the average olivine has settled 20 m. If now the contents of the chamber down to a depth of 100 m is erupted, calculate:

   (a) the MgO contents and phenocryst contents of
       (i) the first magma erupted,
       (ii) magmas derived from depths of 10, 20 and 100 m;
   (b) the volume proportion of magma erupted with an MgO content of 16%.

12

Petrographic aspects of plutonic rocks

Introduction

Compared with volcanic rocks, plutonic rocks have cooled more slowly, under greater pressure and in the more protracted presence of volatiles. All three factors combine to influence the structure, texture and mineralogy of plutonic rocks but it is usually assumed that rate of cooling exerts the more important influence on the development of primary structure and texture while pressure and volatile content predominate in determining mineralogical features. Thus we may observe that in a body of magma of sufficiently low viscosity, the slower the cooling rate the greater will be the opportunity for crystals of different densities and sizes to move relative to one another and for that movement to be influenced by currents within the magma or flow of the magma itself. On the other hand we may note that increased pressure imposes different sets of mineral equilibria leading, for instance, to the suppression of the incongruent melting of enstatite, and that the maintenance of a high volatile content (under elevated pressure) will favour the stability of hydrous minerals both as primary phases and secondary replacements of other primary phases, and will also be a control in sub-solidus changes such as exsolution.

Chilled margins

A familiar example of the importance of cooling rate in determining texture in the plutonic environment is well provided by the so-called chilled margins which many intrusions display. A chilled margin is generally envisaged as an envelope of fine-grained rock enclosing and grading into the relatively coarser inner parts of an intrusion (Fig. 12.1). Chilled margins are commonly used to determine the relative age of intrusions and adjacent rock.

Figure 12.1 Chilled margin of Marsco epigranite (bottom) against gabbro (top) of the Cuillin Complex, Druim Eadar da' Choire, Isle of Skye, Scotland. Gabbro shows coarse, even-grained plagioclase and clinopyroxene displaying cracks and alteration; epigranite shows alkali feldspar and quartz with slight turbid alteration of the former.

However, within this broad description numerous significant variations can occur. Chilled margins may be glassy, devitrified, porphyritic, ophitic or **allotriomorphic-granular** in texture; the latter, consisting of even-sized small, shapeless grains, develops when a cooling liquid reaches the stage of peak crystal nucleation rate, having passed rapidly through the peak crystal growth rate so inhibiting extensive growth of earlier crystal nucleii. The fall in temperature also increases the viscosity of the magma and this in turn impedes diffusion of appropriate components to the crystal nucleii. A significant point here is that all the grains are roughly equal in size, a contrast with the **ophitic** texture in which, typically, large numbers of small plagioclase crystals are partially enclosed by relatively few pyroxene grains. The chilled margin of the Skaergaard basic layered intrusion, East Greenland, is ophitic and an explanation of this has been offered by Carmichael, Turner and Verhoogen (1974). They point out that whereas the nucleation rate of a mineral is *inversely* proportional to the square of its entropy of fusion and the square of the fall in temperature, the crystal growth rate is *directly* proportional to these parameters. Since plagioclase has a significantly lower entropy of fusion than that of pyroxene in a rapidly cooling basic liquid, a large number of slowly-growing plagioclase nucleii should become enclosed by pyroxene crystals growing rapidly from a few nucleii.

Herein may also lie one reason why some chilled margins display one rather than the other of these two contrasting textures. The entropies of fusion of minerals relevant for acid rocks (cristobalite, albite and sanidine) are all lower, and mutually closer in value than those of minerals forming basic rocks (Carmichael, Turner & Verhoogen 1974, Table 4–3, p. 167). Thus we might expect the chilled margins of sizeable granitic intrusions to display allotriomorphic-granular texture (either throughout, or in the groundmass of porphyritic cases) while comparably sized basic intrusions would have chilled margins with ophitic texture.

The chilled margin shown in Figure 12.1 does not persist along the entire contact between the granite and gabbro intrusions. In places the granite shows no diminution in grain size, implying thermal equilibrium between country rock and intruding magma at those points. The explanation of this is found elsewhere in small detached fragments of chilled margin material enclosed in the coarser granite, indicating that in places continued pulsing of magma prised off the chilled skin and came into contact with a pre-heated country rock thus suffering no accelerated cooling. It is also possible for a chilled margin to be at least partially melted by later pulses of magma which are hotter than the initial batch.

These are points of some significance when one considers the assumption made by certain petrologists that the fine-grained margin of an intrusion is compositionally representative of the initial magma before any subsequent *in situ* differentiation. Other snags involved in such an assumption are the possibility of reaction between hot magma and cold country rock and the chilled margin's forming not so much a frozen sample of the pristine magma as a membrane at least semi-permeable to the diffusion of fluids. In the case of the Skaergaard intrusion Wager and Deer (1939) were careful to prefer a specimen of the fine-grained gabbro in contact with Tertiary basalt rather than with Precambrian acid gneiss, thinking so to limit the likelihood of using a contaminated sample. As it happens, the Skaergaard intrusion has experienced some chemical and isotopic equilibration with the country rock through the agency of heated groundwater diffusing through the chilled margin. This diffusion was evidently more effective through the stratified and jointed basalts than it was through the massive gneiss.

Even where the enclosing rock is the same at all points the chilled margin may show variation in chemical composition which is not always petro-graphically evident. The upper and lower chilled margins of the Prospect alkaline diabase sill, New South Wales (Wilshire 1967) are chemically different, suggesting the emplacement of a slightly differentiated magma. Hydrothermal alteration of the upper chilled margin of the Palisades Sill, New Jersey (Walker 1969) may account to some extent for its composition's differing from that of the lower chilled margin.

As described in Chapter 13 it is possible to calculate the bulk composition of a well-exposed intrusion and where this has been done in the case of the two sills mentioned above the bulk composition differs from that of the chilled margin, the latter being based upon averaged analyses. Significantly, perhaps, the much larger Palisades Sill turns out to be a multiple intrusion as shown by the presence of an *internal* chilled contact and reversals of mineral composition trends.

A chilled margin is, of course, similar in several ways to a volcanic rock and is petrologically useful, not least in occasionally preserving quenched high temperature phases, or evidence of their original presence in the form of paramorphs. An example of this is seen in the Coire Uaigneich granophyre intrusion in the Isle of Skye, Scotland. The chilled margin contains phenocrysts of quartz paramorphing tridymite (Fig. 12.2). Further into the intrusion discrete primary quartz crystallised after tridymite.

Finally we may note that larger intrusions may display a chilled margin which is only one part of a relatively rapidly cooled facies forming an extensive outer part of the whole mass with its own distinctive textures and mineralogy. The term **border group** was applied to such a set of rocks by Wager and Deer (1939) in their study of the Skaergaard intrusion. Here parts of the border group contain primary orthopyroxene which is absent from rocks of the layered series of the intrusion, its place being taken by orthopyroxene inverted from pigeonite. From the presence of primary orthopyroxene in the quickly cooled rocks, it was deduced that the mineral should be a significant phase in the unexposed layered rocks presumed to form lower parts of the intrusion.

Cumulates

The thermal conductivity of a crystalline rock is only marginally lower than that of a liquid of the same composition and therefore a chilled margin will provide only slight extra thermal insulation for the still fluid magma contained within it. This will, nevertheless, further retard the cooling rate, if only fractionally, permitting more extensive growth of crystal nucleii since diffusion is not being impeded by increased magma viscosity. What would be a state of equilibrium crystallisation, contrasting with the conditions in the chilled margin, is, however, disturbed by

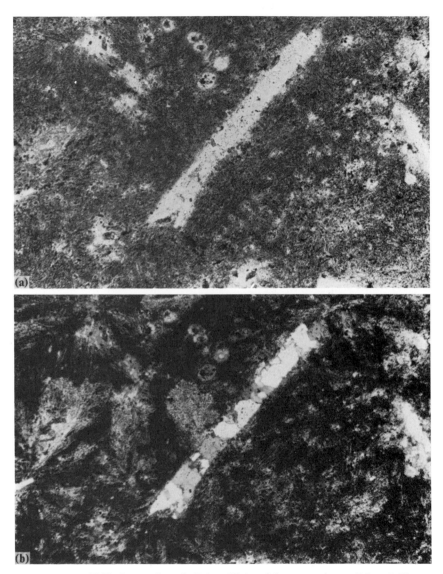

Figure 12.2 Chilled margin of Coire Uaigneich granophyre, Isle of Skye, Scotland. (a) Plane-polarised light. Elongate section of a quartz paramorph after tridymite set in fine-grained groundmass. (b) Crossed polars. Shows the mosaic texture of the quartz paramorph and more clearly the felsitic/spherulitic texture of the (? devitrified) groundmass.

what must be – at least in basic magmas – the inevitable onset of crystal settling.

Rocks which have formed from the piles of crystals settled out of a magma have come to be known as **cumulates** (Wager, Brown & Wadsworth 1960). In this context the phrase 'settled out of a magma' refers to all crystals which nucleated and grew for a while as **primocrysts** within the magma and were subsequently removed from suspension to become attached to any of the bounding surfaces of the intrusion, that is floor, walls or roof. A crystal once attached to such a surface is called a **cumulus crystal.** At the moment of attachment and for some time thereafter, cumulus crystals remain partially enclosed by **intercumulus liquid** having the composition of the contemporary magma, which eventually solidifies as **intercumulus material**. This intercumulus material may take the form of (a) new minerals nucleated within the intercumulus liquid and/or (b) further additions to the cumulus phases and/or (c) reaction replacement of cumulus crystals. In certain circumstances a portion of the intercumulus liquid may be displaced into the adjacent magma by the growth of cumulus phases. An analogy has been drawn with the relationship between allogenic (detrital) grains and authigenic or cementing material in aqueous sediments. Post-cumulus modifications are analogous to diagenetic processes.

If the temperature of the composite mass of cumulus crystals and intercumulus liquid falls very slowly (as might be predicted in a large intrusion), for a while at least conditions approximating to equilibrium crystallisation should obtain so that cumulus phases of solid solution type should be progressively made over by reaction to compositions appropriate to lower temperatures. This would provide an example of equilibrium crystallisation persisting in spite of the intervention of what is conventionally regarded as a potent disequilibrium mechanism, namely, crystal settling.

Numerous examples of cumulates composed of unzoned cumulus phases have been described and an extreme case, that of a monomineralic rock comprising pure bytownite (An_{86}), which forms layers 5–15 cm thick in the Rhum ultrabasic intrusion (Brown 1956), is illustrated in Figure 12.3. Contemporaneous magma trapped as intercumulus liquid could have provided but a limited addition to the cumulus plagioclase before becoming exhausted in plagioclase components. The continued growth of plagioclase of unchanging composition and the absence of intercumulus phases (such as olivine or pyroxene which do occur in adjacent layers) was attributed by Wager, Brown and Wadsworth (1960) to the simultaneous operation of two processes: diffusion of plagioclase components from the supernatant magma to the cumulus crystals; and the consequent squeezing out of inappropriate intercumulus liquid by the growing cumulus crystals. This type of unzoned crystal growth has been termed the adcumulus process and the rocks thereby formed **adcumulates.** The process as described is one of perfect fractional crystallisation since only those elements appropriate to plagioclase are supplied to the site of crystal growth while all other elements are efficiently removed. Economically important rocks may be formed such as the magnetite adcumulates of the Upper Zone of the Bushveld Intrusion, Eastern Transvaal (cf. Wager & Brown 1968, Fig. 207, and p. 378).

It can be seen that the erstwhile intercumulus liquid, modified in composition by the removal of plagioclase components, on its return to the supernatant magma, will render the immediately adjacent magma zone yet more mafic in composition. Consequent undersaturation in plagioclase and oversaturation in other mineral components is one likely contributory cause of the abrupt change in mineralogy of the rock layer immediately overlying the pure bytownite rock.

The expelled intercumulus liquid may perform another important function, namely that of transferring the latent heat of crystallisation away from the growing cumulus crystals into the overlying magmatic heat sink. Some such removal of heat is essential if crystal growth is not to cease.

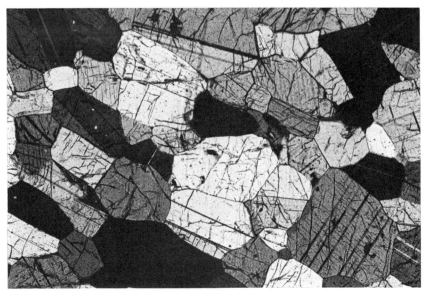

Figure 12.3 Extreme plagioclase adcumulate, Rhum ultrabasic intrusion, Scotland. Field of view shows only unzoned bytownite crystals displaying igneous lamination (see Ch. 13).

The petrographic hallmark of an adcumulate is the absence both of compositional zoning in the cumulus phase(s) and of late stage minerals. This is reasonably easily established with the petrological microscope in the case of plagioclase but less so with other minerals. Confirmation in all cases is achieved only by analytical traverses of crystals made with the electron probe which can reveal variation in element concentration undetected by optical methods. At present, minor element variation can be reliably detected only by the electron probe, and then not in all cases. The possibility exists that a cumulus phase unzoned as far as major elements are concerned may, nevertheless, contain systematic variation in minor element content and this constitutes a virtually unexplored field of research into what might be termed **selective adcumulus** growth.

A common textural feature in layered intrusions is the **poikilitic** enclosure of idiomorphic or subidiomorphic phases by allotriomorphic crystals of a different phase (Fig. 12.4). The well-shaped crystals have been interpreted as cumulus phases and the poikilitic crystals as minerals formed by *in situ* crystallisation of intercumulus or trapped liquid. The cumulus phases frequently show conspicuous normal zoning and the outer zones have also been regarded as intercumulus or pore material. Such rocks have been called **orthocumulates**, the name itself

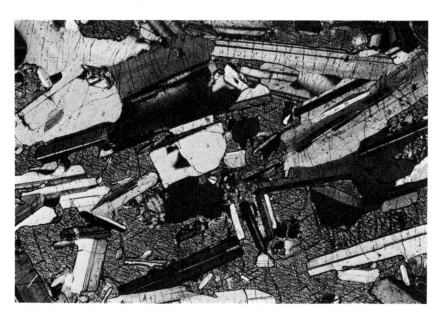

Figure 12.4 Orthocumulate. Poikilitic calcium-rich pyroxene encloses cumulus plagioclase showing normal compositional zoning, and minor cumulus olivine.

reflecting the former belief that the process described to explain the observed features was that normally expected to occur in a cooling magma body undergoing rapid crystal settling.

The pile of cumulus crystals with their enclosing liquid ideally acts as a closed system separated from the rest of the magma body – in contrast to the adcumulus situation described above. Whereas extreme adcumulates have simple mineralogy the more extreme the orthocumulate, the more varied its mineralogy. Cases of plagioclase cumulus enclosed by olivine, pyroxene and opaque oxide are common and the additional presence of late mica, amphibole, chlorite etc. is taken to indicate the ultimate concentration of water in the diminishing intercumulus liquid. It may be said that an extreme orthocumulate displays in one thin section virtually the whole mineralogical range of the layered intrusion.

Appropriately enough rocks intermediate in character between orthocumulates and adcumulates have been called **mesocumulates** and these are likely to be by far the commonest type. In these rocks compositional zoning is subordinate and late stage minerals are but sparingly present.

Poikilitic texture is not confined to orthocumulates. Wager and Brown (1968) have described 'mottled anorthosites' from the Bushveld Intrusion which consist of cumulus plagioclase and occasional chromite poikilitically enclosed by large (up to 25 mm in diameter) unzoned bronzite crystals. Adjacent layers contain bronzite of similar composition but as cumulus crystals. The occurrence of the poikilitic bronzite is ascribed to its nucleation within the intercumulus liquid, and its growth with constant composition to adcumulus diffusion from the overlying magma. The rock is called a **heteradcumulate** and the unzoned poikilocrysts **heterads**. The fact that the intercumulus material (in this case bronzite) does not correspond in composition to that of a reasonable magmatic liquid strongly suggests its derivation from modified intercumulus liquid.

The cumulate types described so far can most conveniently be envisaged as forming from a pile of crystals on the floor of a magma chamber, whatever the height in that chamber at which they may have originally nucleated. This implicitly assumes that the common cumulus phases tend to be more dense than the magma from which they form and that magma viscosity does not significantly impede their downward movement. Little is yet known in this context about conditions in salic magma bodies but in basic magmas certain stages may be reached in which plagioclase particularly may be less dense than the enclosing magma and, consequently, may remain suspended, or rise. In tholeiitic magmas iron enrichment of the fractionated liquid would involve an increase of its density. Goode (1977) has described rocks from the Pyroxenite Zone of the Precambrian Kalka Intrusion, Central Australia, in which cumulus plagioclase occurs as inclusions in pyroxene but not as settled discrete crystals. These do, however, occur in the Norite Zone above. Goode suggests that during the formation of the Pyroxenite Zone the magma density exceeded that of plagioclase (composition An_{76} = density of 2.67) and prevented its settling, except where it was weighted down by enclosing pyroxene. Conditions would then favour the upward rise of discrete plagioclase crystals to form **flotation cumulates** unless, as Goode claims, the plagioclase were to be remelted at higher levels. Perhaps the best-authenticated case of a flotation cumulate is the naujaite of the alkaline Ilimaussaq Intrusion, S.W. Greenland (Ferguson 1964). This rock consists of idiomorphic cumulus sodalite (up to 70 modal per cent) poikilitically enclosed by alkali feldspar and alkali amphiboles and pyroxenes.

In the Marginal and Upper Border Groups of the Skaergaard Intrusion certain rocks consisting of subidiomorphic plagioclase enclosed by other minerals were tentatively identified by Wager and Brown (1968) as **congelation cumulates.** It is possible that the modal content of the plagioclase indicates some crystal concentration but it was felt that the crystals had not effectively settled *out* of their parental liquid and should, therefore, be regarded more strictly as a primocryst (or, perhaps, phenocryst, according to Jackson 1971) than as a cumulus phase. Nearer the margins of the intrusion the magma with its suspended primocrysts kept buoyant by magmatic stirring would be subject to relatively rapid heat loss and the result would be congelation of the liquid around the primocrysts. This explanation superseded an earlier feeling that the rocks in the Upper Border Group might be flotation cumulates because of abundant evidence in the Layered Series of contemporaneous plagioclase sinking.

It has now been established (Naslund 1976) that the Skaergaard magma was laterally inhomogeneous, certainly during formation of the lower part of the Upper Border Group and the upper part of the Layered Series, and probably throughout much of the differentiation process. In the event, it seems quite possible, then, that plagioclase might float in one part of the magma body while it sinks in another.

In Chapter 7 we described the occurrence of skeletal, branching and dendritic crystals in volcanic rocks and attributed these forms to conditions of very rapid crystallisation in the quenching stage. Olivine was picked out for the frequency with which it displays these features. Generally these crystals are small but in one group of volcanic rocks, the Archaean ultramafic lavas known as komatiites (Viljoen & Viljoen 1969a, b) skeletal olivine crystals forming the so-called **spinifex texture** may reach 50 cm in length. Similar textures occur in the olivine-rich rock **harrisite** of the Rhum ultrabasic layered intrusion where the olivines may be up to 15 cm long, and in the Rognsund, Norway, layered gabbro which contains olivines up to 1 metre in length (Robins 1973).

Closer examination of these textures (Donaldson 1974) has detected plate, randomly-oriented, porphyritic and branching varieties within the general heading of 'skeletal/dendritic' in both harrisite and spinifex growths. Whereas all agree that the spinifex olivines grew rapidly, the prevalent interpretation of harrisite (Wager & Brown 1968) has been that this rock is a **crescumulate**, that is, that the olivines have grown slowly upwards into overlying, supercooled basaltic magma from cumulus olivine nucleii at the top of a crystal pile, forming a temporary floor of the magma chamber, growth taking place along concentration gradients to produce unzoned crystals. In other words, a crescumulate is a special form of adcumulate. However, because of similarities with spinifex texture, it appears more likely that harrisitic olivines grew rapidly.

The ratio of crystal growth, Y, to the melt diffusion coefficient, D, is an important constraint on crystal morphology (Lofgren 1974). Y/D is proportional to the extent of undercooling of a melt. At low values of Y/D equant crystals tend to form and as the value increases crystals become skeletal, dendritic, and ultimately spherulitic. The presence in harrisites of porphyritic (equant) olivine implies, then, the existence of high Y/D (Donaldson 1974) in which case the olivine content of harrisite should be similar to that of its parental liquid. Since harrisites always have at least 30% modal olivine it follows that the liquid from which they formed was more basic than basalt.

All that we have said so far about cumulates makes it clear that the bulk chemical composition of such rocks is very unlikely to match that of their parent liquid. Nevertheless, the composition of cumulus minerals does provide important evidence of the nature of their source material since they are liquidus phases and hence allows some deductions about magma composition, temperature, pressure and fugacities to be made. However, one of the principal problems facing a student when examining supposed cumulates is how to decide, first, what is a cumulus phase and, secondly, how much material has been added to the cumulus since it settled.

Ideally cumulus crystals of the same mineral occurring at a given level in an intrusion tend to be uniform in size and shape and, if inequant, often display some degree of preferred orientation. A zoned cumulus mineral may display an unzoned inner part which is taken to be the cumulus as it was at the point of settling. This inner part may also be distinctly idiomorphic even if the whole crystal is much less so. This criterion of shape may in some cases be more trustworthy than that of zoning since clearly a suspended primocryst may suffer zoning as a result of, say, hydrostatic pressure variations as it circulates within convection currents in a large magma body, or of vapour pressure changes within the magma. The frequently observed zoning of volcanic phenocrysts is proof of this. It may be noted that this zoning is usually of the oscillatory type, that is of alternating low- and high-temperature compositions, which is a less likely, if not impossible, result of continued crystallisation after settling. In unzoned cumulates the nature of the cumulus phase is not in doubt but the amount of original cumulus is difficult, if not impossible, to assess petrographically.

Estimates, some based on packing experiments, of the porosity of a crystal pile range from 6% to 50%. The higher values in such a range permit considerable overgrowth of cumulus crystals which may be mapped by zoning; but in favourable circumstances other petrographic evidence may be available.

Figure 12.5 is a photomicrograph of a basic cumulate from the Cuillin intrusion on the Isle of Skye, Scotland. A twinned poikilocryst of

Figure 12.5 Twinned poikilocryst of calcium-rich pyroxene (dark) enclosing idiomorphic cumulus plagioclase crystals. Note difference in size between the enclosed plagioclase and the plagioclase crystals, upper right, enlarged by post-cumulus growth and displaying igneous lamination. Cuillin Intrusion, Isle of Skye, Scotland.

clinopyroxene is seen to enclose randomly-oriented, idiomorphic laths of twinned plagioclase, the size of which differs markedly from that of other, oriented plagioclase crystals outside the clinopyroxene crystal. It appears that enclosure by the clinopyroxene of the small laths isolated them from the effects of the mechanism that oriented the larger plagioclase crystals. Since this mechanism could have been effective only at the time of deposition of all the crystals, the pyroxene with its included plagioclase must have settled as a composite cluster. Indeed, the larger plagioclase crystals appear to be packed tangentially about it. We have, then, a likely sample of the size and shape of plagioclase primocrysts and by comparing them with the larger, oriented plagioclase crystals, may estimate the extent of post-depositional overgrowth in these.

A superficially similar but differently interpreted case permitting the same sort of estimate has been described by Cameron (1969). The bronzite poikilocrysts (or heterads – see above p. 291) in the 'mottled anorthosites' of the Bushveld Intrusion enclose plagioclase crystals with, on average, one-third of the mean diameter of plagioclase crystals in the main body of the rock. Where these latter show preferred orientation so do the plagioclase crystals enclosed by the pyroxene. This indicates that the pyroxene grew quickly after the plagioclase had settled and rapidly insulated a group of plagioclase crystals from further enlargement. The difference in size between the two sets of plagioclase crystals is a measure of the overgrowth of the original cumulus.

Having now listed certain categories of cumulates and, more to the point, dealt in some detail with interpretations of their features, it is time to make a brief cautionary statement about the subjectivity with which some of the nomenclature tended to become invested. For example, an 'adcumulate' need not have formed by diffusion and crystal extension after shallow burial involving physical expulsion of intercumulus liquid with the mechanical difficulties that might involve, but by growth of cumulus crystals remaining at the crystal–supernatant magma interface. Again, orthocumulates, notwithstanding the etymology of the term, are not all that common in layered intrusions (Wager & Brown 1968). Finally, as will have become apparent, there are practical difficulties in distinguishing to what extent the material formed after settling of the cumulus is the product of direct consolidation of intercumulus (trapped) liquid or modified by selective (adcumulus) diffusion from the supernatant magma.

For such reasons cumulate nomenclature has become simplified and more descriptive and it is now customary to use the word 'cumulus' for the settled crystal and the word 'postcumulus' to describe all the material which formed in the place it now occupies within the cumulate, whether it was added to the cumulus, formed additional primary minerals or resulted from reactions (Jackson 1961, 1971; Cameron, 1969).

Sub-solidus textures

The textures we have been describing so far form between magma liquidus and solidus temperatures. A distinctive and important further set of textures develops in several mineral groups as the primary phases continue cooling below the solidus. We may conveniently divide these features simply into exsolution textures and inversion textures.

Exsolution textures. Exsolution, or unmixing, is a process whereby an initially homogeneous solid solution separates into two or more crystalline phases with no change in bulk composition. Most commonly we consider the process to act during a fall of temperature but in some cases it occurs as the temperature rises. Conventionally we think of a **host**, or **parent** crystal, that is the end-member most abundant in the solid solution, as exsolving **guest**, or **daughter lamellae** (if regular in shape) or 'blebs' or 'patches' etc. (if irregular in shape) of the less abundant end-member. The exsolved phase is usually oriented parallel with one or more particular crystallographic planes of the host.

Whether or not exsolution takes place is determined to some extent by the rate of cooling, as is shown by the alkali feldspar series, $NaAlSi_3O_8–KAlSi_3O_8$. In rapidly-cooled rocks alkali feldspars rarely, if ever, reveal exsolution petrographically, although X-rays may detect some separation of phases, whereas in slower-cooled equivalents, perthite, a two-phase assemblage, is common. The term perthite strictly implies the unmixing of daughter sodium-rich feldspar from a potassium-rich parent to give parallel growth, while antiperthite describes the reverse relationship. However, perthite has also been used as a general descriptive term for silicates showing exsolution textures. It should be remembered that certain two phase feldspars, particularly in plutonic rocks, are 'replacement' perthites in which the generally patchy or vein-like areas of albitic feldspar are thought to have formed by reaction between late permeating fluids and primary feldspar.

Although perthites are perhaps the most familiar example of exsolution in silicates, certain members of the pyroxene group provide a petrographically more useful range of exsolution features. As with perthites, pyroxenes showing exsolution are absent from quickly cooled volcanic rocks. Slowly cooled, strongly fractionated bodies of tholeiitic magma such as the Skaergaard and Bushveld Intrusions, are characterised by two compositionally and structurally different pyroxene trends which co-exist through a large section of the fractionated series. The occurrence of co-existing pyroxenes each of which may display exsolution features is a diagnostic feature of tholeiitic intrusions and contrasts with the single pyroxene trend lacking exsolution of alkali basic intrusions. Figure 12.6 summarises in simplified fashion the features of calcium-rich and calcium-poor pyroxene trends as exemplified by the rocks of a tholeiitic intrusion.

The calcium-poor pyroxene trend is represented by primary orthopyroxene in the early stages, this giving way to monoclinic pigeonite which, however, almost always inverts to orthorhombic pyroxene on slow cooling, being then referred to as **inverted pigeonite**. In essence, the exsolution story in these pyroxenes is that a calcium-rich parent exsolves a calcium-poor daughter, and vice versa. Complications arise because whereas the calcium-rich phase is always monoclinic, the calcium-poor phase may be monoclinic or orthorhombic. It can be seen from Figure 12.6 that exsolution features have certain characteristics that are fixed and others which vary systematically with the structure and composition of the pyroxenes and their fractionation stage. Thus, for example, a monoclinic parent always exsolves an orthorhombic daughter $\parallel$ (100) and a monoclinic daughter $\parallel$ (001) but the form of the last may vary from blebs earlier in the fractionation series to broader and more regular lamellae in later stages. For a given layered intrusion this provides a means of broadly locating a rock 'stratigraphically' within a layered series.

It will be noted in Figure 12.6 (stage B, B') that calcium-rich and calcium-poor parents each display bleb exsolution of the other approximately $\parallel$ (001) accompanied by fine, regular lamellae $\parallel$ (100). A section cut $\parallel$ (010) will, then, reveal two sets of exsolution lamellae. The relative lack of orientation of the blebs has been attributed to the unmixing's coinciding with inversion of the calcium-poor phase from monoclinic to orthorhombic symmetry. The difference in shape of the exsolved phase is more likely to reflect changes in Y/D (see above p. 293). At higher temperatures Y/D would be relatively lower and thus favour the more equidimensional shape shown by the blebs, whereas at the lower temperatures of the post-inversion stage, Y/D would increase with consequent development of lamellae which though more regular in shape are less equant.

Pyroxenes in smaller tholeiitic intrusions may show exsolution textures, although less commonly, of the broad, regular lamellae variety. A typical example is the graphic-like exsolution of calcium-rich pyroxene shown by inverted pigeonite in the hypersthene dolerite of the Palisades Sill (Walker 1969). The same texture (Fig. 12.7) is seen in

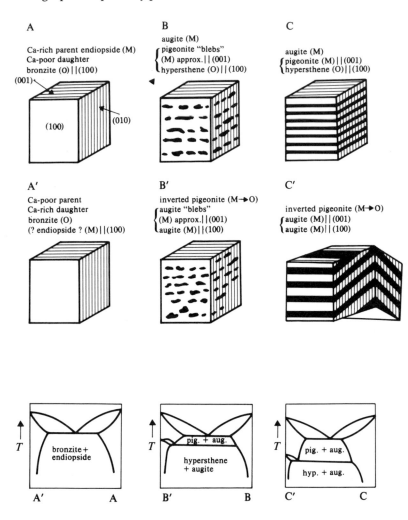

Figure 12.6   Principal sub-solidus relationships of pyroxenes in an idealised slowly-cooled tholeiitic intrusion. M: monoclinic, O: orthorhombic; T: triclinic. Reading from left to right fractionation is proceeding, the iron : magnesium ratio is increasing and the solidus temperatures are decreasing. The general rule for orientation of exsolution lamellae in pyroxenes is that a monoclinic parent exsolves monoclinic daughter lamellae ‖(001) and orthorhombic daughter lamellae ‖(100); an orthorhombic parent exsolves monoclinic lamellae ‖(100). The diagram shows the nature and exsolution and/or inversion features of co-existing or single (calcium-rich) pyroxenes at various fractionation stages (A to F) indicated in the inset diagram of the common pyroxene quadrilateral. For each stage a hypothetical or simplified phase diagram is also shown to illustrate the origin of the exsolution or inversion characteristics. Note that the diagram for stage E shows a theoretical ‘magma’ solidus intersecting the inversion

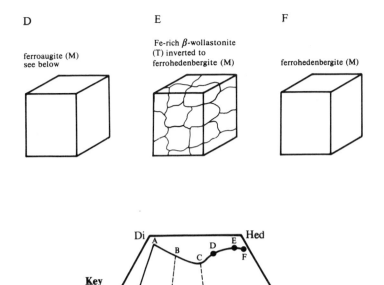

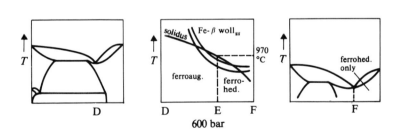

interval of ferriferous β-wollastonite solid solution ⇌ ferroaugite or ferrohedenbergite solid solution in two places according to composition; at 970 °C and 600 bar pressure (dashed lines) ferriferous β-wollastonite will crystallise and as the temperature falls will eventually invert to pyroxene retaining a mosaic texture as described in the text. This phenomenon has been observed only in rocks of the Skaergaard Intrusion. From stage D onwards a single (calcium-rich) pyroxene is found: exsolution is not normally detectable except on the X-ray scale at stage D. At stages B and C *exsolved* pigeonite does not normally invert to orthopyroxene, even when co-existing *parent* pigeonite has inverted, possibly because the monoclinic augite host exerts a stabilising influence on its exsolved calcium-poor (pigeonite) lamellae. Stage C' shows a case of 'herringbone' or chevron exsolution as described in the text (pp. 304–5).

Figure 12.7   Calcium-poor pyroxene (dark) in ferrodiorite from Marscoite Suite, Western Redhills Complex, Isle of Skye, Scotland. Inner part of pyroxene is inverted pigeonite showing blebby but regular exsolution of calcium-rich pyroxene. Surrounding this is calcium-poor pyroxene without exsolution but in optical continuity, presumably nucleated onto the original pigeonite but at a temperature below the inversion temperature and, therefore, formed as primary orthopyroxene.

inverted pigeonite in a much smaller intrusion, the ferrodiorite member of the marscoite suite ring-dyke in the Isle of Skye (Wager *et al.* 1965). The occurrence of such diagnostic exsolution in a pyroxene of this intrusion which is only some 10 m wide, compared with the 300 m or more thickness of the Palisades Sill, suggests that it is worth seeking in other small intrusions where it may not have been expected, or perhaps have been overlooked, or mistaken for some alteration effect.

The most iron-rich pyroxenes in Figure 12.6 that is, those with no co-existing calcium-poor pyroxene, show no exsolution effects that are detectable with the petrological microscope. X-ray investigation, however, has revealed the occurrence of a separate iron-rich pigeonitic phase in ferrohedenbergite (of stage E) in the Skaergaard Intrusion. Pyroxenes of this composition occur also in certain high-level granites where they are usually accompanied by iron-rich olivine. The Loch Ainort granite of the Western Redhills, Isle of Skye (Bell 1966) contains ferrohedenbergite which has a faint schiller texture. Chemical analysis of this pyroxene has revealed a higher than usual ferric iron content and X-ray analysis has detected a magnetic phase exsolved $\parallel$ (100) and $\parallel$ (001).

Before ending this brief discussion of pyroxene exsolution, it needs to be said that examination with the electron microprobe has revealed many complexities not clearly detectable with the petrological microscope. Perhaps most important is the discovery that broad regular exsolution lamellae may themselves include a smaller set of exsolution lamellae. A case in point is a Bushveld Intrusion inverted pigeonite whose augite daughter phase contains a further (?grand-daughter) calcium-poor phase forming very fine lamellae (Boyd & Brown 1968). According to the probe scans these fine lamellae are not as calcium-poor as the host inverted pigeonite, which is, perhaps, not what one would expect. However, this may be because the probe lacks sufficient resolving power in this size range. Another intriguing feature is that if the very fine lamellae are orthorhombic, according to our general rule (Fig. 12.6) they should be oriented ‖(100) of their immediate host (the augite lamellae). In fact they seem to have the same orientation as the augite lamellae which lie ‖(001) of their inverted pigeonite host.

One might expect that certain members of the other major group of chain silicates, the amphiboles, should show exsolution features and, indeed, this is the case. As with the pyroxenes described above a calcium-rich host may exsolve calcium-poor lamellae and vice versa. However, with rare exceptions the described instances occur in metamorphic rocks.

Particularly fine examples of exsolution are provided by the so-called 'ore minerals' in igneous rocks. In this category are included such minerals as ilmenite, magnetite and various sulphides. Titaniferous magnetite exsolves box-like patches of ulvöspinel (Fe_2TiO_4) while ilmenite may exsolve thin lamellae of magnetite which in turn exsolve small amounts of ulvöspinel. Among the sulphides which occur in basic igneous rocks as immiscible globular patches, bornite (Cu_5FeS_4) exsolves lamellae of chalcopyrite ($CuFeS_2$). **Myrmekite** (Fig. 12.8) is another possible example of exsolution found in plutonic rocks. This is an intergrowth of vermicular quartz and (usually sodic) plagioclase which may occur near the rim of a plagioclase crystal, or between grains of perthitic feldspar or, most noticeable, as lobate projections formed where alkali feldspar and plagioclase crystals are in contact. It is not to be confused with **granophyric intergrowth** which involves a geometrically similar relationship between quartz and alkali feldspar and is generally regarded as the product of eutectic crystallisation. The exsolution hypothesis suggests that calcium in alkali feldspar occurs as a high silica molecule (Schwantke's molecule), which unmixes as follows:

$$CaAl_2Si_6O_{16} \quad \rightarrow \quad CaAl_2Si_2O_8 \quad + \quad 4SiO_2$$

Schwantke's molecule anorthite + quartz = myrmekite

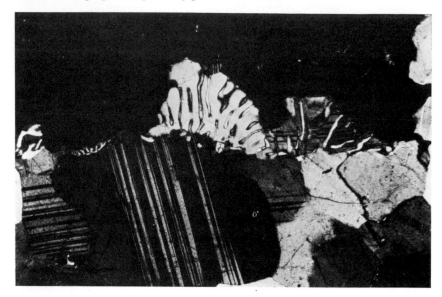

Figure 12.8   Myrmekite. Lobate variety at contact of plagioclase (twinned) and alkali feldspar (in extinction). In the myrmekite quartz forms curved tube-like areas. In the myrmekite mass on the right the plagioclase shows lamellar twinning.

It follows that the higher the content of original Schwantke's molecule the greater will be both the anorthite content of the plagioclase and the amount of quartz in the myrmekite formed. Accurate determination of this proportionality between anorthite content and modal quartz forms a crucial petrographic test of the exsolution hypothesis but one which because of the technical problems of sectioning and measuring these small and irregular intergrowths is so far unresolved.

As in the case of perthites, some of which, incidentally, also contain quartz blebs, alternative hypotheses involving metasomatic replacement reactions are applied to explain the origin of myrmekite.

Myrmekite occurs typically in acid plutonic rocks and is commoner in those with higher calcium contents, such as granodiorites. Quartz–plagioclase intergrowths do occur, however, in basic rocks in which no alkali feldspar is present. As such they fall outside the definition of myrmekite advocated by some workers (Phillips 1974). In certain Bushveld Intrusion cumulates myrmekitic intergrowth appears in the outer, reverse-zoned (i.e. more calcic) parts of cumulus plagioclase (Atkins 1965; Wager & Brown 1968). It is noticeable that in these cases the plagioclase crystal containing the myrmekite retains euhedral outlines, in contrast with the characteristic wart-like projections of myrmekite in acid rocks.

Inversion textures. When a substance is capable of existing in two or more crystalline modifications which, although possessing identical chemical composition, have different physical and chemical properties it is said to exhibit **polymorphism**. The change from one polymorph to another, which is a reaction to establish a new equilibrium stable under a new set of conditions, is termed **inversion**. If the inversion is a reversible reaction it is called **enantiotropic**, and, if irreversible, **monotropic**. We have earlier mentioned two cases of enantiotropic inversion, namely, tridymite to quartz and pigeonite to orthopyroxene, which we will now discuss further in terms of the textural aspects of the reaction.

In the case of the chilled margin of the Coire Uaigneich granophyre, quartz forms **paramorphs** after primary tridymite, that is, it is preserved as pseudomorphs without any change in external form. The original tridymite had a plate-like (hexagonal) habit and this outline is retained. Viewed in plane-polarised light the crystals display nothing unusual but under crossed nicols they reveal a mosaic texture comprising several sub-units of quartz in differing optical orientations (Fig. 12.2).

A similar mosaic texture is shown by certain green Ca–Fe-rich cumulus clinopyroxenes (ferrohedenbergites) in the uppermost rocks of the Layered Series of the Skaergaard Intrusion (Fig. 12.9). In this case

Figure 12.9 Mosaic texture of β-ferriferous wollastonite now inverted to ferrohedenbergite; the darker and lighter patches have different extinction positions. Other minerals present are plagioclase, apatite and opaque oxide. Skaergaard Intrusion, East Greenland.

the original phase was a pyroxenoid, triclinic ferriferous β-wollastonite. Using the natural material Lindsley, Brown and Muir (1969) established experimentally the *P–T* condition for the inversion reaction. Happily quartz inverted from tridymite occurs in closely associated rocks and a petrogenetic grid combining the two sets of *P–T* data has permitted estimates of 600 ± 100 bars and 980–950 °C for the pressure and temperature limits of crystallisation in the later stages of the Skaergaard Intrusion to be made. This is the only case of ferriferous β-wollastonite – ferrohedenbergite inversion recorded from natural rocks. Its absence in otherwise closely similar rocks of the Bushveld Intrusion is attributed to higher pressures during crystallisation of that much larger mass.

The distinctive textural feature of the examples just described is the mosaic appearance under crossed nicols of the differently oriented patches, or **domains**. An approach to this is seen in the case of certain inverted pigeonites, for example, in the Palisades Sill, which although now wholly orthopyroxene retain shadowy patchy extinction features. This appearance is not to be confused with the shadowy extinction of quartz, commonly seen in coarse-grained granites, or the more regular but still shadowy lamellae seen in olivine crystals, both of which are attributed to strain of the crystal lattice, not consequent upon inversion.

Several factors determine whether or not inversion will occur. Inverted pigeonite has not been described in volcanic rocks whereas, although all pigeonites have inverted in Bushveld Intrusion rocks, uninverted pigeonite is not infrequently preserved in other tholeiitic layered intrusions and sills. In the Skaergaard Intrusion uninverted cumulus pigeonite first becomes noticeable at the relatively advanced differentiation stage of the Middle Zone. In the Palisades Sill it is most frequent at a similar stage but also occurs sporadically at other levels. Speed of cooling and load pressure clearly exert some control as may a large interval between solidus and inversion temperatures. The presence of a fluid phase would also assist the reaction.

Inverted pigeonite not infrequently displays a diagnostic combination of inversion and exsolution textures. Figure 12.10 shows an inverted pigeonite in which the exsolution lamellae of calcium-rich pyroxene are oriented in a distinct chevron or 'herring-bone' fashion. The host inverted pigeonite extinguishes in a single position as do the exsolution lamellae. The explanation is that an original (monoclinic) pigeonite twinned on (100) exsolved calcium-rich lamellae with chevron orientation and subsequently inverted, thus eliminating its twinned structure but without affecting the orientation of the exsolution lamellae. Fine lamellae of calcium-rich pyroxene may also appear ‖(100) of the inverted pigeonite.

Figure 12.10 'Herringbone' exsolution in pyroxenes. Centre of the field of view shows, extending to the right, a calcium-rich pyroxene with simple twinning (one side slightly further into extinction than the other) and fine exsolution lamellae of calcium-poor pyroxene disposed in chevron style. Adjacent to the left is an inverted pigeonite crystal almost in extinction with broader lamellae of calcium-rich pyroxene arranged in chevron style, the apices of the chevrons centred on the relict twin plane ((100)) of the original pigeonite. Note the equal illumination of the inverted pigeonite. Skaergaard Intrusion, East Greenland.

Fine chevron exsolution lamellae of calcium-poor pyroxene are also often seen in calcium-rich hosts but these can never be confused with inverted pigeonite 'herring-bone' textures, although they may occur in the same rock, because the former host not only possesses higher birefringence but the twin plane is still in existence and thus the two twin individuals extinguish in different positions.

Partial melting of plutonic rocks

Theories of the origin of igneous rocks are very much concerned with mechanisms of partial melting of solid material, whether it be ultramafic mantle as a source of basaltic magma, or crustal rocks from which acid liquids may be derived. Direct petrographic evidence of partial melting, however, is less commonly recorded. Figure 12.11 illustrates several petrographic features of the process of partial melting and solidification of the liquid so produced.

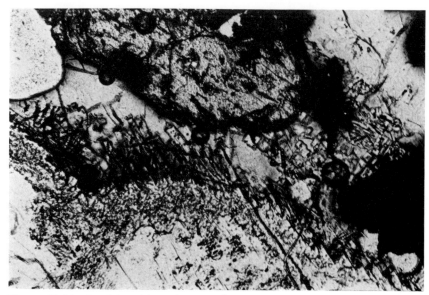

Figure 12.11   Partial melting of a syenitic block forming the core of a volcanic bomb. At the bottom is alkali feldspar showing sieve-texture and a fringe of quench feldspar crystals. The quench feldspars project into glass formed by partial melting of feldspar, pyroxene (large crystal, upper part of picture) and opaque oxide. Note (a) how the tint of the glass varies indicating variation in composition; (b) patchy alteration and replacement of pyroxene. See text for further details. Ascension Island, South Atlantic.

The sample illustrated is a nodule of coarse-grained syenitic rock consisting of alkali feldspar, green pyroxene and an opaque oxide, which forms the core of a volcanic bomb, a not uncommon site for plutonic nodules. The syenite evidently occurs as a sub- or intra-volcanic intrusion which has been disrupted by a later pulse of lava which detached and carried fragments up to the surface.

Petrographic evidence of re-heating of the syenite is present in the cracking and clouding of the feldspar and the occurrence of glass proves that melting took place with the added effect that the edges of the feldspar also show fine-scale **sieve-texture**. The glass varies noticeably in colour, being colourless adjacent to the feldspar but turbid and brown next to the opaque minerals and pyroxene. It must vary, then, in chemical composition and here we have signs that liquids produced by partial melting of a multi-phase rock for a while, at least, are chemically heterogeneous.

The pyroxene is replaced patchily by glass and an opaque mineral and since it is a sodium-rich pyroxene this could be interpreted as evidence of incongruent melting involving an acmitic composition to produce liquid and iron oxide.

The preservation of glass means that the liquid was rapidly quenched and this would be readily achieved during the aerial flight of the volcanic bomb. There is one other feature of quenching to observe, however, and that is the growth of small idiomorphic quench crystals of feldspar which have nucleated on the edges of their parental feldspars, and of acicular opaque crystals, presumably of iron oxide, which concentrate in the areas of darker glass. These, too, have nucleated for the most part on feldspar edges. The sample provides a complete record of the range of events from heating of solid source material to the formation of liquid and its subsequent crystallisation, although this last phase is a rather special case. It is worth noting that all three primary phases were melting at the same time, although elsewhere in the specimen (not illustrated) it is clear that the amount of liquid formed is less at the contact between two crystals of the same mineral (i.e. equivalent to melting a 'pure' end-member) than at the contact between two different minerals (i.e. equivalent to a two-component mixture with lowered melting point).

The amount of liquid formed was relatively small and evidence that it might have been able to concentrate into a discrete and mobile mass (a magma) is lacking. This formation of intergranular films and pockets of liquid is, however, likely to be the sort of thing that occurs in the early stages of partial melting on the grand scale.

13

The interpretation of data for plutonic rocks

Volume of intrusions

An important object of any study of igneous rocks in the field is to gain an idea of the volume of the separate units present. Given adequate maps and/or air photographs, assessing the total volume of a lava flow, particularly if it is recent, is a relatively easy matter. Doing the same for a partially exposed pluton is more difficult and frequently only the area exposed is quoted. The time-honoured method of calculating the volume is to extrapolate the boundaries downward to a reasonable depth and model a three-dimensional shape accordingly. That the data so acquired, especially when reassessed in the light of geophysical measurements, provide surprising results is aptly illustrated by the following two examples.

The sketch map of the Tertiary igneous complex of the Isle of Skye (Fig. 13.1) shows that the outcrop area of the Cuillin Intrusion of ultrabasic and basic rocks is virtually the same as that of the Red Hills granitic rocks: about 77 km². The margins of the Cuillin Intrusion dip generally inwards suggesting a rough inverted cone shape whereas the granitic rocks seem to form a cylindrical mass with vertical or steep outwardly dipping walls. Assuming that each mass projects to the same depth (5 km) and has the same outcrop area we find that the Cuillin Intrusion would have a volume one-third that of the Red Hills granites. (Volume of a cone $= \frac{1}{3}\pi r^2 h$; volume of a cylinder $= \pi r^2 h$ where h = height.) Apparently, then, area exposed is not proportional to volume in such a case. However, gravity work (Bott & Tuson 1973) reveals that the granitic rocks are unlikely to exceed about 1.7 km in thickness. If we recalculate the granite volume using this value we find the intriguing result that the Cuillin Intrusion has a volume of 129.6 km³ and the Red Hills granites a volume of 130.9 km³. Thus surface area turns out to be proportional to volume, in this specific case.

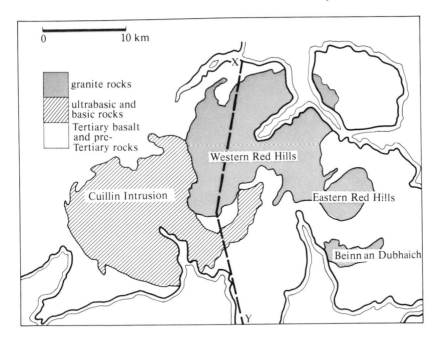

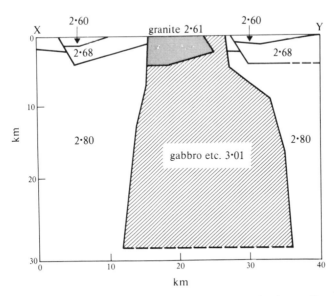

Figure 13.1 The Tertiary igneous complex of Skye. The lower figure shows density model for the section X–Y on upper figure.

One reason for determining the volume of each member of an igneous series as accurately as possible is that no realistic petrogenetic model can be constructed without such data. We should, then, further point out in the case of the Skye igneous complex that the gravity survey demonstrated the existence of a large positive anomaly interpreted as being caused by a subjacent mass of basic or ultrabasic rock with a volume of either 3500 km^3 or 1500 km^3 depending on which rock type was used for the model. Accordingly, in terms of volume proportions alone the Skye granites could be differentiation products of the underlying mass since it is commonly held that up to 10% of acid differentiate may be derived from 100% of basic parental liquid. The Skye granites have not been regarded as differentiates of the Cuillin magma since field evidence shows clearly that they have an intrusive relationship towards the Cuillin Intrusion. As we also see, relative volume considerations would preclude such a relationship.

The Skaergaard Intrusion provides our other example. From careful mapping and downward extrapolation of inwardly dipping margins (correcting for later tectonic reorientation), Wager and Deer (1939) envisaged a funnel-shaped intrusion with a volume of some 500 km^3. Recent gravity and magnetic surveys (Blank & Gettings 1973) indicate a different form, that of a tapering, generally horizontal sheet with two feeder pipes and having a markedly smaller total volume of earlier unexposed differentiates (the Hidden Zone) than that envisaged by Wager and Deer (1939) (Fig. 13.2).

Time for emplacement of plutons

The time any lava flow under observation takes to extrude is another straightforward measurement and when related to the volume of the flow provides the important information of a volcano's productivity. Obviously we cannot time in the same visual way the inflow of magma forming a pluton but an estimate of this may be possible by indirect means provided that we accept some comparability with a volcanic process.

The volcano Kilauea on Hawaii provides our information. Geodetic measurements reveal that the volcano inflates as magma accumulates within the structure prior to eruption. During the 34 days from 27 September to 31 October 1967 part of the summit region of Kilauea expanded more or less symmetrically to a maximum height of 105 mm over a circular area some 11 km in diameter (Kinoshita, Swanson & Jackson 1974). Using the formula for the volume of a cone this gives a figure of 0.03 km^3, an average daily inflation of 9×10^{-4} km^3, or

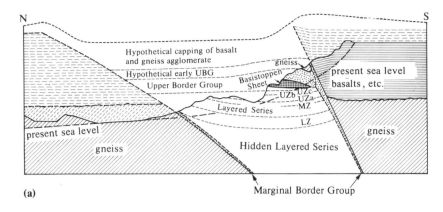

(a)

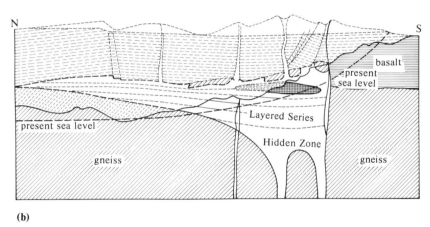

(b)

Figure 13.2 Cross sections of the Skaergaard Intrusion, East Greenland, showing appearance before monoclinal flexuring and erosion. (a) after Wager and Deer (1939), (b) after McBirney (1975).

9×10^5 m³, which we take to be caused by the intrusion of magma into some kind of storage chamber. We may now apply the rate to the infilling of the Skaergaard Intrusion justifying this on the grounds that the magmas involved were compositionally similar (tholeiitic), that the thickness of cover in each case was also approximately the same (see Ch. 11), and that the Skaergaard magma is considered to have been emplaced as a single influx without any lengthy interruptions in the process (Wager & Brown, 1968). Using Wager and Deer's (1939) calculated volume of 500 km³ it would take 5.5×10^5 days or approximately 1500 years for the magma body to accumulate as an intrusion.

This approach involves large extrapolations and no doubt conceals considerable errors but is probably more realistic than using the subaerial extrusion rate of a very large lava flow if only because the latter encounters much less frictional resistance. To illustrate the difference, however, we consider the largest tholeiitic flow ever observed to erupt, the Lakagigar fissure eruption of 1783 in Iceland (Thorarinsson 1970). Some 12 km³ of lava was extruded over a period of 145 days an average daily flow rate of 0.08 km³. At this rate the Skaergaard magma chamber would have been filled in about 17 years. There were pauses in the Lakagigar eruption and the lava may have been extruded in as little as 85 days, that is a daily average of 0.14 km³ which applied to the Skaergaard Intrusion means a little over 10 years for emplacement.

Flow of magma

A more rigorous approach to increasing our understanding of the flow characteristics of magmas is to make use of some of the principles of fluid dynamics. Unfortunately, the rigour diminishes in the face of incomplete information about the essential parameters of specific cases. However, some insight into the processes may be gained from considering simple examples and approximations.

The volume flow rate of a viscous fluid through a cylindrical channel under constant pressure gradient is given by

$$Q = \frac{\pi P r^4}{8 \eta l} \tag{13.1}$$

where Q = volume flow rate in cm³ sec⁻¹; P = pressure drop in bars; r = the radius of the channel in cm; η = the viscosity of the liquid in poises; l = the length of the channel in cm.

Applying this relationship to a large (~200 km³), simple funnel-shaped magma chamber which is filled with basaltic magma (η = 300; Shaw 1969) via a cylindrical feeder pipe at the base having a diameter of 200 m and a length of 3 km, the pressure drop through the pipe being 1000 bar, we find

$$Q = \frac{3.14 \times 1000 \times 10^{16}}{8 \times 300 \times 3 \times 10^5} = 4.36 \times 10^{10} \text{ cm sec}^{-1}$$

or 3.76 km³ per day.

This volume flow rate is of a very different order from that deduced above for Kilauea but this is a very simple case in which the magma pours into an empty cavity and has no other work to perform such as lifting rock cover, encountering obstacles, stoping its way into place and so on. The important point of the illustration is that the movement of large quantities of magma in short periods of time is an entirely feasible process.

The type of flow imposed upon a magma, that is laminar or turbulent, is of interest, for example, in the case of initially heterogeneous magma which would be more efficiently homogenised by turbulence; or, again, in view of the effects on the orientation of suspended particles such as phenocrysts or xenoliths. To determine whether the conditions indicate laminar or turbulent flow we calculate the dimensionless Reynolds number, *Re*, which in terms of the average volume flow rate is given by

$$Re = \frac{2\rho Q}{\pi r \eta} \tag{13.2}$$

where $\rho =$ the density of the fluid. Turbulent flow occurs when $Re > 2000$. In our example above, with $\rho = 2.6$,

$$Re = \frac{2 \times 2.6 \times 4.36 \times 10^{10}}{3.14 \times 10^4 \times 300} = 2.39 \times 10^4$$

hence the flow would be turbulent. Note that the much higher viscosities of acid magmas ($\eta = 10^5 - 10^8$ poise) render turbulent flow unlikely.

A porphyritic chilled margin to an intrusion indicates that the magma was emplaced as a suspension of phenocrysts in liquid. In such a case η in 13.2 above becomes the *effective* viscosity, η_e, which increases as the proportion of suspended particles, C, is increased and ρ is the mean density. Then

$$\eta_e = \eta_f (1 - 1.35C)^{-2.5} \tag{13.3}$$

where $\eta_f =$ viscosity of the fluid fraction (Roscoe 1952). Hence the phenocryst content has an important bearing on the nature of flow in a magma. This subject is discussed in some detail by Shaw (1965) who also gives expressions for flow in channels of rectangular cross section and applications to dyke formation.

Some thermal features of intrusions

In a volcanic eruption the effects of kinetic energy expenditure, such as the ejection of pyroclastic material, may be more spectacular but the major part of the energy budget is thermal. In the eruptive part of a volcanic event most of the thermal energy eventually dissipates into the atmosphere and hydrosphere but it must be remembered that a large proportion of the magma available for eruption never breaks surface so that much of its heat is transferred into the enclosing rocks. A certain amount may, of course, be removed rapidly in gas or fluid phases subsequent to the main eruptive episode.

The thermal features of intrusions pose very complex problems which are only slowly responding to mathematical and experimental analysis employing equations of the sort derived from heating engineering studies. For detailed treatments we will refer the reader to particular texts and confine ourselves to a few outlines.

The intrusion of hot magmas into higher and cooler levels of the crust (penetrative convection) provides one source of heat which affects the surrounding rocks. At the time of intrusion, the temperature at the contact, Tm, is given by

$$Tm = \tfrac{1}{2}(Tc + Tl) \tag{13.4}$$

where Tc = the temperature of the country rock and Tl the magma temperature. We can calculate the amount of heat per unit mass ideally transferable from an intrusion from

$$C\Delta T + Lx \tag{13.5}$$

where C = the specific heat of the intruded material (cal $g^{-1} K^{-1}$), ΔT = change in temperature (°C), and Lx = latent heat of crystallisation (cal g^{-1}).

Applying this expression to a given intrusion requires a figure for its mass which in the case of our earlier example of a 200 km³ basaltic intrusion is $2 \times 10^{17} \times 2.9 = 5.8 \times 10^{17}$ g (using 2.9 as the density). If $\Delta T = 200 \,°C$, $C = 0.33$ cal g K^{-1} and $L = 90$ cal g^{-1}, then the heat available for transfer from the intrusion during the first two hundred degree fall in temperature, i.e. to virtual solidification, is

$$[0.33 \,(200) + 90] \times [5.8 \times 10^{17}] = 9.04 \times 10^{19} \text{ cal}$$

We may assess the effect of such a heat source on its country rock by calculating the heat required per unit mass to melt the latter from

$$c_l\Delta T_l + L_fc_s\Delta T_s \tag{13.6}$$

where
    c_l = specific heat of the country rock in the liquid state
    c_s = specific heat of the country rock, solid
    L_f = latent heat of fusion of country rock
    ΔT_l = change in temperature from beginning of, to completion of, melting of country rock
    ΔT_s = change in temperature of country rock up to its beginning of melting.

Hence it is an easy step to calculate the maximum amount of melt which could be produced from known country rock by a given intrusion, a useful constraint when one is pondering the origin of spatially associated igneous rocks (Bell 1966). We must stress that such a calculation sets only an upper limit since it takes no account of important factors such as thermal conductivity of the country rock.

A number of authors have attempted to estimate the time taken for an intrusion to cool through a specific temperature interval. Larsen (1945) gives a detailed calculation of the time (70 million years) required for crystallisation of the Southern and Lower Californian bath-olith, treated as a dyke 100 km wide which cooled by conduction through the walls. An amplified description of principles and methods with applications to various types of igneous body is given by Jaeger (1968). A graphical method which provides approximate solutions of non-steady state heat conduction problems, known as the Schmidt approximation method, is described by Ingersoll, Zobel and Ingersoll (1954). These various approaches allow estimates to be made of the temperature distribution as a function of time in both the intrusion and country rock and so provide some constraint to set upon the progress of crystallisation in the former and metamorphism in the latter (cf. Dunham 1970; Irvine 1974).

On this subject we may make the following generalisation. While the total heat flux, q_w, from a magma body varies according to size, shape, depth of emplacement etc., a range of q_w from 1×10^{-5} to 1×10^{-3} cal cm^{-2} sec^{-1} covers most cases at magma liquidus temperatures, with smaller bodies tending towards the larger values of q_w (Shaw 1965; Bartlett 1969). Reverting once more to our 200 km^3 basaltic intrusion we may model this as both cylinder and inverted cone to illustrate the difference in cooling time, t, for the first 200 °C temperature fall, heat ideally transferable being 9.048×10^{19} cal. For the cylindrical model when $q_w = 1 \times 10^{-3}$, $t = 1.9 \times 10^3$ years; with $q_w = 1 \times 10^{-5}$, $t = 1.9 \times 10^5$ years. For the conical model $t = 2.85 \times 10^3$ and 2.85×10^5 years, respectively, for the high and low rates of q_w. (For both models heat transfer is assumed to be from the lateral and upper surfaces only; lateral area of a cylinder (of radius = height = 4 km) = $2\pi rh$; lateral area of a cone of the same dimensions is πrh.)

Convection in magmas

One consequence of the loss of magmatic heat by conduction from the margins of an intrusion is the establishment of a temperature gradient within the magma. Local differences in density leading to gravitational instability ensue and natural convection movements might consequently be expected to begin. Factors other than the temperature gradient are, however, important in determining the thermal stability of a fluid such as magma and these are taken into account in calculating another dimensionless ratio, the Rayleigh number, *Ra*.

The Rayleigh number, Ra_h, for a horizontal layer of viscous fluid having an upper and a lower surface is given by

$$Ra_h = \frac{L^4 \alpha_T g \beta}{\nu K} \qquad (13.7)$$

where
   L = the height of the layer (cm)
   α_T = the thermal expansion coefficient, $\dfrac{1}{\rho}\left(\dfrac{\partial \rho}{\partial T}\right)_\rho$ (K^{-1})
   g = gravitational acceleration (980 cm sec^{-2})
   β = the vertical temperature gradient (K cm^{-1})
   ν = the kinematic viscosity = $\dfrac{\eta}{\rho}$ where η = magma viscosity
   K = the thermal diffusivity of the magma (cal g^{-1} K^{-1})
   ρ = density

The Rayleigh number, Ra_t, for a vertical tube heated from below is given by the same expression except that L^4 is substituted by R^4 where R is the characteristic radius of the tube in centimetres. Bartlett (1969) has shown that the critical Rayleigh number above which convection will take place is ~1700. Clearly, from 13.7, the larger the intrusion and the lower the viscosity the more likely is convection to occur; but quite small bodies, having high heat flux values, should also be unstable as shown in Figure 13.3.

For low Rayleigh numbers, then, that is, $Ra < 10^3$, transfer of magmatic heat is predominantly by conduction; steady convection sets in at intermediate (10^4) values of Ra and strong, eddying motion obtains when $Ra \sim 10^5$ (Elder 1976).

Crystal settling

A number of textural and structural features displayed by layered intrusions testify to the effect of magmatic movement on crystal orientation. Figure 13.4 is an example of igneous lamination, that is, the orientation of tabular minerals parallel with the plane of layering. This uniform arrangement of mineral grains could not be produced by random crystal settling in a stagnant magma. Stokes' Law, which strictly,

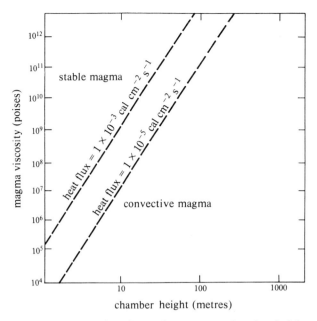

Figure 13.3   Plot of magma viscosity against magma chamber height to delimit stable and convective fields (after Bartlett 1969).

relates to the settling velocity of a single spherical particle in a viscous fluid whose extent greatly exceeds the diameter of the particle, reasonably describes the motion of crystals in a magma, when the Reynolds number is of the order of 0.05 to 0.1:

$$V = \frac{2gr^2(\Delta\rho)}{9\eta}$$ (13.8)

where
   V = terminal velocity (cm sec^{-1})
   r = particle radius (cm)
  $\Delta\rho$ = density contrast between particle and fluid medium
   η = viscosity (poises)
   g = gravitational acceleration.

Several factors influence the straightforward predictions from Stokes' Law. For example a crystal in a magma body of considerable height may grow during movement and its motion may be affected by other crystals. In the first case Shaw (1965) has shown that with particle radius

Figure 13.4   Photomicrograph of igneous lamination.

assumed as a function of time, the distance, d, travelled is given by

$$d = \frac{2gc^2\Delta\rho t^3}{27\eta} \qquad (13.9)$$

where c = the radial growth rate constant (cm sec^{-1}), t = time, and other symbols are as in 13.8.

As for the influence of other crystals, which could become important after supersaturation or mechanical crystal sorting has led to marked crystal concentration, the sedimentation rate, V_s is given by

$$V_s = V_0(1 - \phi)^{4.65} \qquad (13.10)$$

where V_0 = particle settling velocity at infinite dilution, and ϕ = the volume fraction of particles (Lewis, Gilliland & Bauer 1949; Shaw 1965). Thus the free settling velocity is markedly modified by increasing concentration of primocrysts in the magma. In a convecting magma the net settling velocities would be further modified, that is, decreased, in a rising current (although a prolongation of crystal growth could to some extent counteract this) and increased in downward moving currents. Lastly, one should add deviation from the ideal spherical shape as a factor that will exert a small but consistent influence upon the settling velocity of crystals in natural melts.

An idea of the velocity of the current itself may in certain circumstances be gained using the same expressions. The Marginal Border Group rocks of the Skaergaard Intrusion contain blocks of acid gneiss derived from the Precambrian basement. Taking an average radius of 25 cm for these blocks and assuming a magma viscosity of 3000 poises and $\Delta\rho = -0.1$, Wager and Brown (1968) quoted 160 m/hr for the rate at which the blocks should rise through quiescent basaltic magma. Since the blocks are frozen in place at various levels in the marginal rocks it was suggested that downward moving currents of at least the same velocity prevented the otherwise buoyant acid gneiss xenoliths from rising. The estimated velocity of these currents (4 km/day) is very high and they were regarded by Wager and Brown as sporadic events superimposed on a continuous slow convective overturn. If the same reasoning is to be applied to xenoliths in other intrusions it is worth noting that for particles to conform to the Stokes' relation their maximum radius, r, should be such that

$$r = \left(\frac{9\pi^2 Re}{4g\rho e\Delta\rho}\right)^{1/3} \qquad (13.11)$$

where ρe = the density of the liquid, and Re = the Reynolds number (Shaw 1965).

Current and crystal settling velocities calculated for one intrusion should not be indiscriminately applied to others, as we may now show. Particularly variable has been the estimate of magma viscosity. Wager and Brown (1968) followed Hess (1960) in the choice of 3000 poises for basaltic magma whereas Jackson's (1971) curves for crystal settling velocities assumed a viscosity of 300 poises and Goode (1976) presented values of settling velocity based on viscosities of 100 and 1000 poises, pointing out that for deeper-seated magmas higher liquidus temperatures and hence lower viscosities might be expected. Estimated settling velocities for a plagioclase crystal of $r = 1$ mm presented by three of these authorities are as follows: 23 m/year (Wager & Brown); 50 m/year (Jackson); 173 or 17.3 m/year, depending on viscosity (Goode). What is significant in terms of the mechanism of differentiation in the basic and ultramafic layered intrusions considered is that reasonable estimates of crystal accumulation time deduced are short compared with the cooling time available.

For methods of calculating the density and viscosity of magmatic silicate liquids the reader is referred to Bottinga and Weill (1970, 1972).

Four mineral groups whose members are common to both granitic and basaltic rocks, namely plagioclase, olivine, pyroxene and iron–titanium oxide, all show a greater density contrast with granitic than with basaltic magma yet the considerably higher viscosity of granitic magma, not less than 10^6 poises (Shaw 1965), greatly lowers crystal settling velocities. For example, a plagioclase crystal of $r = 1$ mm settles less than 10 cm/year in granitic magma. However, pyroxene and magnetic crystals of the same size would settle at 20 and 40 cm/year, respectively. Since large granitic plutons have very long cooling times even these rates should permit significant crystal settling, yet textural evidence of this has but rarely been reported. We will consider this apparent anomaly further in the next section.

Layered intrusions

In Chapter 12 we considered the textural features and possible modes of origin of cumulates. The characteristic structural feature of these rocks is layering, corresponding to bedding or stratification in sediments and any pluton which includes as a significant integral part a series displaying this phenomenon is by definition a layered intrusion. Layering may, however, be absent from sizeable portions of any layered intrusion. By reason of more rapid cooling rather than of size, sills rarely develop all the interrelated textural and structural characteristics of layered intrusions and are conventionally placed in a separate category.

Conditions are favourable for the formation of a layered intrusion when a large quantity of magma with relatively low viscosity cools through the liquidus temperature of several minerals undisturbed by contemporaneous tectonic movements. Layering may be found in rocks of any common composition but layered intrusions which show optimum development of the characteristic features are predominantly ultrabasic (including ultramafic), basic, or alkaline intermediate (especially undersaturated) in type.

Layered intrusions may be conveniently divided into two categories according to the mechanism of magma supply and retention:

(a) Closed system: magma is emplaced in a single, relatively rapid injection and crystallises without loss.

(b) Open system: magma is emplaced in several pulses separated by time intervals lengthy enough to permit significant cooling and formation of cumulates and intermittent leakage of supernatant magma is possible.

While accepting that natural phenomena rarely conform entirely to artificial classifications we may cite the Skaergaard Intrusion and the Rhum ultrabasic intrusion as examples of the first and second types, respectively (Wager & Brown 1968). The observable part of the Stillwater Intrusion, Montana, U.S.A., appears also to be a closed system type. The Bushveld Intrusion, South Africa, seems to have undergone some replenishment of magma but it is rash to attempt to fit this huge intrusion into any simple category: as with other now continuous masses, such as the Great Dyke of Rhodesia, it may be the end product of the confluence of several initially separate intrusions.

Figure 13.5 depicts an ideal layered intrusion modelled on the Skaergaard closed system type (Wager & Brown 1968). The first surge of magma cools quickly to form an enveloping Border Group which includes a chilled margin grading inwards into less quickly cooled rocks. Within this inwardly accreting shell crystals accumulate on the floor of the slowly cooling magma to form a Layered Series. Solidification downwards of an Upper Border Group accompanies the inward growth of the Marginal Border Group and the accumulation upwards of the Layered Series so that compositional correlation of units within each of these parts of the intrusion is possible. Eventually a portion of liquid becomes trapped between the advancing crystal fronts and solidifies in place as a Sandwich Horizon. The rocks of this horizon and of the chilled margin are likely to be the only ones in the intrusion representing liquid compositions unmodified by crystal settling and are also the only ones unlikely to display some sort of layering.

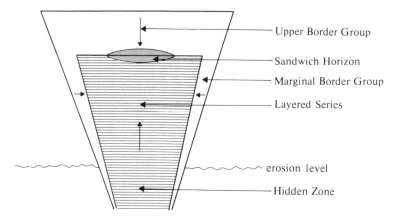

Figure 13.5 Idealised layered intrusion. Not to scale. Hidden Zone is unexposed part of intrusion and may lack layering at least in part. Arrows within the intrusion indicate direction of solidification.

In the most cogent discussion of the nomenclature available, Jackson (1967) has defined a layer as 'a continuous sheet-like cumulate that is characterised by uniform or uniformly gradational properties' and has distinguished two main categories of layers based upon the proportions of cumulus minerals and physical or compositional properties of cumulus minerals, respectively. In the first category a layer may be **isomodal,** that is, have a uniform proportion of one or more cumulus minerals, or **mineral-graded,** that is, display a gradual vertical change in the proportions of two or more cumulus minerals. In the second category are **size-graded** layers in which gradual vertical changes in the grain size of one or more cumulus minerals are found, and **chemical-graded** layers in which gradual vertical changes in the chemical composition of cumulus minerals take place.

A layered series comprises a stack of any combination of these types often displaying much repetition of a particular type to which the term **rhythmic layering** has been applied (Wager & Deer 1939). In the field what emphasises the layering is the type of contact between superposed layers. This, following Jackson (1967), may be a **phase contact** (marked by abrupt appearance, or disappearance, of a cumulus phase); a **ratio contact** (marked by sharp change in the proportions of two cumulus minerals); or a **form contact** (picked out by an abrupt change in, say, size or habit of one or more cumulus minerals). Clearly types of contact may coincide.

The sequence and nature of the layers result from the interplay of crystal sorting by gravity and magma currents, nucleation processes, and crystal fractionation. The best index of the last is the chemical-graded layering, perhaps better known as **cryptic layering** (Wager & Deer 1939) in which cumulus phases which are members of a solid-solution series change composition continuously with fractionation and in accordance with the predictions of the phase equilibria of the relevant systems. For example, plagioclase becomes more sodic and olivine and pyroxenes more iron-rich.

Cryptic layering, so called because the compositional change of the cumulus phase is not detectable by eye in the field, is the key to the 'stratigraphy' of a layered intrusion, the cumulus phases being analogous to zone fossils in a sedimentary sequence. Continuous changes in cumulus mineral composition consistent with down-temperature trends of the parental magma indicate closed system crystallisation: the sequence in the Layered Series of the Skaergaard Intrusion is a prime example. Reversals of a trend suggest a rise in the temperature of the magma most plausibly explained by influx of new, hotter material, in other words, an opening of the system. The eastern part of the Rhum ultrabasic intrusion (Brown 1956) comprises fifteen thick layers each with the same mineralogical character and totalling some 800 m in thickness. Yet this pile of 'macro-rhythmic units' has no detectable cryptic layering. The repetition of an effectively identical unit has been explained as the result of periodic magmatic replenishment coupled with extrusion to the surface of supernatant liquid which would otherwise have crystallised phases of different, lower temperature composition. Thus the Rhum magma chamber is thought to have had inlet and outlet valves which operated in conjunction at least fifteen times during the formation of the layered intrusion.

A combination of mineral-grading and cryptic layering data permits the construction of what is, in effect, a stratigraphic column for a layered intrusion which may be used to identify the structural level of any rock within the sequence. Wager and Brown (1968) provide these for a number of well-investigated intrusions. In some cases the same device is appropriate for thick sills and Figure 13.6 is an example for the Palisades Sill compiled from the data of Walker (1969). We may note the reversals in mineral composition trends and recall that the Palisades Sill is now adjudged to be a multiple intrusion.

Certain diagnostic mineralogical features of fractionated tholeiitic intrusions are also brought out in Figure 13.6, namely the coexistence of two pyroxene trends, the eventual development of interstitial quartz feldspar intergrowth and the occurrence of an 'olivine gap'. The last named feature, referring to the absence of cumulus olivine in a section of the intrusion between a lower section containing more magnesian cumulus olivine and an upper section in which more iron-rich cumulus

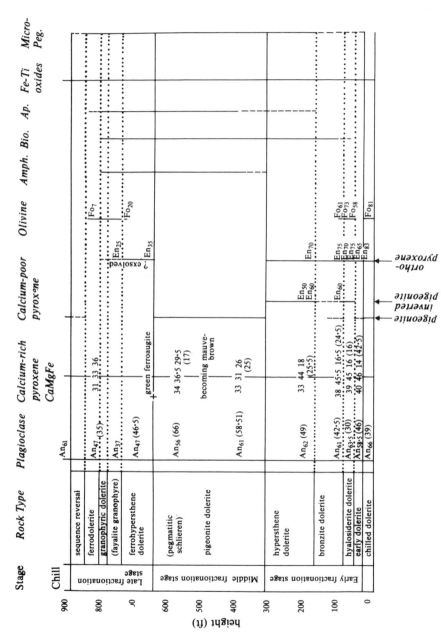

Figure 13.6 Mineral variation as a function of stratigraphic height in the Palisades Sill (data from Walker 1969). Continuous vertical line indicates mineral present in appreciable amount; dashed line, mineral present in trace amount. Numbers in parentheses indicate volume modal per cent.

olivine is found, is an important case of phase-layering and formed the basis for dividing of the Layered Series of the Skaergaard Intrusion into Lower, Middle (cumulus olivine-free) and Upper Zones. Further subdivision of these three zones is based upon the appearance in quantity of other cumulus minerals such as augite, magnetite, apatite and inverted ferriferous-wollastonite (Wager & Brown 1968, Figs 14, 15).

Exceptions prove, that is test, rules. A case in which one cumulus mineral group (calcium-rich pyroxene) shows regular cryptic variation while an accompanying cumulus mineral lacks it is the Layered Series of the Kap Edvard Holm Intrusion, East Greenland (Elsdon 1971). The discrepancy is explained according to the differing susceptibilities of the crystallisation of the two minerals to increasing water vapour pressure which is indicated by the presence of hydrous minerals. From a liquid of given composition plagioclase will crystallise with an anorthite content that is proportional to increase in the water vapour pressure of the liquid. Thus as plagioclase separation proceeds if the water content of the remaining liquid increases owing to upward diffusion of water the calcic composition of successive plagioclase batches will be more or less maintained. The same will not apply as distinctly to pyroxene compositions which in the Kap Edvard Holm Intrusion undergo normal iron-enrichment.

Repetition of layers indicates some interruption of the slow process of continuous gravitational settling of crystals since early-formed slowly settling grains will be caught up by later formed, more quickly settling grains producing uniform textures and mineralogy. We referred earlier to fast and slow currents in the Skaergaard magma. Crystal settling in the latter produced 'average rock', that is, rock of uniform texture. The fast currents have been likened to turbidity currents and have been cited as the mechanism responsible for the type of mineral-graded layer which displays **gravity-stratification**. Such a layer has a sharp base rich in denser, usually darker minerals, and grades upwards into a rock richer in less dense, usually lighter-coloured minerals. A fast current is envisaged as starting as a crystal-charged denser mass which plunges downwards at tne edge of the magma chamber and on meeting the base spreads laterally across the floor depositing its load of crystals, the denser and larger first, and eventually coming to rest. Other sedimentary features such as scour and fill structures lend strength to the suggestion that rapidly moving currents are a feature of some layered intrusions.

The velocity of such laterally-moving currents in aqueous environments is readily calculated from the formulae of fluid mechanics. Application of these formulae to magma currents is a step that cannot be justified in many cases because of lack of critical data. Usually unknown factors, such as the direction and gradient down which the current flowed are, however, measurable in the case of slump and **trough-banding** structures occurring in some layered sequences. Trough-banding (Wager & Deer 1939) is an arrangement of lenticular, concave-upwards rhythmic layers in channels within which magma currents were confined and whose plunge can in some cases be mapped in the field. The velocity of flow in an open channel may be calculated from

$$U^2 = \frac{2gmS}{C_f} \tag{13.12}$$

where
  U = the mean velocity
  g = gravitational acceleration
  m = the hydraulic mean depth = the ratio of the average cross-sectional area of the channel to the wetted perimeter
  S = the slope of the channel floor
  C_f = the friction coefficient (Kay & Nedderman 1974).

C_f is a coefficient related to the Reynolds number and is smallest at large values of Re; tables and formulae for its determination are provided in standard texts on civil engineering and fluid mechanics. Goode (1976) has suggested that for erosion and movement of sediment in lower flow regime conditions the current velocity would be approximately the same as the settling velocity of the sedimentary material.

Returning now to the question of crystal settling in granitic magmas we observe that although directional textures, such as the alignment of feldspar phenocrysts, are not uncommon in larger plutons, layering in which mineral grains seem to display gravity stratification has been reported in very few cases (Harry & Emeleus 1960) and then when mineralogical evidence such as high modal fluorite contents points to unusually low magma viscosity. We have seen (Fig. 13.3) that magma bodies with low heat flux and the typical viscosities of granite are likely to undergo natural convection. The type of flow in liquids is characterised by the Reynolds number (13.2) which we may also write as

$$Re = \frac{\rho UL}{\eta} \tag{13.13}$$

where L = a length dimension, e.g. of a channel and another dimensionless ratio, the Froude number, Fr, given by

$$Fr = \frac{U^2}{gL} \tag{13.14}$$

The Reynolds number and the Froude number are, then, measures of the ratio of inertial force to, respectively, viscous and gravitational forces. The high viscosities and low particle settling velocities typical of granitic melts necessitate low *Re* and *Fr*, hence particles entrained in convection currents in these melts will be well dispersed.

Such are the observations and theoretical predictions but a further petrographic test might be applied to granitic plutons of considerable vertical extent. The Stokes' relation means that a doubling of particle radius, other coefficients remaining the same, can lead to a quadrupling of the settling velocity. The evidence to seek is not so much for larger single crystals, however, as for composite clusters of several crystals which reveal primocryst features. As we have seen (Ch. 12) primocrysts tend to be idiomorphic so that a composite cluster should reveal interior contacts that preserve these idiomorphic outlines after settling while exterior surfaces will grow freely and allotriomorphically by orthocumulus or adcumulus processes interfered with by the growth of other grains. If such clusters are detected, both their frequency at given levels and any change in chemical composition with height in the intrusion would be guides to crystal sorting and cryptic variation in granitic intrusions. In other words such a test is one way of detecting whether or not certain granitic masses are layered intrusions.

Some chemical features of layered intrusions

In Chapter 12 we noted that the bulk chemical composition of a cumulate is unlikely to match that of its parental liquid. In spite of this and the virtual absence of any other rock which could be regarded as an unmodified liquid composition, Wager and Deer (1939) and Wager and Brown (1968) were able to derive compositions of the successive liquids of the evolving Skaergaard intrusion which have become a virtually universal standard for calibration and comparison in igneous petrology. The method used depended upon a knowledge of the initial composition of the original magma and of the volume of the various zones of the layered rocks. The initial composition was assumed to be that of a particular rock in the chilled margin. From this composition were subtracted the compositions of the various zones in proportion to their volume as deduced from field mapping. Part of the intrusion, the Hidden Zone, could not be mapped and had not then been drilled and its volume had to be estimated by other means. Figures 13.7 and 13.8 depict the graphical method employed.

Figures 13.7a and 13.7b are variation diagrams of any given oxide against any parameter recording solidification of an intrusion. Figure

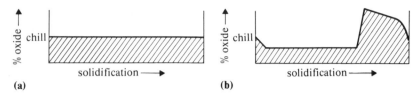

Figure 13.7   Variation diagram of an oxide against solidification. (a) Totally undifferentiated body of solidified magma (e.g. homogeneous glass). (b) Differentiated magma body. The shaded sections are equal in area.

13.7a shows the oxide curve for a totally undifferentiated body of magma, for example, a homogeneous glass, and Figure 13.7b the curve for the same body of magma differentiated by crystal settling. The areas below the curves on the two graphs are equal. Figure 13.8 develops this reasoning. In this case the variation of a given oxide is plotted against 'percentage solidified', that is, the percentage of the total volume of the cumulate rock series which had accumulated at a given time. We arbitrarily set the volume of the Hidden Zone as 60% solidified in Figure 13.8.

A curve for the oxide in question in the exposed rocks of the intrusion is constructed from chemical analyses of exposed average rocks (those occurring between the strongly crystal-sorted horizons). One other known point on the curve is the content of the oxide in the chilled margin rock. Now the amount of our oxide in the whole intrusion is proportional to the area made up by sections B + C + D (i.e. as if the magma were undifferentiated). The amount of our oxide in the exposed rocks is proportional to the area comprising A + B. The oxide curve must continue to the left of the 60% solidified point since the Hidden Zone rocks must contain a positive amount of the oxide: this amount is proportional to the area D. Thus the area of A + B + D is also proportional to the total amount of oxide in the intrusion, hence A + B + D = B + C + D and therefore A = C. The area of A is known from the

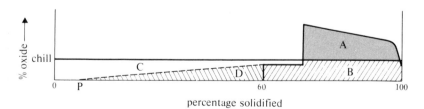

Figure 13.8   Graphical method for determining relative volume of unexposed rocks of layered intrusion (after Wager & Brown 1968).

chemical analysis plots and the area of C is fixed by moving point P until the area of C equals the area of A. If, when this is done, point P, as in Figure 13.8, indicates that the oxide percentage falls to zero somewhere within the Hidden Zone, then a different volume (percentage solidified) for the Hidden Zone is selected and the process repeated until a satisfactory complete curve is derived.

This operation provides graphs for each oxide in the various *rocks* but we have now to deduce the oxide content of the corresponding *liquids* from which the rocks were derived. The method is similar in conception to that above and Figure 13.9 exemplifies its use.

The first step is to plot an average rock composition curve for a given oxide as described above. In Figure 13.9 we assume the Hidden Zone forms 70% of the intrusion. If we take a point at 35% solidified to represent the average of the Hidden Zone we know that the amount of the oxide in all the rocks between 35% and 100% solidified is proportional to the shaded area A. The amount of the oxide in the liquid existing at 35% solidified is given by point L which is fixed by lowering it, or raising it, so that the rectangle LMNO which it defines equals the shaded area of A. Other liquids at different percentages solidified are fixed in the same way defining a series of rectangles each equalling the area marked off by the corresponding rock composition curves. The process is repeated for all the relevant oxides and thus full compositions for the successive liquids are provided.

The curves provide a clear illustration of the effect of crystal fractionation in that the amount of oxide in the liquid exceeds that in the corresponding average rock until a cumulus phase containing a significant amount of that oxide begins to separate in quantity.

The bulk composition of a differentiated intrusion may be straightforwardly calculated if the relative volume, average chemical

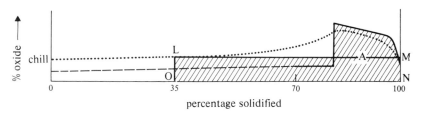

Figure 13.9 Graphical method for determining oxide content of liquid from which cumulates form. L is liquid existing at 35% solidified. Shaded area A is proportional to oxide content of rocks between 35% and 100% solidified. Point L is fixed by defining rectangle LMNO to equal the area A. Defining successive similar rectangles produces locus of liquid points (dotted curve) (after Wager & Brown 1968).

composition and rock density of its several units are known. The relative weight proportion of each unit is calculated by multiplying its volume percentage by the rock density. In each unit each oxide in the average composition is multiplied by the weight percentage of the unit to give a set of 'raw' weight percentages of oxides for the unit. The figures for the various units are summed and each recalculated as a corrected percentage of the sum. The list of corrected figures then represents the bulk composition of the whole intrusion.

The analysis so gained may be compared with that of the chilled margin, when available. Assuming each to be a valid composition, any significant differences point to some kind of 'opening of the system' such as loss of liquid from the magma chamber, or multiple intrusion or hydrothermal alteration. Depending on which elements are involved, and in what proportions, and on petrographic evidence, any of these and other possibilities may then be further investigated. Critical to any such calculations are accurate field data and chilled margin rocks which truly represent initial magma. We have noted in Chapter 12 and earlier in this chapter how uncertain some assumptions have proved to be.

We will conclude this section by taking brief note of three other techniques concerned with measurements of the liquid from which cumulates have formed.

Assuming that the trapped liquid in an orthocumulate is a sample of the contemporaneous magma it should be possible to estimate the magma composition by subtracting from a whole rock analysis the composition of the cumulus assemblage as derived from modal proportions and constituent mineral composition. After determining the modal weight percentage of each cumulus mineral (from volume and density) the weight percentage of an oxide in the material remaining after extraction of the cumulus assemblage is given by

$$\frac{(W_R) - \dfrac{MW_M}{100}}{100 - M} \tag{13.15}$$

where W_R, W_M = the weight percentage of the oxide in the whole rock and the extracted mineral respectively, and M = the modal weight percentage of the extracted mineral.

This 'orthocumulate method' was used by Brown (1956) on a plagioclase–olivine orthocumulate from the Rhum ultrabasic intrusion. The principal factor affecting accuracy is the difficulty of determining precisely the amount of cumulus material.

Henderson (1970) has described a different approach, that is, to determine the proportion of mesostasis, the last-formed material in a cumulate which can, in part, (Ch. 12) represent reaction between cumulus phases and intercumulus liquid. The method makes use of (a) the existence in the rocks of elements with high and low crystal/liquid partition coefficients, referred to as high-k and low-k elements, respectively; and (b) the occurrence of rhythmic layering in which two virtually monomineralic layers, A and B, of contrasting mineralogy (e.g. plagioclase and pyroxene) are closely associated. The very restricted mineralogy eases the problem of determining accurate modal proportions. Then the volume proportion of mesostasis in rock A, V_A, is given by

$$V_A = \frac{(R_A - C_A) - W(R_B - C_B)}{(C_B - C_A)} \tag{13.16}$$

where

R_A, R_B = the concentration of high-k element in whole rocks A, B, respectively

C_A, C_B = the concentration of high-k element in the cumulus assemblages of rocks A, B, respectively

W = the ratio of the whole rock concentrations of low-k element.

Henderson gives examples of the calculation using uranium as a low-k element and strontium as a high-k element and finding values of 15% and 24% for the proportion of mesostasis in olivine + pyroxene-rich and plagioclase-rich rocks, respectively.

The future for this kind of quest lies in determining the conditions of crystallisation from the mineral assemblage (such as temperature and oxygen fugacity from co-existing iron-bearing phases or P–T conditions from co-existing silicates which exhibit different polymorphs as described in Ch. 12) and then experimentally reproducing liquids which under these conditions give rise to the cumulus phase assemblage observed (McBirney 1975).

14

Trace elements in igneous processes

Introduction

Several previous chapters have been concerned with the way in which major element compositional data can be put to work in the formulation and testing of petrogenetic hypotheses. Trace elements have not so far been considered in detail because the way in which they are best handled is often substantially different from the treatment for major elements.

The term 'trace element' cannot be rigidly defined but is usually taken to mean those elements present in rocks in concentrations of a few thousand parts per million, or less, by weight. Trace elements of particular interest to igneous petrologists include Rb, Ba, Sr, Zr, Y, Nb, Th, the rare earth elements (La to Lu), Ni, V, and Cr, though there are many others. Major elements such as K and P may often profitably be considered along with the trace elements.

Different minerals may incorporate or exclude trace elements with even greater selectivity than they do major elements. These preferences exert such a critical influence on trace element distributions during igneous processes that their analysis can lead to constraints on the nature and composition of the mineral assemblages with which a magma may have previously equilibrated, or vice versa. The use of trace element data has increased greatly with the development of accurate and easily applied determinative techniques (especially X-ray fluorescence, neutron activation and mass-spectrometric isotope dilution) and today many long-standing petrogenetic arguments centre on the trace element contents of magmas and their possible source regions. This chapter is concerned with the approach to quantitative treatment of trace element behaviour in simple models of magmatic evolution and the way in which these may be applied to observed data.

However, before proceeding it is worth making a small digression concerning the composition of the Upper Mantle because this is a subject of frequent reference in the following pages. Firstly our direct knowledge of Upper Mantle rocks is derived very largely from the ultramafic nodules brought to the surface by kimberlite pipes and alkali basalts. The reader is referred to Dawson (1971) and Forbes and Kuno (1965) for details of these occurrences. Briefly, the commonest type of ultramafic nodule from kimberlite is garnet lherzolite consisting of abundant olivine and orthopyroxene with subordinate amounts of chrome diopside (Cr-rich clinopyroxene) and pyrope-rich garnet. Possible melting relationships of such rocks at depth have already been discussed in Chapters 9 and 10. Spinel-bearing lherzolites are formed at shallower depths and are more characteristic of the nodule suites found in basalts. With regard to the trace element contents of mantle rocks geochemists frequently use the concept of the earth model based on element abundances in chondritic meteorites (see Mason 1966). This is particularly useful as a help in displaying analytical results for the rare earth elements where so-called 'chondrite-normalised' data are used (see, for example, Fig. 14.2). In this technique rare earth analyses are expressed as the analysed concentration of the element in the rock (or mineral) divided by its average concentration in chondrites. This has the effect of ironing-out the saw-tooth distribution which the raw data normally show since rare earth elements with odd atomic numbers are less abundant than their even-numbered neighbours. The smooth or more-or-less regularly sloping patterns shown by normalised data are much more readily interpreted.

Distribution coefficients. When a mineral is in chemical equilibrium with a liquid, elements are partitioned between the two phases according to their chemical activity in each. For an element whose concentration is low in both (generally speaking less than 1% or 10 000 ppm)[1] application of Henry's Law leads to the relationship

$$\frac{\text{Conc. in mineral}}{\text{Conc. in liquid}} = K_D$$

where K_D is a constant known as the 'distribution' or 'partition' coefficient for the given crystal–liquid equilibrium. In principle K_D can be measured either from synthetic crystallisation experiments or from phenocryst–matrix relations in glassy volcanic rocks. Some typical values are summarised in Table 14.1, but it must be emphasised that estimates vary quite widely in practice. This is especially true of the very low values for olivine and orthopyroxene and some of the higher values for acidic compositions. To date, most are based on phenocryst–matrix analysis and undoubtedly include some cases where equilibrium was not

Table 14.1 Typical trace element distribution coefficients

Mineral	Liquid composition*	K	Rb	Sr	Ba	Ce	Sm	Eu	Eu†	Yb	Ni	Cr
Olivine	B	0.001	0.001	0.001	0.001	0.001	0.002	0.002	0.002	0.002	10	0.2
Orthopyroxene	B	0.001	0.001	0.01	0.001	0.003	0.010	0.013	0.013	0.05	4	2
Clinopyroxene	B	0.002	0.001	0.07	0.001	0.10	0.26	0.20	0.30	0.28	2	10
	A	0.04	0.03	0.5	0.13	0.5	1.7	1.6	1.8	1.6	–	–
Garnet	B	0.001	0.001	0.001	0.002	0.02	0.22	0.32	0.35	4.0	0.4	2
	A	0.02	0.01	0.02	0.02	0.4	2.7	1.5	6.5	40	–	–
Spinel	B	0.01	0.01	0.01	0.01	0.08	0.05	0.03	0.03	0.02	5	10
Plagioclase	B	0.2	0.07	2.2	0.2	0.10	0.07	0.3	0.06	0.03	0.01	0.01
	A	0.1	0.04	4.4	0.3	0.3	0.1	2.1	0.1	0.05	–	–
Amphibole	B	1.0	0.3	0.5	0.4	0.20	0.52	0.59	0.58	0.49	3	12
	A	0.08	0.01	0.02	0.04	1.5	7.8	5.1	8.9	8.4	–	–
Fe-mica	B	2.7	3.1	0.08	1.1	0.03	0.03	0.03	0.03	0.04	3.5	7
	A	–	2	–	10	0.32	0.26	0.24	0.27	0.33	–	–
K-feldspar	A	–	0.4	4	6	0.04	0.02	1.1	0.01	0.01	–	–
Apatite	A	–	–	–	–	35	63	30	60	25	–	–
Zircon	A	–	–	–	–	2.5	3	3	8	300	–	–

* B = basaltic; A = intermediate or acidic.
† Eu interpolated from adjacent trivalent rare earths.

In this table, incompatible element (K to Yb) distribution coefficients for olivine and orthopyroxene are taken from mineral comparisons in nodules (Hanson 1977) and may be appropriate to prolonged equilibration within the mantle and zone-refining. Typical mineral–matrix determinations give values approximately one order of magnitude higher for these and are certainly more realistic for fractional crystallization of magmas. It must be emphasised that all K_Ds show a wide range in practice and are only nominally represented by the values in this table.

established. Nevertheless, there is clear evidence that distribution coefficients are strongly composition-dependent, e.g. for most elements in garnet and clinopyroxene, and for rare earths in most minerals, there is a ten-fold increase in K_Ds from equilibrium with basaltic liquids to that with acidic ones. In part this simply reflects composition-dependence of the lattice parameters for the minerals, especially those which form solid solution series. Here it is often helpful to relate the behaviour of the trace element to that of the major element for which it substitutes by dividing the distribution coefficient of the trace element by that of the 'carrier' element. The compound distribution coefficients so obtained for Sr/Ca in plagioclase and Ni/Mg in olivine show much less compositional dependence. However, use of compound K_Ds is often inconvenient since it requires full chemical analyses of both major and trace elements. The observed variability can also to some extent be expressed as one of temperature dependence, since crystal–liquid equilibria in more differentiated liquids invariably relate to lower temperatures. Thus Drake and Weill (1975) have shown that the distribution of Sr in plagioclase is a simple logarithmic function of temperature over the range 1400° C to 1100° C. This of course raises the possibility of using phenocryst–matrix data as geothermometers (see O'Nions and Powell 1977 for limitations in this respect). There is also some evidence that distribution coefficients of K, Rb, Sr and Ba for clinopyroxene vary with pressure, but the effect seems too small to have practical value as a geobarometer. A further difficulty arises with elements which have more than one oxidation state, since their distribution coefficients then become strongly dependent on the oxygen fugacity pertaining during equilibrium. Thus the behaviour of Eu is frequently anomalous relative to the other rare earth elements since at low oxygen fugacity it exhibits a divalent state which is preferentially incorporated into plagioclase (Drake 1975) and rejected from ferromagnesian minerals, especially clinopyroxene (Grutzeck *et al.* 1974). Thus in applying distribution coefficients we need access either to such a complete experimental study (and knowledge of P–T–f_{O_2} conditions in our rocks) that we can calculate appropriate values, or else we must simply choose values from a table such as given here for similar liquid compositions to that which we are studying.

A simple and direct use of distribution coefficients is the reconstruction of trace element concentrations in a magma from analysis of minerals which equilibrated with it. Thus Hart and Brooks (1977) have presented trace element data for unaltered clinopyroxene phenocrysts in Archaean greenstones from the Superior Province, Canada and applying distribution coefficients as in the above equation calculated Sr (140 ppm), K/Rb (470) and K/Ba (16) for the parent magma. This shows that the greenstone volcanics, which in bulk are considerably modified with respect to alkali contents, were essentially similar to modern island-arc tholeiite magmas. Similarly Hamilton (1977) used cumulus clinopyroxene from the Bushveld Mafic Phase to show that the original magma of this layered intrusion had a light rare earth-enriched abundance pattern; and Shimizu (1975) demonstrated that the lithophile element contents of diopside in the inclusions of some South African kimberlites were too high for any reasonable crystal–silicate liquid equilibrium and necessitated metasomatic enrichment of parts of the Upper Mantle.

In order to consider the evolution of a magmatic liquid we need to quantify its equilibrium with more than a single mineral phase. In this case, partitioning is described by the bulk distribution coefficient (D) which is calculated from the weight proportions (w) of each mineral in the assemblage:

$$D = \Sigma_{i=1}^{n} \, w_i K_{Di}$$

For example, for a hypothetical garnet peridotite consisting of 60% olivine, 25% orthopyroxene, 10% clinopyroxene and 5% garnet:

$$D_{Ce} = 0.6 \times 0.001 + 0.25 \times 0.003 + 0.1 \times 0.1 + 0.05 \times 0.02 = 0.012$$

and similarly, $D_{Yb} = 0.244$. Clearly, one phase with a distribution coefficient very different from the remainder, especially a very high one, can have a dominating effect on the concentration of a trace element even when it is present in quite small amounts. In the above example, garnet is responsible for the fact that the distribution coefficient for Yb is more than twenty times higher than that for Ce. In a more extreme case, the separation of small amounts of minor phases such as zircon and apatite could effectively control the behaviour of rare earths (and Zr) and Sr respectively. It also follows that occlusion or entrapment of the liquid itself in the solid assemblage (e.g. as mesostasis) – equivalent to a phase with $K_D = 1$ for all elements – can cause an appreciable change in effective distribution coefficients. This may be another reason for the compositional dependence of phenocryst–matrix estimates for distribution coefficients, since crystals in acid volcanics tend to contain many inclusions of accessory minerals.

With these qualifications in mind we may recognise two general types of behaviour. Elements with $D \ll 1$ are termed **incompatible** – they will be preferentially concentrated in the liquid phase during melting and crystallisation. In contrast, those with $D > 1$ are called **compatible** and these will be preferentially retained or extracted in the residual or crystallising solid phases respectively. Of course, if the equilibrium phase assemblage of the solids changes during magmatic evolution, as different minerals are consumed during melting or as new minerals appear on the liquidus during crystallisation, certain elements may change from compatible to incompatible, or vice versa. For example, Sr would be incompatible in ultramafic compositions but would become compatible in the presence of significant amounts of plagioclase, and K would be compatible if the mineral assemblage included much mica, amphibole or K-feldspar but would otherwise be incompatible. Thus it is sometimes helpful to specify elements which are incompatible with respect to the common minerals involved in the production and evolution of magmas in the mantle – olivine, pyroxenes, spinel and garnet – as **lithophile** trace elements (e.g. K, Rb, Sr, Ba, Zr, Th and the light rare earths). Some authors have used the more descriptive term 'large-ion lithophile' (or LIL) elements for this purpose, and others 'hygromagmatophile'.

Two of the principal factors controlling trace element entry into the lattices of major mineral phases are ionic radius and charge (see Mason 1966 for further reading). Nickel, for example, enters readily into olivine ($K_D \simeq 10$, see Table 14.1) because its ionic radius is similar to that of Mg^{2+} and its charge is the same. Table 14.2 shows in these terms why the incompatible elements have low K_Ds. Compared with most major elements their ionic radii are too large, or too small, or their valencies

Table 14.2 Ionic radius and charge of some important major and trace elements

	Major elements		*Minor and trace elements*
Ion	*Radius* * (Å)	*Ion*	*Radius* (Å)
Si^{4+}	0.48	P^{5+}	0.25
Al^{3+}	0.61	Rb^{+}	1.57
Fe^{2+}	0.69	Sr^{2+}	1.21
Mg^{2+}	0.80	Ba^{2+}	1.44
Ca^{2+}	1.08	Zr^{4+}	0.80
Na^{+}	1.10	Nb^{5+}	0.72
K^{+}	1.46	La^{3+}	1.13
Ti^{4+}	0.69	Y^{3+}	0.98

*Radii given are for 6-co-ordination, although some elements more commonly occur in other co-ordination states e.g. Si^{4+} in four-fold co-ordination has a radius of 0.34 Å. See Whittaker and Muntus (1970) for a compilation.

(charges) are too high. Several of these elements (for example, Zr and the minor element P) are relatively abundant and are commonly found forming their own minerals (zircon, apatite).

As will be shown in the following section, incompatible elements may become highly concentrated during melting and crystallisation processes. During such processes major elements may show comparatively little change so that rocks which arc broadly similar in major element composition may be widely variable in terms of incompatible elements. Rocks enriched in these elements are often recognised even in the absence of trace element data by their high potassium contents. Although they are not common, as already pointed out in Chapter 2 the effect is sometimes so extreme as to result in obvious mineralogical peculiarities (e.g. presence of leucite) and the necessity for a special system of nomenclature (see Table 2.1).

Melting and crystallisation models

Simple models have been developed to quantify the changes in trace element concentrations which occur during partial melting and fractional crystallisation. The reader is referred to Wood and Fraser (1976, Ch. 6) for the derivation of the basic equations – here we shall simply quote the results and discuss their significance. It is important to remember that they are based on highly idealised assumptions and at best can only be regarded as crude approximations to reality. They do, however, serve to illustrate the general principles and limiting effects of simple igneous processes.

Partial melting. *Batch melting.* The simplest model for the partial melting of a complex mineral assemblage is one in which the liquid remains at the site of melting and is in chemical equilibrium with the solid residue until mechanical conditions allow it to escape as a single 'batch' of primary magma. Under these circumstances the concentration of an element in the liquid (C_L) is related to that in the original unmelted source (C_0) by the expression:

$$\frac{C_L}{C_0} = \frac{1}{F + D - FD}$$

where F is the weight proportion of melt formed and D is the bulk distribution coefficient for the residual solids *at the moment when the melt is removed from the system*. The original source mineralogy and the changes which it undergoes during melting are thus immaterial except in so far as they determine, respectively, the value of C_0 and the mineralogical composition of the final residue. (Nevertheless, in order to define D it is necessary to calculate the proportions of residual minerals, so that some authors prefer to formulate the above expression in terms of their initial proportions and relative melting rates – see Wood and Fraser, 1976, p. 220 for details.) Clearly with increasing degrees of partial melting different minerals may be progressively consumed, causing discontinuous changes in the value of D. This is particularly important in the melting of possible mantle materials, since minor phases such as garnet, amphibole and plagioclase would in general complete their melting significantly earlier than olivine and orthopyroxene. Thus their effect of depressing enrichment in the melt of heavy rare earths, K and Sr respectively will only be apparent for small degrees of melting (generally less than about 10%). For simplicity, this complicating factor has been ignored in Figure 14.1a, which simply shows the ratio of concentrations in melt and source versus F for the case of constant D. It can be seen that enrichment of highly incompatible elements in the melt approaches $1/F$ and can reach very high values for small degrees of partial melting, though always with a maximum limit of $1/D$. This can be used to constrain the degree of melting in the production of primary magmas and incompatible trace element contents in the mantle. Thus, in the partial melting of the garnet peridotite for which we calculated $D_{Ce} = 0.012$, the maximum enrichment of Ce in the melt would be just over 80. Since alkali basalts generally have light rare earth contents greater than 100 times those of the average chondrite meteorite, these can only be produced directly from a mantle chemically analogous to meteorites if the degree of melting is extremely small. Using this type of argument Kay and Gast (1973) proposed that the mantle must have two to five times chondritic rare earth abundances and that alkali basalts represented about 2% partial melting, nephelinites about 1% and ultra-potassic basalts 0.5% or less. Other authors have considered that such small partial melts could not be efficiently extracted and have argued for enrichment by crystal fractionation of primary magmas, or for more lithophile-enriched mantle sources, in order to produce such rock types with larger degrees of melting.

Returning to Figure 14.1a, we see that the concentration in the melt relative to the source for compatible elements with $D > 1$ approximates to $1/D$ throughout much of the melting range, especially for very high values of D, and only increases towards the limit of unity for very high values of F. It should be noted, however, that special physical conditions are required in order to approach total melting – most systems would probably become mechanically unstable, due to the buoyancy of liquids

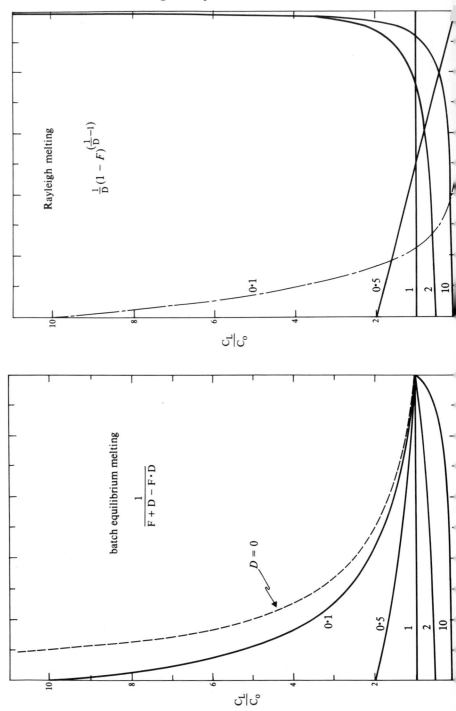

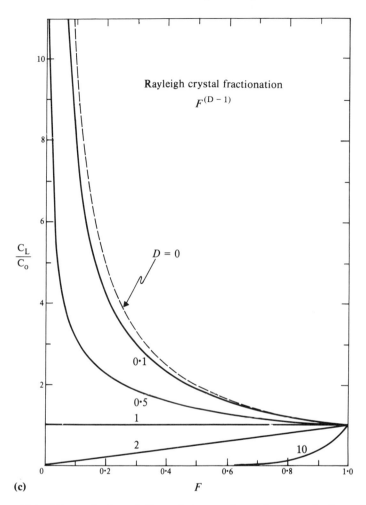

Rayleigh crystal fractionation

$$F^{(D - 1)}$$

$D = 0$

0·1

0·5

1

2

10

(c) F

Figure 14.1 Theoretical variation of trace element concentrations during melting and crystallisation, versus the melt fraction (F). Value marked on each curve is the bulk distribution coefficient (D). (a) Batch equilibrium melting. Concentration is relative to that in the source. The dotted curve shows the limiting enrichment case of $D = 0$. Note that all curves tend to an enrichment (or depletion) ratio of $1/D$ as F tends to zero, i.e. small partial melts. (b) Rayleigh melting. Concentration at any value of F is relative to that in the original (unmelted) source. Although an improbably extreme model this does demonstrate some characteristics of incremental or 'dynamic' batch melting, notably the rapid decrease of incompatible element concentrations in successive melts to values below those in the original source. (c) Rayleigh crystal fractionation. Concentration is relative to that in the unfractionated liquid ($F = 1$). Comparison with (a) and (b) clearly shows the rapid removal of highly compatible elements for small degrees of crystallisation and the unrestricted enrichment of incompatible elements for extreme crystallisation. (These properties only apply for a closed system – see text.)

less dense than the residuum, after about 30% melting. The concentration of elements (both compatible and incompatible) in the refractory residue (C_R) is always given by $D.C_L$ and it follows that even for small degrees of melting the residue becomes severely depleted in incompatible elements, whereas compatible elements remain close to their initial concentrations. These relationships enable us to estimate the average Ni and Cr contents of the Upper Mantle and primary magmas derived from it. Ultramafic nodules, e.g. in kimberlites, are undoubtedly samples of mantle material from around 100 km depth. There is some uncertainty concerning the extent to which mantle at this depth is 'barren', i.e. has already yielded partial melts, or whether it has been artificially enriched in lithophile elements, e.g. as a result of volatile metasomatism or during incorporation of xenoliths into the host magma. Regardless of these points, the average contents of Ni and Cr (ca. 2400 ppm and 2700 ppm respectively) must approximate those in unmelted ('fertile') mantle. If we assume $D_{Ni} = 7$ and $D_{Cr} = 3$ for the residual mineralogies, then primary magmas formed by up to 50% partial melting should contain, within a factor of 2 say, Ni = $2400/7 = 350$ ppm and Cr = $2700/3 = 900$ ppm. Silica-undersaturated volcanic rocks generally have much higher contents of these elements than tholeiites and andesites, so that alkali basalts with ~200 ppm Ni can on these grounds be regarded as a close approach to direct mantle melts.

Small degrees of partial melting can also cause significant changes in the ratio of two incompatible elements with different Ds (e.g. K/Ba, Ce/Sm and Ce/Yb). At constant D the limiting enrichment factor for one element relative to another is D_2/D_1. Returning to our hypothetical garnet peridotite as an example, if D_{Yb}/D_{Ce} is 20 then for 1% melting the Ce/Yb ratio of the liquid would be twelve times that in the source and for 10% melting it would be three times. The effect is less marked for the smallest values of D. Thus for the same assemblage $D_{Rb} = 0.001$ and $D_{Sr} = 0.01$, but the increase in Rb/Sr is only a factor of 1.8 for 1% melting and 1.1 for 10% melting. K and Rb, with very similar K_D values for most minerals, are scarcely fractionated from each other at all except in the presence of amphibole.

Incremental batch melting. A second type of melting model involves repeated extraction of melts, each formed by an equilibrium process, from the same source (sometimes referred to as 'fractional melting'). For defined melting intervals the above formula can be applied provided that the initial composition for each stage is calculated as the appropriate residual concentration from the previous stage. For example, in a two-stage melting process with melt fractions F_1 and F_2 respectively, the concentration of a trace element in the second liquid is given by:

$$\frac{C_{L2}}{C_0} = \frac{D_1}{(F_1 + D_1 - F_1 D_1)(F_2 + D_2 - F_2 D_2)}$$

and that in the second residue by $C_{R2} = D_2 C_{L2}$.

The rate of depletion of incompatible elements in successive residues and the liquids derived from them is more rapid than in single batch equilibrium melting, and later liquids may actually have incompatible element concentrations and incompatible/compatible element *ratios* lower than in the original source material.

This is important in the assessment of the petrogenesis of Mid-Ocean Ridge tholeiites. Such rocks (and some ophiolites) are characteristically depleted in lithophile trace elements (Gast 1968; Schilling 1975). In particular, their light rare earth contents (about three to eight times that in the average chondrite meteorite) are almost always less than their heavy rare earth contents (about seven to twelve times chondrite). Since $D_{Ce} < D_{Yb}$, the Ce/Yb ratio in the source region must be even lower than in these rocks, and obviously cannot be reconciled with the idea of a mantle whose chemistry is broadly similar to meteorites. Thus Gast (1968) suggested that the source of oceanic tholeiites (perhaps the low-velocity zone) had previously been depleted by the loss of a small partial melt. To test this hypothesis we can calculate the Ce and Yb enrichments in the second melt from the above equation. For the first stage of melting we can take our previous data for garnet peridotite and $F_1 = 0.01$ (1% melt), whereas for the second stage we shall recalculate D_{Ce} and D_{Yb} on a garnet-free basis, partly because garnet is consumed more rapidly on extensive partial melting and also because it is not stable at the depth of the low velocity zone. The second stage values of D_{Ce} and D_{Yb} become 0.012 (as before) and 0.045 respectively. Substituting these and $F_2 = 0.15$ (15% melt), we get the enrichments of Ce and Yb in the second melt, relative to the *original* source concentration, as 3.4 and 5.0 respectively. This shows that the rare earth data are consistent with Gast's hypothesis if the original source had unfractionated rare earth abundances about twice those in chondrite meteorites and if the first (very small) partial melt were removed by melting at a depth where garnet was stable. All the other conditions specified would have to be met and it is worth noting that some authors believe that the level of K, Rb and Ba abundances in Mid-Ocean Ridge tholeiites militates against this simple conclusion (e.g. Hanson 1977).

Rayleigh melting. All Rayleigh processes are named after the famous physicist who formulated an expression to describe mass fractionation effects during distillation. In the present context we simply mean melting in which an infinitesimally small amount of liquid is formed in equilibrium with the residue and then removed from the system. Although this may be physically unreasonable as a model for magma production, it serves to illustrate the limiting extremes of changes in the trace element contents of melts and residues where the melt migrates rapidly in comparison with diffusion rates in the solid phases. It could be pertinent to mantle metasomatism and highly fluid CO_2-rich melts, for example.

Since distribution coefficients from earlier melt residues enter the calculation (as in the previous example), minerals originally present can influence the final melt composition even though they may have been totally consumed. Because of this, and the fact that D may change continuously during melting, proper formulation of the concentration relationships is complex. In this case it is more convenient to work with Ds for the starting composition and the relative rates of melting for each phase. The reader is referred to Wood and Fraser (1976, pp. 220–222) for the exposition of this approach. Here, for simplicity, we shall again simply give the result for constant D:

$$\frac{C_L}{C_0} = \frac{1}{D}(1 - F)^{\left(\frac{1}{D} - 1\right)}$$

The changes in concentrations are more extreme than in batch melting (see Fig. 14.1b). The limiting enrichment of incompatible elements in the melt is still $1/D$, but that of compatible elements tends to infinity as F approaches 1. The melts are still in equilibrium with the residue (if only momentarily) so that $C_R = DC_L$ once more. It follows that incompatible elements are yet more efficiently purged from the source than in incremental batch melting.

A model sometimes used in the literature is Rayleigh melting in which the increments are collected into a common reservoir where they are perfectly mixed. This is in practice numerically indistinguishable from batch melting except for the behaviour of compatible elements at very large degrees of melting.

Zone refining. Once a body of magma begins to move it comes into contact with solid material with which it may interact. Where the solid is of completely different composition from the magma source region we would class any such interaction as contamination, but initially the magma would have to pass through higher levels of the source region itself which may not have undergone prior melting. The extent to which interaction may proceed is arguable, but a limiting case is represented by complete equilibration. This occurs in true zone refining (actually based on analogy with an industrial process for removing unwanted impurities from metals) by total melting in front of the advancing magma and deposition of crystalline cumulates behind it. Since compatible elements are essentially buffered at a concentration of $1/D$ during re-equilibration (see Fig. 14.1a) they are relatively unaffected by this process. Incompatible elements, however, are continuously enriched in the magma according to the equation:

$$\frac{C_L}{C_0} = \frac{1}{D} - \left[\frac{1}{D} - 1\right]e^{-nD}$$

when n is the number of equivalent volumes processed by the liquid. It should be noted that this enrichment again has a limiting value of $1/D$ though this may not be achieved until the original liquid has passed through tens or hundreds of times its own volume of solid. For a magma to advance in this way requires the continual supply of sufficient superheat to assimilate solid material and its geological feasibility must be doubted. On the other hand some authors have contested that the same effect could be accomplished without assimilation, as a result of diffusion of incompatible elements into a magma from the walls of the conduits through which it ascends (e.g. Green & Ringwood 1967). This could be important for the volatile elements such as K and Rb but it is difficult to envisage high diffusion rates for, e.g. the rare earth elements. It could also be argued that the immediate environs of the magma conduit would rapidly be purged of incompatible elements and that this would itself limit further operation of the process.

Crystal fractionation. We have already implied (Ch. 9) that the major element compositions of erupted basalts show that they are in many cases not likely to have been in equilibrium with the supposed mineralogy of their mantle source regions but that they must have undergone considerable modification during ascent, by removal of crystalline solids. We now turn to the general problem of quantifying the effects of this on trace elements. Since it is the evolution of the *magma* with which we are most concerned, we shall redefine C_0 as the *initial* concentration of an element in the primary magma (i.e. the same as C_L during the partial melting process which produced it).

If *all* the crystalline products were to remain in chemical equilibrium with the magma this would simply be a reversal of batch melting and the equation given above for this would apply (F now being the proportion of original liquid remaining). However, this seems an unlikely situation – the growth of zoned crystals would probably proceed faster than they could re-equilibrate with the changing liquid composition and gravity settling would remove the denser phases as cumulates as the liquid advanced. Thus the Rayleigh law probably provides a better, if extreme, model for crystal fractionation just as an equilibrium model is generally most appropriate for partial melting.

The simplest case to consider is a closed system in which a body of magma is isolated in a magma chamber and undergoes continuous crystal fractionation. The solution differs from that for Rayleigh melting in that this time the liquid is a uniform reservoir and the crystals are removed as fast as they are formed. This gives

$$\frac{C_L}{C_0} = F^{(D-1)}$$

Moreover the total residue is never in equilibrium with the liquid (except for the first infinitesimal increment of crystallisation) so that the expression for C_R must be derived from first principles, yielding

$$\frac{C_R}{C_0} = (1 - F)^{(D-1)}$$

This solution for the liquid composition is illustrated in Figure 14.1c. The enrichment of incompatible elements closely resembles that of the equilibrium solution until more than about 75% of the magma has crystallised, when there is a rapid approach to $1/F$ – this time without the restriction of a theoretical maximum limit. Very high degrees of crystal fractionation are, however, unlikely to be accompanied by efficient separation of melt from crystals, so that this portion of the diagram may only be practically applicable to the composition of trapped interstitial liquid (mesostasis) during the growth of zoned crystals or to liquids expelled from the cumulate pile by filter pressing. The case for highly compatible elements differs significantly from the equilibrium model, these being removed even more rapidly in the early-formed solids so that their concentration in the liquid soon falls below $1/D$. Thus if we take $D_{Ni} = 7$, a reasonable value for crystallisation of an olivine–orthopyroxene–spinel–clinopyroxene assemblage, the initial Ni concentration will fall by about one-half for 10% crystallisation and over 98% for 50% crystallisation. Under such conditions, even cumulates which crystallise late in the sequence will be strongly depleted in compatible elements. An example is given by the Ni contents of olivines from the Skaergaard intrusion (see Wager & Brown 1968, p. 179), which decrease from 2000 ppm in the Marginal Border Group to 400 ppm in the Lower Zone cumulates and to 5 ppm or less in the Upper Zone (equilibrium concentrations in the magma would be 10 times less). Analogous behaviour is exhibited by Cr in the pyroxenes and V in ilmenite and magnetite.

On the other hand, closed system crystal fractionation is far less effective than equilibrium partial melting for concentrating one incompatible element relative to another. Taking $D_{Ce} = 0.10$ and $D_{Yb} = 0.15$, fairly reasonable values for the average cumulate composition of the Skaergaard intrusion, the relative enrichment given by $(F^{(D_{Ce} - D_{Yb})})$ is only 1.1 after 85% crystallisation and 1.2 after 97% crystallisation. As indicated previously though, in actual cases the whole character of the magma would have changed so much during this degree of solidification that there would be considerable uncertainty over D values. K/Rb and K/Ba ratios are likely to be affected even less than Ce/Yb, but Rb/Sr ratios will increase sharply once the crystallisation of plagioclase results in Sr becoming compatible. The late (filter-pressed) granophyre magmas at Skaergaard had Rb/Sr ratios more than fifty times that in the initial liquid.

More complex models

The models investigated above are useful for illustrating general principles but cannot be expected to describe real geological processes adequately. In particular, neither partial melting of up-welling mantle material, nor high level fractional crystallisation are likely to occur entirely within closed systems. As examples of more complex models which attempt to accommodate open system behaviour, the reader is referred to Langmuir *et al.* (1977a) and O'Hara (1977). The first describes a continuous partial melting process in which a proportion of the melt at any stage is retained mechanically trapped within the source material and mixes with the liquid formed next (**dynamic melting**). Broadly speaking the result is similar to that described above as incremental melting, especially as regards incompatible element *ratios* but without the severe depletion in their contents in later liquids associated with the closed system case. O'Hara's model is for open system crystal fractionation, in which periodically a small proportion of the liquid is drawn off as a volcanic magma while the magma chamber is replenished with further unfractionated liquid. The predicted effects have some features in common with zone refining in that high degrees of enrichment of incompatible elements may be achieved without recourse to improbably high degrees of crystallisation, and significant variation in the ratios of such elements could result (but see Pankhurst 1977a for some notes on limitations in this regard). Conversely the concentration of compatible elements is maintained at a reasonably high value by the fresh addition of primary magma. In some respects this is a quantification of the model first suggested by Brown (1956) to explain the cyclic repetition of cryptic layering in the Tertiary intrusion of Rhum, Inner Hebrides.

Application of trace elements to petrogenesis

Partial melting and crystal fractionation. So far we have seen how trace element data may be used to reconstruct and characterise magma compositions and how the effects of simple models of partial melting and crystal fractionation may be recognised in individual cases. We now turn to the more general problem of interpreting data for a whole suite of rocks and the extent to which trace elements can constrain possible petrogenetic relationships.

The first approach to such a body of data is to look for linked variation in trace elements corresponding to other petrological or geochemical variability. For example, either sympathetic or antipathetic covariation of trace element contents with differentiation indices (such as SiO_2 or Fe/Mg) or with one another. Once such linked variations are found then the obvious question to ask is 'are they consistent with the rocks being **cogenetic** – related by crystal fractionation to a common parent magma or by partial melting to a common source?' To answer this question confidently requires recourse to a much wider range of petrological information such as field relations, mineralogical and petrographic variation or major element chemistry and even isotope geochemistry (see Ch. 15). For the moment we shall assume that any such information is consistent with the hypothesis but not conclusive, so that we wish to use the trace element data as a final test. Clearly it is necessary that trace element contents and ratios should vary consistently throughout the set, the enrichment in incompatible elements and the depletion in compatible ones increasing uniformly with smaller degrees of melting or more advanced crystal fractionation. The most highly incompatible elements such as Rb, Nb, Ta and Th should increase more rapidly than rare earths and Sr (and light rare earths more rapidly than heavy). Of course the mineral phases likely to be involved must be known in order to assess *how* incompatible different elements may be and distribution coefficients for the crystal–liquid equilibrium should be reasonably well known, at least for one trace element relative to another. Our simple models would then lead us to expect that the overall range of an incompatible element contents should not greatly exceed $1/D$ and that of incompatible element ratios D_2/D_1. These restrictions, of course, concern the evolution of *magma* compositions and so can only be applied directly to volcanic rocks which are believed to represent liquids. This, and the uncertainty of appropriate distribution coefficients, is at present a limitation to the use of trace element models in the petrogenesis of granites. An example of progressive trace element variation in a group of cogenetic volcanic rocks from the island of Réunion in the Indian Ocean is illustrated in Figure 14.2.

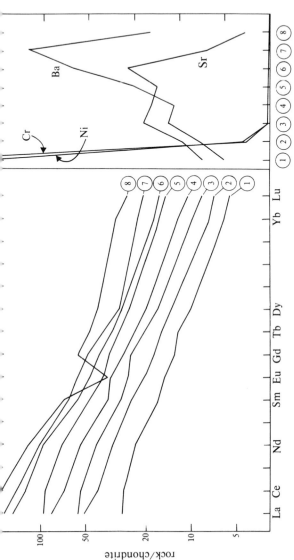

Figure 14.2   Chondrite-normalised rare earth patterns and trace element variation for volcanic rocks from the island of Réunion, Indian Ocean (after Zielinski, 1975). The numbered samples range from olivine basalt (1) through hawaiite and mugearite to trachyte (7 & 8). Overall there is a seven-fold increase in REE abundances through the series, with very little change in the ratio of light-to-heavy REE. Sample 1 has higher Ni and Cr than expected from a primary magma (see text), presumably due to cumulus olivine and pyroxene, so that the true variation in incompatible elements for *liquid* compositions may be only a factor of four. This could be explained by partial melting of a garnet-free peridotite source (residual garnet would result in relatively constant Yb and Lu concentrations – even decreasing in the smallest partial melts – see Shimizu and Arculus (1975) for an example). On the other hand, the fall in Ni and Cr to vanishingly low concentrations suggests that crystal fractionation may be a more important control. Zielinski (1975) demonstrated consistency with a model involving 70–80% fractional crystallisation of a parent magma such as sample 2. The negative Eu-anomaly and fall in Sr content for the trachytes indicate the removal of plagioclase (and possibly apatite) in the final stages of this model. The fall in Ba in sample 8 also suggests the involvement of K-feldspar.

Often we should like to go further than this and distinguish *between* crystal fractionation and differential partial melting as the major factor. This is a far more difficult problem. As our closed system Rayleigh solution indicated, the anticipated effect of crystal fractionation would be extreme depletion of compatible elements in the later liquids, but very little change in incompatible element ratios. The Réunion data fit this expectation very well, in particular the early decrease in Ni and Cr (although the initial values are high enough to suggest that the most primitive rock contains cumulus olivine and spinel or pyroxene) and the subparallel rare earth element patterns. Thus Zielinski (1975) was able to account for these variations by crystal fractionation of the observed phenocrysts in the rocks (mainly olivine, clinopyroxene, plagioclase and titanomagnetite) from parent magmas resembling the second most primitive rock in the series. The fall in Sr and Ba in the trachytes and the simultaneous appearance of a negative Eu-anomaly (see below) reflect the dominance of plagioclase in the late stages of differentiation.

If trace element data are sufficiently complete, it may be possible to use co-variation plots to infer bulk distribution coefficients for any element by reference to one known to be highly incompatible. A mathematical model treatment of Rayleigh crystal fractionation based on this approach and methods for deducing parent magma compositions therefrom has been given by Allegre *et al.* (1977). However, considering that real processes will only at best correspond approximately to mathematical models, we should in general expect that differential partial melting and crystal fractionation will offer complementary solutions to any particular problem. For Réunion, Zielinski (1975) preferred the latter on the basis of well-known field observations, but it would be possible to model most aspects of the trace element data in isolation according to a model in which the more evolved rocks were partial melts of older basalts, for example. On the other hand, trace element geochemistry can be an extremely simple and powerful tool for distinguishing between high-pressure and low-pressure environments where distinctive mineralogies are involved. Thus we are much more likely to get a definite answer if we rephrase the question to 'are the observed variations caused by deep processes, perhaps close to the magma source region, or by high level modification or melting?'

In the deep continental crust and below about 100 km in the mantle, garnet and clinopyroxene will be expected as important residual phases after small degrees of melting (garnet would persist up to 15–20% melting of a typical garnet–peridotite). This results in bulk distribution coefficients approaching unity for heavy rare earth elements and Y (which behaves geochemically very much like Yb). These elements are consequently buffered at a low concentration in the melt, whereas light rare earths are enriched along with other incompatible elements. As the degree of melting increases above about 10%, the proportion of residual garnet falls and heavy rare earths and Y should become progressively more incompatible. Figure 14.3 shows the predicted type of co-variation for K, Sr and Y in a series of alkali basalts, hawaiites and mugearites from Mull, North-West Scotland, which has been ascribed to differential partial melting of a garnet–peridotite source. (Curiously, the same classification in terms of normative mineralogy was applied to the Réunion lavas – a warning not to be too ready to attach petrogenetic significance to rock series.)

At depths of less than 40 km, plagioclase can be expected to be a stable phase and will exert significant control on trace element patterns because its distribution coefficients for rare earths, Sr and Ba differ markedly from those of other common igneous minerals. The tendency to incorporate light rare earths rather than heavy rare earths is generally counteracted in practice by the simultaneous crystallisation of pyroxenes or amphibole but the most distinctive effect of feldspars is their preference for divalent Eu compared to the remaining rare earths. K_Ds for Eu depend strongly on oxygen fugacity, but in most natural silicate liquids there is an appreciable Eu^{2+}/Eu^{3+} ratio and this increases at low temperatures. Thus differentiation involving plagioclase results in depletion of Sr (and to a lesser extent Ba, which only becomes truly compatible once K-feldspar crystallises) and a progressively negative Eu-anomaly (cf. Fig. 14.2) together with enrichment of incompatible elements. Another example of this type of low pressure modification of magmas is illustrated for an island-arc tholeiite suite in Figure 14.4. Conversely, high Sr contents and positive Eu-anomalies are found in rocks formed by plagioclase accumulation, especially feldspathic gabbros, but some large stratiform anorthosites which occur in Archaean shield areas are a curious exception to this (e.g. O'Nions & Pankhurst 1974a). Small positive Eu-anomalies found in some basaltic rocks are more probably primary features resulting from exclusion of Eu^{2+} by residual clinopyroxene in the source region (Leeman 1976).

In many instances, of course, partial melting will be a high-pressure phenomenon and crystal fractionation a low-pressure one; but the reader should guard against the automatic assumption that this is so. Firstly, crystal fractionation of eclogitic cumulates at mantle depths could lead to strong fractionation of rare earth patterns and incompatible element ratios. It must also be remembered that very few of the magmas which reach the surface are likely to be primary in the true sense (i.e. in full chemical equilibrium with mantle compositions), having lost at least 10% olivine (O'Hara 1968). This will often only result in slight enrichment of incompatible elements and no additional change in their inter-element ratios, so that it is quite feasible for the products of a single volcano to exhibit compatible element depletion and high-pressure melting rare earth patterns simultaneously. The more alkaline rocks of the southern part of the Lesser Antilles island arc are a good example of this apparent ambiguity (Shimizu & Arculus 1975; Brown *et al.* 1977). On the other

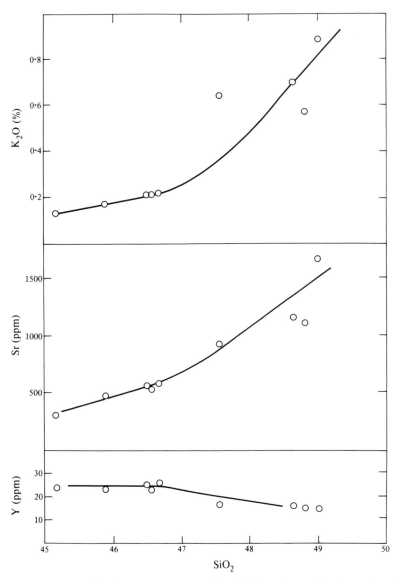

Figure 14.3   Co-variation of K, Sr and Y with SiO₂ for selected Tertiary volcanics (alkali basalt, hawaiite and mugearite) from the Isle of Mull, N.W. Scotland. The fall in Y coinciding with increasing K, Sr and SiO₂ can only be explained by the separation of progressively greater amounts of garnet. Since garnet is not a phenocryst phase in these rocks, generation by differential partial melting of garnet peridotite is probably the process most responsible for their chemistry. The behaviour of Sr as an incompatible element (paralleling K) also rules out significant low-pressure fractionation since plagioclase *is* a phenocryst phase. Data from Beckinsale *et al.* (1978).

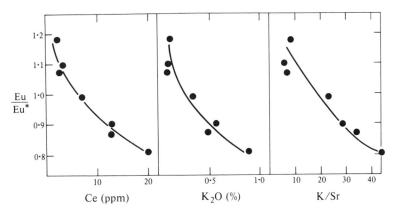

Figure 14.4 Co-variation of Eu/Eu*, Ce, K and K/Sr for arc tholeiites from the South Sandwich Is. (Eu* is the expected Eu concentration obtained by interpolating between Sm and Gd as though there were no anomaly – cf. Figure 14.2 – so that Eu/Eu* is less than 1 for negative anomalies.) The development of increasing negative anomalies in these rocks with increasing incompatible elements must be due to the progressive separation of plagioclase at low pressures. The relative compatibility of Sr is indicated by the increase in K/Sr ratio, Sr contents being fairly constant but somewhat erratic. Data from Hawkesworth *et al.* (1977).

hand, the separation of plagioclase need not be evidence of crystal fractionation from an initially more basic magma. Thus the low-pressure trace element characteristics of many granites (e.g. those of Skye and Mull, Thorpe *et al.* 1977) might equally be interpreted in terms of partial melting of feldspathic material within the crust. Each case should be treated on its merits in the light of as many additional data as possible, as indicated at the outset of this section.

Source region inhomogeneity. In some cases it is found that trace element data for a suite of obviously related rocks cannot be fitted by any single model of crystal fractionation from a common parent magma or differential partial melting of a homogeneous source, either because the range of incompatible element contents or ratios exceeds theoretical limits or else because major element or mineralogical constraints rule out the inferred model. Naturally the behaviour of compatible trace elements is important in establishing that the range is not artificially extended by cumulus minerals (as in the case of the entire suite shown in Fig. 14.2). Once all other possibilities are exhausted, we have to conclude that at least some of the observed variation is a primary feature of the parent magmas, which could be ascribed either to contamination or to inheritance from the source region.

The effects of contamination are considered in the next section but there are now well established grounds for interpreting the overall variation in the trace element geochemistry of oceanic basalts as due to in-homogeneous distribution in the Upper Mantle. As a specific example, Figure 14.5 shows rare earth patterns for volcanic rocks from Iceland. These range from low abundance patterns with light rare earth depletion, characteristic of abyssal basalts from mid-ocean ridges, through flat or

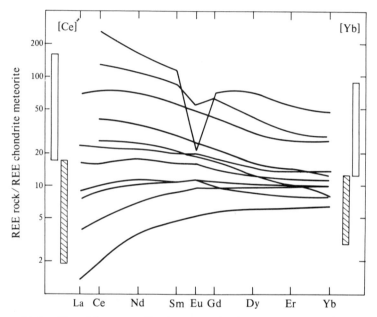

Figure 14.5 Chondrite-normalised rare earth element concentrations in some Recent volcanic rocks from Iceland (after O'Nions and Gronvold 1973 and O'Nions *et al.* 1976). The generally smooth changes in the shape of the patterns as the total concentration increases is consistent with at least some of these rocks being related by partial melting of a common source or crystal fractionation. The vertical bars at the sides of the diagram show the maximum ranges in enrichment feasible for 1–30% batch melting (shaded bar) and 0–90% Rayleigh fractionation (open bar) for the elements Ce and Yb, assuming $D = 0.02$ and 0.07 respectively. These values correspond to solid phases in equilibrium with each liquid corresponding to about 60% olivine, 22% orthopyroxene and 18% clinopyroxene. Distribution coefficients for these minerals were taken from Table 14.1, although these may be underestimated for the case of fractional crystallisation, where phenocryst–matrix values are several times higher. Nevertheless, it is still apparent that the total range of Ce concentrations ascribable to either process individually cannot encompass all the observed values. In combination they could just cover the Ce variation, but not the range of Ce/Yb ratios, i.e. Yb would have to show much more sympathetic variation than it does. It is concluded that the source region of Icelandic volcanics is heterogeneous, as also indicated by isotope data (see Ch. 15).

slightly light-enriched patterns for more Fe-rich tholeiites, to strongly fractionated patterns for alkali olivine basalts, and high abundance patterns with negative Eu-anomalies for low-K andesites, dacites and rhyolites. Ce abundances vary through a factor of 125 and Yb a factor of 8. As indicated on the diagram, this cannot be explained either by 1–30 melting of peridotite or by 90% crystal fractionation of a basalt, nor even by a combination of these two processes. The more evolved rocks could be differentiates from a light rare earth-enriched basalt, or crustal remelts of such rocks (O'Nions & Gronvold 1973). But when the range of major element chemistry and Sr-isotope compositions of the basalts alone are considered it becomes clear that it is necessary to postulate a range of rare earth depletion in the source (Schilling 1973; O'Nions *et al.* 1976).

It is obvious that melting and crystallisation within the mantle must result in inhomogeneous distribution of trace elements. This could arise during a discrete igneous cycle (e.g. incremental melting), or over very long periods of time. Neither major nor trace element chemistry can, however, tell us anything about the time factor – for that we are entirely dependent on isotope systematics (see Ch. 15).

Contamination and mixing. The idea of a magma of deep origins changing its chemical character as a result of reacting with crustal material is one of the oldest in igneous petrology. Specific contamination hypotheses have always evoked strong opposition and few have survived the test of time. However, modern geochemistry, which has proved the downfall of many past ideas, has in turn produced new contamination hypotheses.

Certain trace elements and ratios may be strongly affected by weathering processes and the treatment of this book naturally refers throughout to the study of fresh rocks. But other forms of modification to low temperature and pressure equilibria can apply during the emplacement of magmas (indeed, crystal fractionation may be viewed in this way, but here we mean processes involving external media, e.g. hydrothermal interaction). The clearest evidence for such effects usually comes from isotope geochemistry but one specific case is particularly relevant to trace element studies – the reaction between oceanic magmas and seawater. The elements most affected, becoming enriched in the altered igneous rocks, are those which are most mobile (i.e. soluble) during weathering and so become relatively concentrated in seawater, i.e. Na, K, Rb, Cs, Sr, Ba and P. The effect is greater for the larger ionic species so that K/Rb, Rb/Cs and Sr/Ba ratios are generally lowered with alteration (Hart 1971). On the other hand Ti, Zr, Nb, Y and the heavy rare earths are extremely resistant to alteration (and are relatively depleted in seawater anyway) so that petrogenetic arguments and classifications based on these elements may still be valid even for altered and metamorphosed basalts (Pearce & Cann 1973;

Winchester & Floyd 1976). There is some uncertainty over the behaviour of the light rare earths, Hellman and Henderson (1977) claiming that these are enriched relative to heavy rare earths during spilitisation, whereas other authors have found no such effect in abyssal basalts and ophiolites which show other geochemical and isotopic evidence for seawater interaction (Kay & Senechal 1976; Smewing & Potts 1976; O'Nions & Pankhurst 1976).

Contamination, in the more common usage of the word, means bulk assimilation of crustal rocks by a magma and this should be amenable to a mixing model. Other forms of mixing which can be so treated are those involving two magmas (hybridisation) or two or more solids (which would also describe source region heterogeneity). The object is to test for compatibility with a mixing hypothesis and to identify the end-members. Tests may be based on two-dimensional plots of either element concentrations or ratios (inter-element or isotope ratios). Trace elements are generally preferred for this purpose since they frequently show a greater range of variation than major elements. When one element is plotted against another, mixing data will define a straight line joining the two end-member compositions, and samples representing different proportions of these should plot in the same order regardless of the elements chosen. However, other igneous processes such as partial melting and fractional crystallisation can also result in approximately linear plots between pairs of trace elements, so that this is not generally a conclusive test. Plotting ratios is more diagnostic, but these usually show less variation and are sometimes subject to larger analytical errors (though not always so). For plots of ratios, a straight line is no longer the general solution. Langmuir *et al.* (1977b) have shown that all mixing models can be described by a single equation which is hyperbolic in form:

$$Ax + Bxy + Cy + D = 0$$

where x and y are the plotted parameters. If we let $x = p/q$ (an inter-element or isotope ratio) and $y = r/s$, the coefficients of the hyperbola are given by:

$$A = q_1 \cdot s_2 \cdot y_2 - q_2 \cdot s_1 \cdot y_1$$

$$B = q_2 \cdot s_1 - q_1 \cdot s_2$$

$$C = q_1 \cdot s_2 \cdot x_1 - q_2 \cdot s_1 \cdot x_2$$

$$D = q_2 \cdot s_1 \cdot x_2 \cdot y_1 - q_1 \cdot s_2 \cdot x_1 \cdot y_2$$

where the subscripts refer to two points on the curve (for example, but *not* only, the two end-member compositions). Specific simplifications follow by substituting $q_1 = q_2 = 1$ when x is an element rather than a ratio and $s_1 = s_2 = 1$ when y is an element. A straight line results when $B = 0$, which is generally only true for an element–element plot.

In practice we may not know the composition of the end-members; indeed this may be something which we wish to infer from the data. The procedure is to plot suitable parameters and assess by eye whether a hyperbolic curve may provide a reasonable fit (if this is obviously not so for several pairs of parameters, then a mixing model of any sort can be rejected). Taking two well-separated and representative points on the diagram, the coefficients are determined as above and the hyperbola plotted (Fig. 14.6). Provided a reasonable fit is still indicated, some

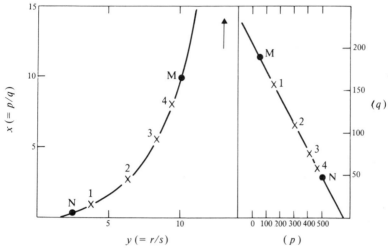

Figure 14.6 The left-hand side of the figure shows a hypothetical x–y (ratio–ratio) plot for a mixing series with actual end-members N and M as marked – e.g. x could be Rb/Sr and y could be Ce/Yb. Knowledge of the end-member compositions is not essential in principle since the curve is uniquely defined by any two data points, although this would not be precisely so in practice owing to experimental error and non-ideal behaviour of natural systems. Curvature of the hyperbola can be in the opposite sense, according to the relative values of q (Sr) and s (Yb) in N and M. In this example, limiting (minimum) values for the ratios in end-member N are given by the intersection of the hyperbola at $x = 0$, $y = 1.6$. For M, x is indeterminate and a long extrapolation to the asymptote gives a maximum value of 14 for y – a rather poor approximation to the true value of 10. The right-hand plot shows p versus q for the same hypothetical case. Here a straight line relationship results, as in all element–element mixing plots. The low value of q in end-member M serves to give a close constraint on the maximum limit of p by extrapolation to zero. Similar constraints would be found for r and s by plotting these against q. (The parentheses are intended to indicate that any of the element concentrations p, q, r, and s can be used in this way according to convenience.)

constraints on possible end-member parameters may result from the approach of the hyperbola to intersection points with the axes or to asymptotic values. However, one a mixing model has been justified, actual element concentrations in end-members are more easily obtained from linear element–element plots involving one which tends to zero in one end-member, e.g. Rb in basalts.

Proper testing of mixing models should be based on several such plots chosen to complement each other (see Langmuir *et al.* 1977b for further details). In practice very few convincing mixing models based on trace element data have been established, with the exception of a few which depend heavily on isotope ratio plots (e.g. Vollmer 1976; Faure *et al.* 1974). Differentiation between liquid and solid end-members is theoretically possible, since compatible elements will not behave systematically if a solid end-member is present, but liquid–liquid mixing will produce hyperbolic relationships for all elements. One final point is that even satisfactory plots can only constitute admission of a feasible mixing model, not absolute proof. Interpretation must always depend on both geochemical and *geological* identification of possible end-members. A clear example of this is given by Langmuir *et al.* (1977b) who show that solid–solid mixing is a possibility for the trace element variation of Icelandic and Reykjanes Ridge basalts, a conclusion which in effect simply confirms previous ideas of source region heterogeneity.

Note

1. B. O. Mysen (*Geochim. Cosmochim. Acta*, 42, 1253–63, 1978) gives detailed experimental evidence for REE partitioning under mantle conditions. He shows that Henry's Law breaks down at much lower concentration limits than previously supposed, e.g. approximately 100 times chondritic abundance for Sm in the liquid. Above these levels, K_Ds increase markedly with REE concentration even at constant P.T. and bulk composition, although the ratio of K_Ds for any pair of incompatible elements remains more constant.

Exercises

1. Trace element analyses of minerals separated from an olivine clinopyroxenite (80% olivine, 20% Cpx) nodule and of the host basalt in which it occurs yield the following results:

	Ce_N	Yb_N	K	Rb	Sr
Cpx	1.9	3.9	15	0.03	21
Olivine	0.1	0.1	25	0.15	0.9
Basalt	21	12	3900	10	305

(Ce and Yb are chondrite-normalised, others in ppm.) To what extent do the data support the idea that the nodule is a xenolith derived from the source region of the basalt? On this basis estimate the degree of partial melting, assuming overall rare earth abundances in the source of one to two times those in chondrite meteorites, and hence constrain its K, Rb and Sr contents.

2. The following trace element data refer to two basalts typical of the range observed in tholeiites from the Neovolcanic Zones of Iceland (see Fig. 14.5). As an alternative to the idea of mantle inhomogeneity, it is suggested that the more evolved basalt could be derived from the other via two stages of crystal fractionation of its parent magma, the first being eclogite fractionation at depth during which the light rare earths are enriched relative to the heavy rare earths, and the second stage being a normal high-level process resulting in enrichment of total rare earth contents to the observed level. Estimate the amount of crystallisation necessary in each stage to account for the rare earth concentration and then test the model against the remaining data. (N.B. assume the eclogite consists of 50% clinopyroxene, 50% garnet and that the high-level cumulates are 50% olivine, 20% clinopyroxene and 30% plagioclase.)

	Ce_N	Yb_N	K	Rb	Sr	Ni	Eu/Eu*
Basalt 1	2.5	7.5	300	0.5	50	100	1.0
Basalt 2	18	12	900	1.5	170	75	1.0

3. Samples from an eroded volcanic cone range from basalt (both tholeiitic and alkaline) to trachyte, with the following trace element analyses.

Sample	Rb	Sr	Y	Zr	(ppm)
1	15	500	16	152	
2	25	470	17	154	
3	50	425	20	160	
4	75	365	24	170	
5	125	250	30	180	

Investigate the variation of these elements in terms of a mixing model, attempting to define (and interpret) possible end-member compositions.

15

The use of isotopes in petrology

Isotope geochemistry must now be regarded as an essential branch of modern igneous petrology. Natural processes lead to variations in the isotopic composition of certain elements, which may become characteristic of distinct geochemical environments. The importance of isotope variations in petrogenetic studies is that they frequently survive the chemical fractionation which accompanies the formation and evolution of magmas. Thus isotopes may often be interpreted as geochemical tracers of magma origins even where trace element concentrations and ratios have been extensively modified.

We may divide isotope studies into two groups, one based on variations caused by the decay of radioactive elements, the other based on variations caused by mass fractionation in chemical reactions. These are conventionally referred to as **radiogenic** and **stable** isotope studies respectively, although the radiogenic isotopes are of course also stable – it is only their radioactive parent isotopes which are unstable. It is now known that a third major cause of isotope variations exists, which affects both radiogenic and stable isotopes – that is, primary heterogeneity in the solar system. This is recognised in extra-terrestrial material, especially certain phases of meteorites, but so far there is no definite evidence of any primordial isotopic heterogeneity within the Earth (or the Moon). Nevertheless, it should be remembered that many bulk geochemical and isotopic characteristics of the Earth are only inferred from analogy with meteorites.

Radiogenic isotope variations

The naturally occurring, long-lived radioactive decay schemes (summarised in Table 15.1) are mostly well known from their application to geochronology. Of these, the K–Ar scheme is of little

Table 15.1 Radioactive decay schemes used in petrology

Radioactive parent isotope	Natural abundance (atoms %)	Half-life (years)	Decay type	Decay constant λ ($\times 10^{-11}$ y^{-1})	Radiogenic daughter isotope	Aprox. abundance (atoms %)	Ratio measured
^{40}K	0.0118	1.25×10^9	(i) K capture (10.5%)	5.81	^{40}Ar	99.7*	^{40}Ar/^{36}Ar
			(ii) beta (89.5%)	49.62	^{40}Ca	97	^{40}Ca/^{44}Ca
^{87}Rb	27.85	4.88×10^{10}	beta	1.42	^{87}Sr	7	^{87}Sr/^{86}Sr
^{147}Sm	14.97	1.06×10^{11}	alpha	0.654	^{143}Nd	12	^{143}Nd/^{144}Nd (or ^{143}Nd/^{146}Nd)
^{232}Th	100	1.40×10^{10}	chain	4.95	^{208}Pb	52	^{208}Pb/^{204}Pb
^{235}U	0.72	7.04×10^8	chain	98.485	^{207}Pb	22	^{207}Pb/^{204}Pb
^{238}U	99.28	4.47×10^9	chain	15.5125	^{206}Pb	25	^{206}Pb/^{204}Pb

*In the atmosphere.

value as a petrogenetic tracer because Ar, being a gas, is too easily lost or gained by a magma and the K–Ca scheme has yet to be developed. All the remaining elements concerned are trace metals in igneous rocks. Moreover their atomic masses are relatively high so that their measured isotopic compositions are normally considered to be immune to the mass fractionation which affects the stable isotopes described later (i.e. *all* observed variations are ascribed to radioactive decay).

The isochron method for dating co-magmatic suites of igneous rocks, which may be applied to all these decay schemes, depends on the fact that in a given sample the rate of accumulation of the radiogenic isotope relative to a non-radiogenic isotope of the same element is related to the concentration ratio of parent and daughter elements. Thus the 'isochron equation' for the Rb–Sr scheme is

$$\left[\frac{^{87}Sr}{^{86}Sr}\right] = \left[\frac{^{87}Sr}{^{86}Sr}\right]_0 + \left[\frac{^{87}Rb}{^{86}Sr}\right](e^{\lambda t} - 1)$$

in which t is the time since the rocks crystallised from an isotopically homogeneous magma and λ is the decay constant.[1] The subscript '0' refers to the initial (magmatic) Sr-isotope composition and all the other ratios are as measured today. Since the amount of radiogenic ^{87}Sr present in rocks is generally very small (though it can be appreciable in old minerals with very high Rb/Sr ratios), the atomic ratio $^{87}Rb/^{86}Sr$ may be regarded as a simple multiple of the Rb/Sr weight ratio – about 2.9 times this quantity. Exactly analogous equations may be written for the Sm–Nd, Th–Pb and U–Pb schemes. Each represents a single straight line (isochron) through all the data points for a single co-magmatic suite of rocks. In geochronology we are chiefly concerned with derivation of the age (t) from the slope of the isochron (see Moorbath 1971) but the value of the initial $^{87}Sr/^{86}Sr$ (or $^{143}Nd/^{144}Nd$ etc.) ratio at the time of crystallisation is also easily derived by extrapolation to the intercept at Rb/Sr = 0. It is this *initial ratio* which is used as a petrogenetic indicator since its value in a magma source region will depend on the prior Rb/Sr ratio of the source, integrated back through time to the formation of the Earth when isotopic ratios are assumed to have been completely uniform. We shall continue to illustrate this basic principle with respect to the Rb–Sr system since this is the one most widely used in petrological studies.

It is not always necessary to plot an isochron in order to obtain an initial $^{87}Sr/^{86}Sr$ ratio; if the age of the rock is known it can be calculated from the measured parameters in the above equation. For maximum precision the age correction (i.e. $^{87}Sr/^{86}Sr - {}^{87}Sr/^{86}Sr_0$) should be as small as possible and this is achieved by analysing samples with the lowest Rb/Sr ratios. For very young basaltic rocks the correction is often insignificant.

Strontium isotopes

Consider the Sr-isotopic composition of a segment of terrestrial material (a portion of the mantle, say) starting with a primordial value of $[^{87}Sr/^{86}Sr]_I$ at the time of formation of the Earth. The simplest case, that of a closed system with constant Rb/Sr ratio, is illustrated by the 'single-stage' growth line (O–R) of Figure 15.1, the gradient of which depends upon the $^{87}Rb/^{86}Sr$ of the mantle segment. An apparently reasonable value for $[^{87}Sr/^{86}Sr]_I$ is the initial $^{87}Sr/^{86}Sr$ ratio of many types of meteorites at the time of *their* formation, generally about 4600 Ma ago. A very precise estimate, insensitive to the exact age, comes from those meteorites with the lowest Rb/Sr ratios – the basaltic achondrites, whose 'best' initial ratio (usually referred to as BABI) is 0.69898 ± 0.00003 (Papanastassiou and Wasserburg 1969). For most terrestrial applications it is sufficiently precise to adopt a value of 0.699.

It is a fundamental tenet of isotope geology that isotopes of heavy trace elements are not fractionated in any way during ordinary chemical processes. Although there may be rare circumstances, particularly in the melting of complex continental crustal material, where isotopic disequilibrium could pertain, it is generally accepted that mantle-derived magmas will initially have the same isotopic composition as their particular source region. Thus estimates of the present-day $^{87}Sr/^{86}Sr$ ratio of our closed system mantle segment may be obtained from those rocks most likely to have been derived directly from such magmas, i.e. oceanic basalts. Many recent oceanic tholeiites have $^{87}Sr/^{86}Sr$ ratios close to 0.704; in a comprehensive survey Faure and Powell (1972) report an average of 0.7037 ± 0.0002. If we take this as the end-point of the single-stage growth line of Figure 15.1 we have of course implicitly defined the Rb/Sr ratio of our mantle segment – it works out at 0.024 ± 0.001. The continental crust, on the other hand, which has differentiated from the mantle by the combined (and often repeated) processes of juvenile magmatic addition, metamorphic consolidation, erosion and sedimentation, should be characterised by higher Rb/Sr ratios and consequently more radiogenic ^{87}Sr than the mantle. Taylor (1964) estimated the mean Rb/Sr ratio of the crust to be about 0.24. A still higher value of 0.35 was calculated for the mean Rb/Sr of the exposed rocks of the Canadian shield by Shaw *et al.* (1967). Freshwater run-off from shield areas has $^{87}Sr/^{86}Sr$ in the range 0.712–0.726, which would imply Rb/Sr = 0.18 ± 0.06 averaged over the

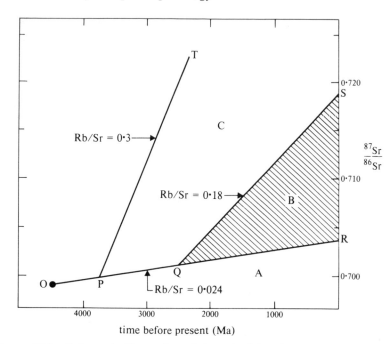

Figure 15.1   Schematic illustration of the principle of Sr-isotope evolution in the Earth with time. Point O represents the primordial (homogeneous) $^{87}Sr/^{86}Sr$ of the mantle, which grows with time to its present day value (R) represented by the average composition of oceanic basalts. To all intents and purposes, the rate of growth is linearly dependent on Rb/Sr ratio. Thus Sr in continental crust with higher Rb/Sr ratios will evolve along much steeper paths diverging from OR at the time of its effective differentiation from the mantle. In practice, such crustal growth curves are merely notional – the Rb/Sr ratio and age of crustal rocks varying very widely indeed; but two are shown for illustrative purposes: PT represents some of the oldest known crustal rocks, the 3800 Ma-old Amitsoq gneisses of W. Greenland, with an average Rb/Sr of about 0.3. QS represents the 'average' composition of the North American shield areas (Faure & Powell 1972). The latter has been used, with the mantle growth curve OR to divide the diagram into fields A, B (shaded), and C in order to assess the relative contributions of crustal and mantle Sr to igneous magmas (see text).

past 2500 Ma (Faure and Powell 1972, p. 25). (Seawater has a lower $^{87}Sr/^{86}Sr$ ratio of 0·709 at the present day, largely due to isotopic exchange with basaltic rocks of the ocean floor (Spooner 1976).)

The observations lead to two fundamental principles of Sr-isotope geology:

(a) Rocks derived from a common (homogeneous) source or parental magma should all possess identical initial $^{87}Sr/^{86}Sr$ ratios.
(b) Magmas derived from sources which have always had low Rb/Sr

ratios (e.g. the mantle) should have correspondingly low initial $^{87}Sr/^{86}Sr$ ratios, whereas those derived from, or contaminated with, long-lived sources with high Rb/Sr ratios (e.g. the upper continental crust) should have higher initial $^{87}Sr/^{86}Sr$ ratios.

The petrogenetic application of these principles can be made both in general terms to the broad classes of igneous rocks, and more specifically to the rocks of an individual igneous province, intrusion or volcano.

A summary of the available data for some distinctive classes of igneous rocks is given in Table 15.2. Island-arc tholeiites and carbonatites show restricted ranges of initial $^{87}Sr/^{86}Sr$ ratios, similar to those of oceanic basalts, confirming their igneous derivation from low Rb/Sr sources. This is particularly striking when the carbonatite data are compared with those for sedimentary limestones (mostly 0.706–0.712) which clearly play no part in their petrogenesis (see Faure and Powell 1972, pp. 56–9). In the case of island-arc tholeiites,

Table 15.2 Generalised initial $^{87}Sr/^{86}Sr$ ratios of igneous rocks

Rock type	$(^{87}Sr/^{86}Sr)_0$	*Comments*
Oceanic basalts	0.7023–0.707	Mean of ocean island basalts = 0.7037[†] Abyssal tholeiites generally ≤ 0.7030
Island-arc volcanics	0.703–0.707	Andesites from New Zealand and South America tend to be higher (often > 0.705) than the remainder, which average 0.7037[†].
Continental mafic rocks	0.700[*]–0.718	Majority <0.710. Jurassic volcanics of Gondwanaland typically high (e.g. ca. 0.712)
Alkalic volcanics	0.704–0.714	E.g. Italian and East African volcanics.
Carbonatites	0.702[*]–0.707	Mean 0.7034[†]
Granites	0.700–0.740	Phanerozoic orogenic batholiths typically 0.704–0.709
Ultramafic intrusives and nodules	0.703–0.730	Low Sr contents make these rocks prone to post-crystallisation contamination, including laboratory contamination in some early analyses
Kimberlites	0.704–0.718	Lowest values from samples claimed to be least altered

[*]Lowest values in range from Archaean rocks, when mantle $^{87}Sr/^{86}Sr$ would have been lower than today.
[†]Faure and Powell (1972).

however, Sr-isotope data alone cannot support distinction between alternative source regions in the mantle above or below the subducted lithospheric slab or within the slab itself (but see following section on Nd-isotopes).

Granites. A perennial problem in igneous petrology to which isotopic studies have been applied with mixed success is the origin of granites. Are rocks of granitic (*sensu lato*) composition formed by fusion of chemically evolved material within the Earth's crust, or are they the products of extreme differentiation of mantle-derived primary magmas? Granites show a very wide range of initial $^{87}Sr/^{86}Sr$ ratios (Table 15.2) which may of course reflect many different petrogenetic factors. This problem was considered by Faure and Powell (1972) who presented a compilation of data for granites analysed up to that time. They recognised three categories based on low, intermediate and high $^{87}Sr/^{86}Sr$ ratios, a procedure which is also followed here (refer to Figure 15.1). It must be emphasised that this subdivision is somewhat arbitrary and may not have an exact significance in relation to the question posed above.

Firstly, granites within or very close to field A show no isotopic evidence of crustal involvement. They could be derived either from primary mantle magmas or alternatively by fusion of juvenile intermediary sources in which there has been insufficient time for accumulation of significant additional ^{87}Sr. A large proportion of these granites are essential components of Precambrian shield areas, although the errors associated with estimating their initial $^{87}Sr/^{86}Sr$ ratios often do not allow distinction between the arbitrary limits of fields A and C. Nevertheless, it seems that the major crust-building cycles of magmatism and metamorphism seen in such areas resulted in the accretion of mantle-derived material during relatively short episodes which did not involve significant reworking of older crustal rocks (Moorbath 1975). This in itself is *not* evidence for the mantle origin of granites since the data permit derivation by partial fusion or ultrametamorphism of basaltic and andesitic volcanics formed not much earlier, perhaps during the same major episode. This idea may be seen as more likely in view of the difficulty of producing large quantities of silica-rich magmas from mantle sources and is consistent with the trace element chemistry of Archaean granitic gneisses (e.g. O'Nions and Pankhurst 1978). Another common rock type in this group is that of undersaturated syenites, for which a mantle origin by small degrees of melting may be more plausible. The Proterozoic granites and syenites (both saturated and undersaturated) of the Gardar Province of S.W. Greenland are a typical example for which an origin by differentiation of primary basic magmas has been claimed (Blaxland *et al.* 1978).

At the other extreme, granites in field C must contain a high proportion of Sr from old crustal sources. Significantly these are frequently small, high level intrusions of peraluminous or peralkaline granites with associated pegmatites. Examples are the Heemskirk granite of western Tasmania (Brooks and Compston 1965) and some of the Mesozoic peralkaline granites of Nigeria (van Breemen *et al*. 1975). Different authors disagree over the extent to which even these high initial $^{87}Sr/^{86}Sr$ ratios (over 0.730) reflect a crustal origin for the granite magma itself, some favouring the late introduction of Sr-bearing fluids derived from country rocks during emplacement.

By far the greatest proportion of intrusive granites, including most Phanerozoic orogenic batholiths, have intermediate initial $^{87}Sr/^{86}Sr$ ratios which fall in field B, requiring more complex explanations. Broadly there are three possibilities:

(a) Derivation by fusion of continental crust which either has a lower than average Rb/Sr ratio or else is very much younger than 2500 Ma. Since Rb is generally depleted along with K and other large-ion lithophile elements in high-grade metamorphic rocks it is quite probable that the lower crust has lower Rb/Sr and $^{87}Sr/^{86}Sr$ ratios than the exposed shield areas. The Caledonian granites of N.E. Scotland, with initial $^{87}Sr/^{86}Sr$ ratios of 0.714–0.717, have been interpreted as partial melts from an unseen deep crustal layer formed about 1300 Ma ago (Pankhurst 1974).

(b) Contamination of mantle-derived magmas by assimilation of older crust. Many orogenic granites are xenolithic and seem to have been emplaced by stoping of upper crust. Other contamination mechanisms could include interaction with hydrothermal water containing more radiogenic Sr and mixing of mantle material with crustal sediments entrained with subducted lithosphere (Armstrong 1968).

(c) Derivation from mantle abnormally enriched in lithophile trace elements (especially Rb relative to Sr). This has been suggested for the Mesozoic granites of California by Brooks *et al*. 1976a).

Choice between these possibilities in any individual case must at present be based on other geochemical or geological evidence, such as the feasibility of contaminants of the appropriate composition. However, detailed regional studies of the Cordilleran batholiths of North America (Kistler and Peterman 1973; Armstrong *et al*. 1977) have emphasised that crustal control of initial $^{87}Sr/^{86}Sr$ ratios is fundamental. There is sharply defined geographical variation, high $^{87}Sr/^{86}Sr$ ratios (i.e. ca. 0.709) occurring in areas underlain by ancient basement rocks, and low

^{87}Sr/^{86}Sr ratios (down to 0.703) occurring in areas of Phanerozoic eugeosynclinal deposits (basalts, andesites and greywackes i.e. mostly young rocks with low Rb/Sr ratios). This would appear to favour derivation from, or extensive contamination with, *local* crustal rocks at shallow depths ((a) or (b) above), but it is also possible that old basement is coupled at deeper levels to old lithospheric mantle, as suggested below for the petrogenesis of some continental basic volcanic rocks. Since crustal rocks are characterised by extremely variable Rb/Sr ratios, as well as higher ones than the mantle, old continental crust is exceedingly inhomogeneous with respect to ^{87}Sr/^{86}Sr ratios. It should be expected that mobilisation of such material to form granitic 'melts' might not cause complete isotopic homogenisation. If local source region variations in Rb/Sr also partially survive magma genesis, this could result in a correlation between Rb/Sr and initial ^{87}Sr/^{86}Sr ratios – a pseudo-isochron – which may equally be recognised by the normal isochron plot yielding an age which is significantly older than that of emplacement (Pankhurst and Pidgeon 1976; Roddick and Compston 1977). If the Rb–Sr system were inherited *intact*, an isochron plot would date the last major homogenisation in the source region rather than granite emplacement; but in general fractionation of Rb/Sr ratio would result in an intermediate 'age'.

Contamination of basic igneous rocks. Other rock types listed in Table 15.2 show wide variation of initial ^{87}Sr/^{86}Sr ratios, even those which on the basis of their mineralogy and major element chemistry must originate in the mantle (kimberlites, ultramafic intrusives and most continental basaltic magmas). In many cases, variation occurs not only overall but within individual igneous complexes or lava successions. Usually there is noticeable co-variation with other geochemical parameters, the more acidic rocks having higher initial ^{87}Sr/^{86}Sr ratios. Since Sr contents tend to be lower in the more evolved rocks, this has often been interpreted as an indication of crustal contamination. In a detailed study of the Jurassic basalts and dolerites of the Transantarctic Mountains, Faure *et al.* (1974) demonstrated good linear relationships between initial ^{87}Sr/^{86}Sr ratios (which range from 0.708 to 0.714) and SiO_2, TiO_2, MgO, CaO, alkalis and $1/Sr$. Interpreting the last of these plots (Fig. 15.2) according to a binary mixing model (see previous chapter) they postulated end-members corresponding to an uncontaminated parent olivine-tholeiite (^{87}Sr/^{86}Sr $\simeq 0.706$) and a silica-rich contaminant (^{87}Sr/^{86}Sr $\simeq 0.720$) which might be a partial melt extracted from the Precambrian metamorphic basement.

In many other examples, however, contamination by assimilation conflicts with the low SiO_2 and high MgO of the basic rocks concerned

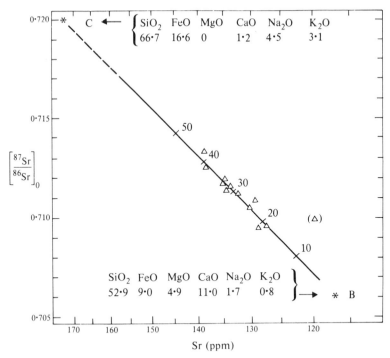

Figure 15.2 Sr-isotope mixing model for the Jurassic Kirkpatrick basalt, Antarctica (after Faure *et al.* 1974). The model is based on a linear fit between initial $^{87}Sr/^{86}Sr$ and $1/Sr$, as plotted here, although the latter has been marked in terms of Sr for convenience (the data point in parentheses was omitted from the fit). End-members B and C were selected according to estimated likely values for their $^{87}Sr/^{86}Sr$ ratios. The major element compositions shown were derived from other diagrams showing co-variation with $^{87}Sr/^{86}Sr$, and they were identified as a parental continental basalt magma (B) and a continental contaminant (C). The numerals marked on the mixing line indicated the percentage of C in the mixing model, which approaches the improbably high value of 50%.

and mechanisms involving *selective* contamination by incompatible lithophile elements or even ^{87}Sr alone have been proposed. These include migration of extremely small partial melts from the wall rocks (Green and Ringwood 1967), akin to zone refining, and transfer of trace elements in a fluid phase derived either from the magma or the contact aureole of intrusions. An example of the latter is the Insch layered basic intrusion of N.E. Scotland in which ultramafic and mafic cumulates have initial $^{87}Sr/^{86}Sr$ ratios of 0.703–0.712 (Pankhurst 1969).

Contamination of a primary magma is not the only feasible explanation for co-variation of $^{87}Sr/^{86}Sr$ with other chemical parameters. Two alternatives which have been discussed recently are disequilibrium melting and gross inhomogeneity of the mantle. The idea of disequilibrium melting is based on the observation that the mineral phases of ultramafic nodules derived from the Upper Mantle by incorporation into basaltic magma are not isotopically equilibrated as would be expected if continuous diffusion of ^{87}Sr were efficient at mantle temperatures. Harris *et al.* (1972) suggested that if such disequilibrium persisted during partial melting, magmas might be produced with different $^{87}Sr/^{86}Sr$ ratios from that of the original source region, particularly since phases with high Rb/Sr ratios (and consequently the most radiogenic Sr) such as phlogopite and amphibole would preferentially enter the first melts. Hofmann and Hart (1975), however, claim that experimental evidence for Sr-diffusion rates in mantle material precludes the survival of heterogeneities on such a small scale at temperatures close to that of melting. Thus the first principle of Sr-isotope geology outlined above may be considered to have been reinstated.

In considering mantle inhomogeneity as a cause of such wide variation of $^{87}Sr/^{86}Sr$ ratios, we are effectively challenging the spirit, if not the letter, of the second of these principles and this requires more lengthy discussion.

Mantle evolution and inhomogeneity. Our previous discussion of the data in Table 15.2 has tacitly assumed that the mantle has uniformly low $^{87}Sr/^{86}Sr$ ratios resulting from low Rb/Sr ratios since its formation. There is, however, overwhelming evidence that a single-stage evolution model cannot be applied to the mantle in general. Firstly there is significant heterogeneity at the present day, represented by the variation within oceanic basalts, which has been emphasised by the growth of high-precision measurement techniques (Fig. 15.3). Thus fresh abyssal tholeiites dredged from the deep ocean basins and ridges mostly have $^{87}Sr/^{86}Sr$ less than 0.7030, down to a minimum of about 0.7023. On the other hand, tholeiites from oceanic islands have values generally above 0.7030 and alkali basalts extend this range up to at least 0.706. An overall correlation with alkalinity (Peterman and Hedge 1971) cannot be explained by seawater contamination since $^{87}Sr/^{86}Sr$ is often constant for the volcanic products of an individual island within close limits (O'Nions & Pankhurst 1974b). The range must thus reflect real variation of trace element and isotope ratios in the mantle. If these heterogeneities had been established at the time of formation of the Earth, each segment with a different Rb/Sr ratio could conceivably be a separate single-stage sub-system or domain. In this case they should now all lie on the same isochron, with a slope age of 4600 Ma and a

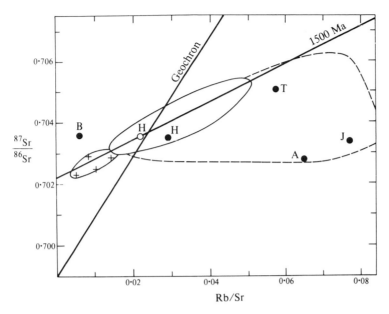

Figure 15.3 Rb–Sr isochron plot for modern oceanic basalts with the ordinate plotted as Rb/Sr rather than ^{87}Rb/^{86}Sr (after Sun & Hanson 1975; Brooks *et al.* 1976a). Fields are delimited for mid-ocean ridge basalts (crosses), ocean island tholeiites (open circles) and alkali basalts (filled circles) B = Bouvetøya, A = Ascension, H = Hawaii, J = Jan Mayen, T = Tristan da Cunha; many others not shown individually. The alkali basalts show considerable scatter, mostly well to the right of the Geochron (a 4570 Ma isochron through BABI), partly due to the increase in Rb/Sr over mantle source regions for very small degree of partial melting. The abyssal tholeiites plot to the left of the Geochron, i.e. their observed Rb/Sr ratios are too low to account for their ^{87}Sr/^{86}Sr ratios in 4570 Ma. Together with the island tholeiites they form a reasonably linear array which could be interpreted as representing mantle differentiation about 1500 Ma ago, with the source region of the abyssal tholeiites being relatively depleted in Rb at that time but see text for alternative explanations.

primordial initial ^{87}Sr/^{86}Sr equal to BABI (called the **Geochron**). There seems little possibility that this can be so, since the oceanic basalts themselves scatter well off the Geochron (Fig. 15.3). The Rb/Sr ratios of the parent magmas need not, of course, be the same as those of their source regions, but the trace element models considered in the last chapter indicate that the fractionation factor is probably in the range 1.0–1.3 (major element compositions suggest a relatively high degree of partial melting of tholeiites, about 10–20%, certainly enough to eliminate residual phlogopite or amphibole, and the absence of negative Eu-anomalies from the rare earth element patterns of such rocks excludes significant low pressure plagioclase fractionation). Thus the fact that some points plot to the *left* of the Geochron (particularly those for mid-ocean ridge basalts) means that some of the radiogenic ^{87}Sr which they contain is 'unsupported', i.e. there is more than can have been produced during the Earth's existence by the inferred Rb/Sr ratio of their source regions. This can only be explained if the Rb/Sr ratio has been reduced from an initially higher value, for example by the extraction of a previous partial melt. This suggests that major parts of the mantle must have undergone multi-stage isotopic evolution.

A generalised equation for a multi-stage isotope evolution model has the form

$$\left[\frac{^{87}Sr}{^{86}Sr}\right]_T = \left[\frac{^{87}Sr}{^{86}Sr}\right]_I + \sum_{i=1}^{n} \left[\frac{^{87}Rb}{^{86}Sr}\right]_{i,T} (e^{\lambda t_{i-1}} - e^{\lambda t_i})$$

where t_i is the time at the end of the ith stage and the ^{87}Rb/^{86}Sr ratio for each stage is calculated as it would be measured today ($T = t_n = 0$) had no further fractionation of Rb from Sr occurred (i.e. allowing for the radioactive decay since t_i). The age of the formation of the Earth is t_0. Unfortunately there is no unique solution to this equation unless we know the number of stages and the duration and Rb/Sr ratio for each. Since Rb is a more incompatible element than Sr and is therefore more efficiently transferred from the mantle to the crust by igneous processes, we may intuitively suppose that for the mantle as a whole, Rb/Sr decreases with each fractionation event, perhaps continuously with time as indicated by the upward convex evolutionary path of Figure 15.4. Attempts to define such a curve by measuring the initial ^{87}Sr/^{86}Sr of mantle-derived rocks of widely different ages (e.g. Faure & Powell 1972, Fig. XII.2) have been notably unsuccessful, presumably because the mantle was also heterogeneous in the past. A simplified model which appeals to some geologists depends on major geochemical differentiation events in the mantle being accompanied by effective isotopic re-equilibration. This results in a 'pseudo-two-stage' evolution in which observed variations are essentially dominated by the final stage. Thus the crude linear alignment of the tholeiite data in Figure 15.4 has been ascribed to a major event about 1500 Ma ago (Brooks *et al.* 1976b) although it is also possible that differentiation has been semi-continuous and that this age does not refer to a discrete event. Data for oceanic alkali basalts scatter rather more about a mean line with a somewhat lower slope, but this may simply reflect a general increase of up to 30% in the Rb/Sr ratios of very small partial melt fractions compared to their source region (Duncan & Compston 1976). These arguments have been extended to the sub-continental mantle, as an alternative explanation to contamination for the variability of

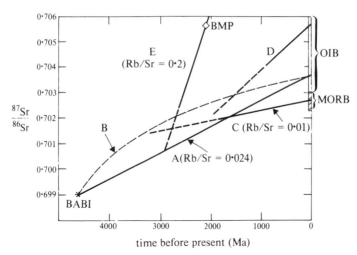

Figure 15.4 More detailed evolution model for Sr isotopes in the Earth's mantle. Line A is the single-stage growth curve of Figure 15.1, which clearly cannot account for the variability of $^{87}Sr/^{86}Sr$ in oceanic basalts, although it provides one possible fit for most ocean island tholeiites with $^{87}Sr/^{86}Sr$ ca. 0.704. Line B is an alternative continuous differentiation model ending up at the same point today, but with initially higher Rb/Sr ratios decreasing throughout the Earth's history to values less than that of 0.024 required by the single-stage model. Line C is a possible final stage of growth for mid-ocean ridge basalts with $^{87}Sr/^{86}Sr < 0.703$ and Rb/Sr $\leqslant 0.01$ (see also Fig. 15.3). The intersection with Line A at about 1500 Ma ago is merely another indication that data for abyssal and ocean island tholeiites approximate to a 1500 Ma isochron (Fig. 15.3). Line D represents a typical alkali basalt source region, which must have Rb/Sr higher than 0.024, even allowing for some increase in this ratio during partial melting. Line E represents mantle exceptionally enriched in lithophile trace elements, such as has been proposed for parts of the sub-continental mantle (BMP = initial magma composition of Bushveld Mafic Phase, Hamilton 1977).

$^{87}Sr/^{86}Sr$ in continental basic igneous rocks (e.g. Brooks *et al.* 1976a; Pankhurst 1977b). As an extreme example, Hamilton (1977) claims that the Proterozoic intrusives of southern Africa (including the Great Dyke of Rhodesia and the Bushveld Mafic Phase) are derived from mantle which began a growth stage with a very high Rb/Sr of about 0.2 close to 2900 Ma ago (see Fig. 15.4). An important problem for the next few years in igneous petrology will be to use intensive isotope and trace element studies in an attempt to discriminate between inheritance of such long-lived heterogeneities from the lithospheric mantle or asthenosphere and crustal contamination mechanisms as the major factor in producing the geochemical characteristics of igneous rocks.

Neodymium isotopes

The application of Nd isotopes in petrology is a relatively recent development (Richard *et al.* 1976; De Paulo and Wasserburg 1976; O'Nions *et al.* 1977). Principles and systematics closely parallel those of the Rb–Sr decay scheme with two important differences. Firstly, the range of Sm/Nd ratios in igneous rocks is very much less than that of Rb/Sr ratios. This, together with the long half-life of ^{147}Sm, means that the natural variability of radiogenic ^{143}Nd is considerably less than that of ^{87}Sr, so that extremely high precision is required to measure it. Geochronological application is virtually restricted to Archaean rocks. Secondly, the daughter product, as a light REE, is generally more incompatible in igneous processes than its parent. Thus ^{143}Nd accumulates relatively faster in light REE depleted environments such as some mafic and ultramafic rocks and parts of the Upper Mantle than in the light REE-enriched sources of alkali basalts and crustal rocks such as granites. This of course is the reverse of the behaviour of the Rb–Sr system and results in an inverse relationship between ^{143}Nd/^{144}Nd and ^{87}Sr/^{86}Sr ratios in oceanic basalts (Fig. 15.5). Since Nd is several orders of magnitude less concentrated in seawater than Sr this anti-correlation seems to be strong confirmation that, at least in the oceanic environments, the isotopic variations in both elements are inherited from long-lived heterogeneities in the source region and are not due to alteration or other secondary processes. Exceptions to this rule are some abyssal basalts which show clear petrographic evidence for alteration and contain leachable Sr with high ^{87}Sr/^{86}Sr ratios (O'Nions *et al.* 1977) and fresh tholeiites from the South Sandwich Is. – an active island arc usually regarded as representing an early stage of development. The latter have ^{143}Nd/^{144}Nd ratios indistinguishable from those of unaltered abyssal basalts but primary ^{87}Sr/^{86}Sr ratios in the range 0.7038–0.7039 (Hawkesworth *et al.* 1977). This suggests that the island-arc magmas contain a significant proportion of Sr derived from seawater altered material, presumably derived from the sea floor subducted along associated Benioff zones.

A further consequence of the inverse behaviour of the two decay schemes in lithophile depleted sources is the fact that ocean-ridge tholeiites with unsupported ^{87}Sr due to past depletion of Rb have an apparent deficiency of ^{143}Nd resulting from increased Sm/Nd ratios and therefore plot to the *right* of the Geochron in an Sm–Nd isochron diagram.

As yet there are few Nd-isotope data for continental volcanics or granites, but this should provide new complementary evidence with

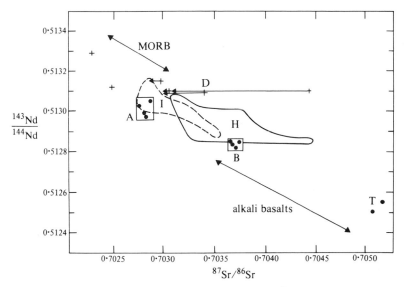

Figure 15.5   Anti-correlation of ^{143}Nd/^{144}Nd and ^{87}Sr/^{86}Sr for oceanic basalts, after O'Nions *et al.* (1977). Individual data points are shown for volcanics from Ascension, Bouvetøya and Tristan da Cunha, as well as fields for Hawaii (H) and Iceland/Reykjanes Ridge (I). This relationship seems to confirm that isotope variations in the mantle result from long-lived trace element heterogeneities, since Rb/Sr and Sm/Nd ratios also show antipathetic behaviour and enrichment in partial melting processes. The only rocks which do not fit are altered abyssal basalts in which ^{87}Sr/^{86}Sr has been increased by seawater interaction, such as those shown here from DSDP Leg 37(D). The arrows indicate the reduction of ^{87}Sr/^{86}Sr for these rocks when alteration products are removed by leaching (O'Nions & Pankhurst 1976), but the Nd-isotopes are apparently resistant to such alteration, a valuable property for petrogenetic applications.

which the controversial interpretations of Sr-isotope data reviewed above may be resolved.[2]

Lead isotopes

It has become standard practice with the U–Pb system to model evolution from a supposedly homogeneous composition at the formation of the Solar System. Thus the expressions parallel that given previously for multi-stage evolution of Sr-isotopes, e.g.:

$$\left[\frac{^{206}\text{Pb}}{^{204}\text{Pb}}\right]_T = \left[\frac{^{206}\text{Pb}}{^{204}\text{Pb}}\right]_I + \sum_{i=1}^{n}\left[\frac{^{238}\text{U}}{^{204}\text{Pb}}\right]_i (e^{\lambda t_{i-1}} - e^{\lambda t_i})$$

A corresponding equation for $^{207}Pb/^{204}Pb$ is obtained by substituting ^{235}U and λ' (its decay constant). The primordial composition of Pb is obtained from the U-free troilite phase (Fe_2S) of meteorites ($^{206}Pb/^{204}Pb = 9.307$; $^{207}Pb/^{204}Pb = 10.294$). T is the time from the formation of the Solar System (at t_0) to the present day, for which the best estimate is now taken as 4570 Ma (also based on meteorite data). The ratio $^{238}U/^{204}Pb$ is referred to as μ, starting with an initial value of μ_1. This is calculated as though it were measured today, so that the $^{235}U/^{204}Pb$ ratio is always given by dividing μ by the observed $^{238}U/^{235}U$ ratio, 137.88.

One of the chief problems with U–Pb studies is the extreme mobility of U once oxidised to the hexavalent state (as during weathering). Thus U/Pb ratios measured in igneous rocks are rarely those acquired during crystallisation. This problem is usually avoided by combining equations for the two U–Pb decay schemes, thus eliminating U contents and forming a relationship between $^{206}Pb/^{204}Pb$ and $^{207}Pb/^{204}Pb$ (see Doe 1970 for details).

Mantle lead. Examples of simple single-stage ($i=1$) Pb-isotope evolution are illustrated diagrammatically in Figure 15.6. As shown, Pb from stratiform ore deposits of various ages lies close to a curve with a μ value of about 8.3. It was originally thought that this indicated that the mantle had undergone a simple evolution and that it should therefore be homogeneous with respect to both U/Pb and Pb-isotopes. Rock Pb data from recent basalts do not, however, confirm this conclusion, considerable isotopic variation being found within oceanic islands (even those such as Ascension I. and Tristan da Cunha in which $^{87}Sr/^{86}Sr$ is essentially constant). Results such as this provided some of the earliest evidence for large-scale heterogeneity of the mantle. Indeed, some authors now question the validity of a single-stage model even for the stratiform Pb ores.

The next-simplest model would be one in which different domains in the source region had the same initial Pb-isotope composition but different U/Pb ratios, either because of primordial heterogeneity in U, or Pb, distribution or as a result of a fractionation event very early in the Earth's history. In this case the Pb-isotope composition of each domain would evolve along separate growth curves as shown in Figure 15.6 but would end up on a single straight line – a 4600 Ma, Pb–Pb isochron called the Geochron. Modern oceanic Pb falls to the right of the Geochron, showing that the overall U/Pb ratio of the mantle has increased during geological time. (Because the half-life of ^{235}U is so short the natural $^{235}U/^{238}U$ ratio must have decreased quite sharply early in

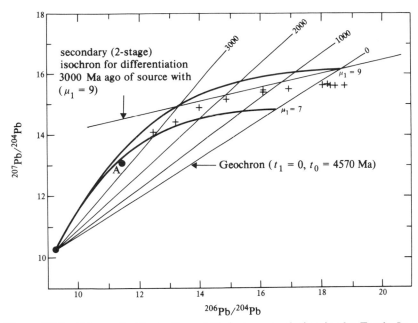

Figure 15.6 Schematic illustration of Pb-isotope evolution in the Earth. In a single-stage model Pb-isotope ratios increase from a primordial value (0) along curved paths according to U/Pb ratios. Two such curves are shown, for μ-values ($^{238}U/^{204}Pb$) of 7 and 9 respectively. The data for stratiform Pb-ore bodies of various ages (crosses) approximate to such a curve with $\mu = 8$, suggesting that they may be a good analogue for the mantle in this respect, although the highest $^{206}Pb/^{204}Pb$ ratios observed are too radiogenic for a single-stage model. Sub-systems with different μ values undergoing single-stage evolution would today lie on a straight line – the Geochron. Other primary isochrons are shown for separation from the mantle (with reduction of U/Pb ratios to zero) at times of 1000, 2000 and 3000 Ma ago. High-grade metamorphic rocks in the crust often have very low U/Pb ratios, so that their Pb isotope compositions are generally much less radiogenic than those of recent volcanics. The least radiogenic Pb observed in terrestrial rocks (3800 Ma-old Amitsoq gneisses of W. Greenland) is plotted as point A. Two stage evolution would result in secondary isochrons of much lower slope. For example, one is shown for a mantle sub-system starting the first stage with $\mu_1 = 9$ until 3000 Ma ago, at which time it underwent differentiation to form separate domains with different μ_2 values. These would now lie on the secondary isochron shown. High μ values late in the Earth's geochemical development can thus produce Pb-isotope compositions which plot to the right of the Geochron, as most ocean basalt data do (see Fig. 15.7).

the Earth's history and so late increases in U/Pb result in relatively more rapid growth of ^{206}Pb compared to ^{207}Pb.) In fact, Pb-isotopes from modern oceanic volcanics exhibit a pronounced linear tendency (Fig. 15.7), indicative of an even more complex history. Differentiation

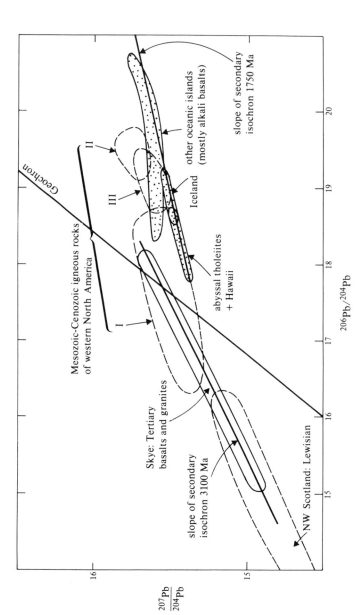

Figure 15.7 $^{207}Pb/^{204}Pb$ versus $^{206}Pb/^{204}Pb$ for some igneous rocks. Ocean basalt Pb falls to the right of the Geochron, indicating an overall increase in the U/Pb ratio of the mantle with time. The more radiogenic compositions of the alkali basalts compared to the tholeiites indicate higher U/Pb ratios in their source regions. Fields I, II, and III are for basalts, andesites and granites from western North America (after Zartman 1974) and have been taken to indicate various proportions of incorporated Pb from old crustal basement. Similarly, the linear array for the Tertiary basalts and granites of Skye (Moorbath & Welke 1969) was interpreted as a mixing line between a recent oceanic-type Pb and unradiogenic Pb derived from the local Lewisian basement gneiss. Because of the systematics of two-stage Pb models, however (see Fig. 15.6), this line also corresponds to a secondary isochron with an age close to that of the formation of the Lewisian, so that an alternative interpretation, i.e. inheritance of mantle heterogeneities which have existed beneath Skye since that time, is also theoretically

of an originally homogeneous mantle at some time in the past, resulting in domains with different U/Pb ratios, constitutes a two-stage model. In this case, present day Pb-isotope compositions would not lie at the end of the primary growth curve but along a **secondary isochron** whose slope and intersection with the curve define the time of differentiation (t_1). According to this type of model, oceanic Pb suggests a major differentiation of the mantle about 1500–2000 Ma ago, with the less radiogenic Pb of abyssal tholeiites being derived from regions with low second-stage μ_2 values and the more radiogenic Pb of alkali basalts being derived from regions with high μ_2.

.Thus interpretation according to a two-stage model is consistent with the similar treatment of Rb–Sr data given on page 372. However, it is important to realise that in neither case is the model a unique solution and that the scatter of real data show that it must in any case be a gross oversimplification. Church and Tatsumoto (1975) argued that the Pb-isotope data could be more closely fitted by a model involving five discrete growth stages and O'Nions *et al.* (1978) have shown that the observed spread of $^{206}Pb/^{204}Pb$ and $^{207}Pb/^{204}Pb$ ratios would result from *continuous* differentiation of the mantle, in which the slope of the Pb-isotope plot would have no time significance, a view shared by Tatsumoto (1978).

Crustal lead. In most igneous processes there is a tendency for U/Pb ratios to increase with differentiation (alkali basalts often much higher than suspected for the mantle source regions). The separation of U from Pb is continued even more efficiently in the oxidising environment of the continental crust, where a very wide range of U/Pb ratios is observed. Sometimes this separation is complete, U and Pb forming their own ore deposits and considerable amounts of U ending up in seawater. Conversely some crustal rocks, especially metamorphic ones, become depleted in U relative to Pb. Thus ancient crustal basement usually contains Pb which has not evolved isotopically very much since its first separation from the mantle. As an example, trace Pb extracted from the 3800 Ma Amitsoq gneiss of W. Greenland has amongst the least radiogenic compositions known for terrestrial rocks, with $^{206}Pb/^{204}Pb$ down to 11.5 and $^{207}Pb/^{204}Pb$ down to 13.2 (see Fig. 15.6). Mineralised sediments from W. Greenland have $^{206}Pb/^{204}Pb$ as low as 11.1. Unusually unradiogenic Pb in Recent igneous rocks could thus be taken as evidence for contamination of the parent magma with, or its derivation from, old crustal material. Unlike the principle application of Sr-isotopes, however, the converse cannot always be assumed – some crustal rocks will contain highly radiogenic Pb and contamination with these may not be detected. Even more than with Sr-isotopes, therefore, each case must be examined on its own merits, with the maximum possible information about Pb-isotopes in possible contaminants.

Lead isotopes in igneous rocks. Pb-isotope compositions for island-arc volcanics and continental basalts generally fall within the field shown in Figure 15.7 for ocean island alkali basalts, though they tend to have rather higher $^{207}Pb/^{204}Pb$ ratios for given $^{206}Pb/^{204}Pb$ ratios. This could be taken as indicating slight contamination with continental sediments in which U is enriched relative to Pb; but oceanic sediments are insufficiently radiogenic to account for the difference (Church & Tilton 1973).

Considerably greater variation is shown by continental orogenic igneous rocks (volcanics and granites). The best studied area for such rocks is undoubtedly the cordillera of western North America, where Pb isotope compositions vary systematically with the sub-surface geology. These results have been summarised by Zartman (1974) as defining three distinct fields, also shown in Figure 15.7. Field I shows a rather wide range in Pb-isotope compositions but all are unradiogenic in comparison with the known age of the rocks (Mesozoic and Cenozoic). Such compositions are confined to the eastern Cordilleran region (i.e. the Rocky Mountains) where extensive Precambrian crystalline rocks outcrop. Zartman concludes that these rocks were derived from lower crust of similar ancient age, or possibly from U-depleted Upper Mantle. Field II has rather radiogenic Pb compositions ($^{206}Pb/^{204}Pb > 19.1$) but is fairly homogeneous. This applies to igneous rocks further west in an area dominated by miogeosynclinal sedimentation throughout the late Precambrian and Palaeozoic and it seems that these represent the U-enriched counterpart to the U-depleted basement from which they were derived. Pb from these sediments must have contributed substantially to the later igneous magmas. Finally, the western coastal batholiths and related andesites have homogeneous Pb-isotope compositions in an intermediate range (Field III), broadly similar to that exhibited by island-arc volcanics. Thus there seems to be a fundamental lithospheric control on the Pb-isotopic compositions of magmas in this region, just as there is for Sr. This implies an origin in the crust or ancient differentiated mantle immediately beneath it, or at least very extensive contamination at these sites.

An even more spectacular example of anomalously old Pb in igneous rocks is displayed by the Tertiary basalts and granites of the Isle of Skye, N.W. Scotland (Moorbath & Welke 1969). These form a regular linear trend (Fig. 15.7) between a modern island-arc type of Pb composition and an unradiogenic Pb typical of the 2700 Ma-old Lewisian gneiss of the area. Not surprisingly, this was interpreted as a mixing line formed by contamination of magmas derived from the mantle in Tertiary times with crustal basement. As expected, the granites showed the most displaced compositions, corresponding to extreme contamination or even origination within the crust, but the basic volcanics also show the effect to lesser degree. On the other hand, Sr-isotope compositions for the basic rocks are all within the expected range for mantle Sr. The slope of the linear trend in Skye Pb compositions corresponds to a secondary isochron, with an age of 3100 Ma, close to the age of formation of the Lewisian gneiss. This is a necessary consequence of mixing if the old Pb component evolved according to a two-stage model (i.e. U/Pb ratios unchanged since the major differentiation from the mantle). Unfortunately, it demonstrates that such linear trends can always be interpreted in two ways: either as mixing lines or as secondary (or higher-order) isochrons, implying a common source region which is heterogeneous with respect to U/Pb ratios.

Some authors have gone some way towards resolving this ambiguity by demonstrating mixing models as outlined in the last chapter. Lancelot and Allegre (1974) showed that Pb-isotope ratios for carbonatites from Uganda fell on a hyperbola when plotted against Pb content and argued for mixing between a Pb-poor component with $^{206}Pb/^{204}Pb$ = ca. 18.5 and a Pb-rich component with $^{206}Pb/^{204}Pb$ = 20.84, but were unable to make a definitive geological interpretation. A similar conclusion was reached by Vollmer (1976) who demonstrated a very convincing hyperbolic relationship between $^{206}Pb/^{204}Pb$ and $^{87}Sr/^{86}Sr$ ratios for the Tertiary and Recent potassic volcanics of Italy. One end-member was characterised by high $^{206}Pb/^{204}Pb$ (ca. 20) and low $^{87}Sr/^{86}Sr$ (ca. 0.7035) whereas the other with about seven times the total Pb content, had low $^{206}Pb/^{204}Pb$ (ca. 18.7) and high $^{87}Sr/^{86}Sr$ (ca. 0.714). These were tentatively identified as a mantle source similar to that of oceanic basalts and moderately old crust respectively. The proportion contributed by each varies systematically from north to south and this was interpreted as being due to variable melting of an inhomogeneous crust–mantle mixture. No simple relationship exists between isotope composition and either Pb or Sr contents, indicating that the mixing process involves solids, or possibly that the erupted magmas have been extensively modified by crystal fractionation.

A more specific form of Pb-isotope mixing which is unrelated to most other geochemical variations but which is petrogenetically important applies to zircon populations in igneous rocks. The systematics of U–Pb dating of zircons is beyond the scope of this book (see Doe 1970 for details) but there are several examples of granites which have been shown to contain mixed populations of two types of zircon – one with isotope characteristics related to crystallisation during granite emplacement, the other having crystallised very much earlier (Pankhurst & Pidgeon 1976). The latter are clearly inherited from older crustal rocks incorporated into, if not the actual source region of, the granite magmas.

Stable isotope variations

Variations in the isotopic composition of elements which consist of stable, non-radiogenic, isotopes are related to mass-dependent differences in nuclear properties, whereas the chemical behaviour of elements is principally dictated by their external electronic configuration. Thus inter-isotope effects are comparatively small except for the light elements, in which proportional mass variations are greatest. The heavier natural isotope of hydrogen has a nuclear mass twice that of the more abundant ^1H isotope and the differences here are sufficiently great to merit a separate name and chemical symbol (deuterium, $^2H = D$). Mass fractionation effects are usually considered to be insignificant above mass 40 (K, Ca) and even, in geological processes, for some much lighter elements (e.g. Li). Those elements for which natural isotope fractionations are most pronounced are listed in Table 15.3. Of these oxygen is by far the most frequently applied to problems in igneous petrology and will be used to illustrate the basic principles.

The primary influence of changes in nuclidic mass on chemical properties occurs as a result of corresponding variation in the vibrational frequency of adjacent atom pairs. More energy is stored in chemical bonds formed by the heavier isotopes of an element, so that such bonds are less easily broken, as exemplified by the smaller ionic product of D_2O (1.6×10^{-15}) compared to that of H_2O (10^{-14}). This leads to differences in the lattice energies of minerals when isotopic substitution of their constituents occurs.

Natural fractionation of stable isotopes is of two types – equilibrium and kinetic. In the former case isotopes are partitioned between two or more phases in equilibrium, e.g. quartz and magnetite:

$$\tfrac{1}{2} Si^{18}O_2 + \tfrac{1}{4} Fe_3{}^{16}O_4 \rightleftharpoons \tfrac{1}{2} Si^{16}O_2 + \tfrac{1}{4} Fe_3{}^{18}O_4$$

As in all reversible reactions, the equilibrium condition is described by a constant, K, which is strongly temperature-dependent. This applies to all igneous processes in which equilibrium exists between solids and/or magma. Kinetic fractionations apply in non-equilibrium cases and are due to faster reaction rates for light isotopes (again, a consequence of differences in bond strength). Kinetic effects are minimised at high temperatures and are only geologically important during open system

Table 15.3 Stable isotopes of light elements

Element	Natural isotopes		Measured ratio	Reference standard	Fractionation range for igneous rocks and minerals
Hydrogen	^1H	99.984%	D/H	SMOW (Standard mean ocean water)	$\delta D = -180$ to $-30‰$
	^2D	0.016%			
Carbon	^{12}C	98.89%	$^{13}C/^{12}C$	PDB (Pedee Formation belemnite)	$\delta^{13}C = -30$ to $0‰$
	^{13}C	1.11%			
Oxygen	^{16}O	99.76%	$^{18}O/^{16}O$	SMOW	$\delta^{18}O = -10$ to $+20‰$
	^{17}O	0.04%			
	^{18}O	0.20%			
Sulphur	^{32}S	95.02%	$^{34}S/^{32}S$	CD (Troilite, Fe_2S, from Diablo meteorite)	$\delta^{34}C = -10$ to $+30‰$
	^{33}S	0.75%			
	^{34}S	4.21%			
	^{36}S	0.02%			

Natural abundances are only approximate.

behaviour at relative low temperatures, e.g. hydrothermal and metasomatic processes.

Notation of isotope fractionations. In order to compare variations in stable isotope ratios (e.g. $^{18}O/^{16}O$) it is necessary to eliminate any laboratory bias introduced during analysis. This is achieved by reporting results relative to an internationally agreed standard substance (see Table 15.3), although in practice measurements are made by comparison with a calibrated secondary standard. Fractionations are usually quantified in terms of 'del'-values, in per mil (‰) e.g.

$$\delta^{18}O = \frac{(^{18}O/^{16}O)_{sample} - (^{18}O/^{16}O)_{standard}}{(^{18}O/^{16}O)_{standard}} \times 1000$$

Thus a sample with $^{18}O/^{16}O = 2.000 \times 10^{-3}$ relative to a value of 2.005×10^{-3} for SMOW is reported as having $^{18}O = -2.49‰$ (SMOW), though in practice a secondary standard would usually be used rather than seawater itself. The definitions for δD, $\delta^{13}C$ and $\delta^{34}S$ are analogous.

Fractionations between two phases, A and B, may also be defined in terms of the **fractionation factor**, $\alpha_{AB} = (^{18}O/^{16}O)_A/(^{18}O/^{16}O)_B$, although this is usually restricted to equilibrium fractionations only. This has the advantage that α is related to the equilibrium constant, K, by the equation $K = \alpha^n$, where n is the number of exchangeable atoms. Thus if the reaction equation is written with only one exchangeable atom in each compound, as above, then $K = \alpha$. For small values of δ ($\leqslant 10‰$), we can further deduce the approximate relationship

$$1000 \ln \alpha_{AB} = \delta_A - \delta_B (= \Delta_{A-B})$$

Figure 15.8   Equilibrium mineral–mineral fractionations for (a) oxygen and (b) sulphur isotopes as a function of temperature. (Qz = quartz, Mt = magnetite, Ab = albite or K-feldspar, An = anorthite, Mu = muscovite, Pl = plagioclase An_{60}, Ol = olivine, Px = pyroxene, Bi = biotite, Py = pyrite, Ga = galena, Cp = chalcopyrite, Sp = sphalerite.) Note that the temperature scale for sulphur is compressed relative to that for oxygen. The fractionations are plotted as 1000 ln α_{AB}, which for most purposes is approximately equal to $\delta_A - \delta_B$, i.e. the difference in $\delta^{18}O$ or $\delta^{34}S$ between the two minerals concerned. Thus a difference of $+5‰$ in $\delta^{18}O$ between co-existing quartz and magnetite would give an O-isotope 'temperature' of about 750 °C. Since $\delta_A - \delta_C = (\delta_A \gtrless \delta_B) - (\delta_C - \delta_B)$, it is also clear that, for example, quartz–feldspar fractionations are always small in an equilibrium assemblage (4‰ or less), whereas feldspar–biotite and quartz–biotite fractionations are usually large. All fractionations tend to very small values at the temperatures of basaltic magmas. Data from the compilation of Friedman and O'Neil (1977).

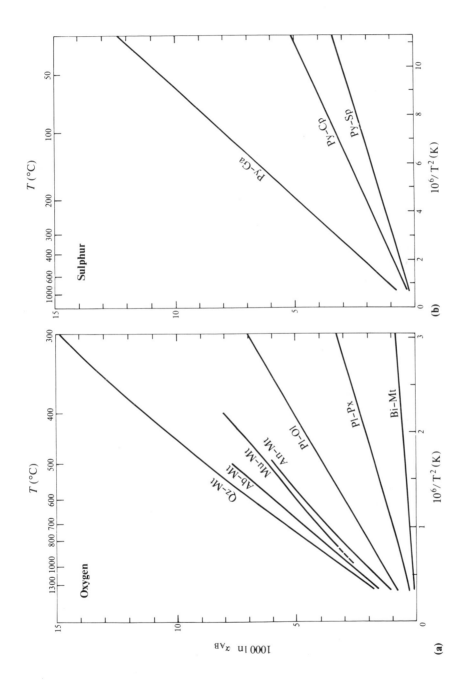

$1000 \ln \alpha_{AB}$

Temperature effects. The form of temperature dependence of the equilibrium constant in exchange reactions is ln K $\propto 1/T^2$ where T is the absolute temperature, so that applying the above restrictions we expect a linear relationship between Δ_{A-B} and $1/T^2$. It follows that all isotope fractionations tend to zero at high temperatures. Figure 15.8 shows O- and S-isotope equilibrium fractionations for the most important mineral pairs plotted against temperature. This leads to the first main application of stable isotopes in igneous petrology – geothermometry. The most useful isotope geothermometers are those involving common minerals with simple chemistry, since fractionation factors for minerals exhibiting a solid solution series, such as the plagioclase feldspars, vary according to the proportions of each end-member. The assumption of equilibrium should be justified by thin-section examination and preferably also by analysing two independent mineral pairs from each rock. Sensible magmatic temperatures ($> 1000\,°C$) have been calculated in this way from phenocryst assemblages in unaltered volcanic rocks, examples of which are summarised by Faure (1977). Geothermometry of plutonic rocks is complicated by a much wider range of crystallisation temperatures, reflected in disequilibrium textures and mineral zonation. Analysis of zoned crystals will generally give a calculated temperature within the crystallisation interval, but values derived from feldspar fractionations are frequently lower than those obtained from other minerals, indicating the continuation of isotopic exchange down to sub-solidus temperatures below 800 °C (e.g. Anderson *et al*. 1971). Because of the occurrence of sulphide minerals, the application of S-isotope geothermometry is virtually restricted to the environment of late- or post-igneous hydrothermal mineralisation, and and here factors other than temperature alone play an extremely important role, notably oxygen fugacity and pH (Ohmoto 1972).

Stable isotopes and magmatic evolution. Because of the wide variety of chemical conditions and compositions which 'magma' may possess, individual mineral–magma fractionation factors, as opposed to those for mineral pairs, are not easily predicted. Thus the behaviour of stable isotopes during partial melting and fractional crystallisation is not amenable to quantitative treatment. Nevertheless we may generally recognise minerals which have relative preferences for either light or heavy isotopes of a particular element, even at high temperatures where the effect may be very small. Thus quartz, and to a lesser extent feldspars and muscovite, preferentially incorporate ^{18}O compared with olivine and, especially, magnetite which contain a greater proportion of ^{16}O. The remaining common igneous minerals show neither preference. Thus we may anticipate that early crystallisation and removal of

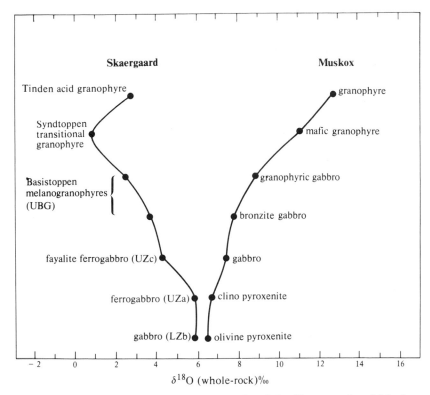

Figure 15.9 Variation of $\delta^{18}O$ in whole-rocks of the Skaergaard and Muskox Intrusion (after Taylor 1968). The vertical scale is a non-parametric sequence of chemical evolution, although it is not certain that the Skaergaard granophyres, for example, were the products of differentiation of the basic magma. Nevertheless, trends towards decreasing and increasing ^{18}O-values respectively are already apparent in data for the layered basic rocks.

magnetite should lead to enrichment of the later magma in ^{18}O whereas removal of olivine and plagioclase with no magnetite might well lead to ^{18}O-depletion. These predictions seem to be borne out by O-isotope data for the rocks of the Muskox and Skaergaard layered intrusions (Fig. 15.9) where the extreme Fe- enrichment trend of the latter may be ascribed to lower oxygen fugacity suppressing early magnetite crystallisation. However, as in other Rayleigh processes, a significant change in $^{18}O/^{16}O$ ratios is only apparent after a very high degree of solidification – at least 95% in these examples. It should also be remembered that the more acid granophyres of these intrusions may not be simple extreme differentiates of the basic magmas, Sr-isotope data for the Skaergaard granophyres suggesting appreciable contamination with the Precambrian country rocks (Hamilton 1963). Thus, on the whole, it seems that only very slight fractionation of stable isotopes can arise during closed system evolution and this should be negligible at the highest temperatures associated with partial melting in the mantle. These conclusions do not apply, as will be seen below, where an aqueous vapour phase is important, i.e. ultrametamorphism and melting of crustal rocks containing hydrous minerals.

Application to petrogenesis – oxygen and hydrogen. Figure 15.10 summarises O- and H-isotope data for various types of igneous rocks and some other geologically important materials. Most fresh volcanic rocks and ultramafic or mafic plutonic rocks fall entirely within the ranges $\delta^{18}O = +5$ to $+10‰$ and $\delta D = -50$ to $-100‰$, designated as 'normal' by Taylor (1968). In fact, primary magmatic values for abyssal basalts, kimberlites etc. are very uniform, especially with respect to $\delta^{18}O$ which averages around $+5.5‰$ – a reasonable estimate for the O-isotope composition of the Upper Mantle. We may recognise three

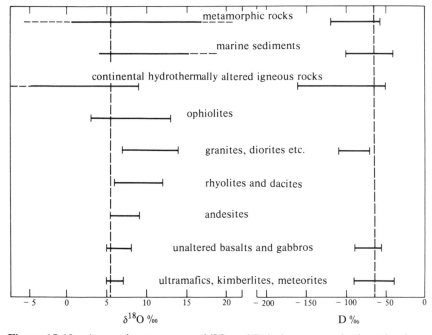

Figure 15.10 Approximate ranges of ^{18}O and D in igneous rocks (mostly after Taylor 1968). The data for hydrogen are much fewer than for oxygen so that δD ranges should be regarded as tentative. The vertical dashed lines represent supposed compositions for mantle oxygen and hydrogen.

characteristic cases of rocks which have isotopic compositions outside the 'normal' ranges: some granites, ophiolites and ^{18}O-depleted igneous complexes.

Granites. Batholiths of intermediate and acid igneous rocks have a wider range of δ^{18}O, extending up to a maximum value of about $+15^{0}/oo$ with a tendency also to more negative δD ($< -100^{0}/oo$). Since marine sediments and metamorphic rocks also exhibit heavy oxygen and, in some cases, light hydrogen (Fig. 15.10), this would be compatible with derivation of some granite magmas by fusion of continental crust, or at least assimilation of crustal rocks. This appears to be confirmed by differences in δ^{18}O between 'S'-type ($+10.4$ to $+12.5^{0}/oo$) and 'I'-type ($+7.7$ to $+9.9^{0}/oo$) plutons of the New England Batholith, Australia (O'Neil *et al.*, 1977), which were classified on the basis of other mineralogical and geochemical criteria to have sedimentary and igneous affinities respectively (Chappell & White 1974).

Ophiolites. Massive ophiolite complexes and altered sea-floor basalts also have a tendency to ^{18}O-enrichment (up to about $+15^{0}/oo$). In this case the major and trace element chemistry, as well as the oceanic environment, rules out assimilation of crustal rocks as an explanation. Concomitant oxidation of iron and development of serpentine minerals indicate that this effect is related to low-temperature interaction with seawater during sea-floor metamorphism (Spooner & Fyfe 1973). Figure 15.11 shows that many igneous minerals (and their alteration products) have positive ^{18}O-fractionations relative to water so that, for example, muscovite or kaolinite in equilibrium with seawater (δ^{18}O = O) would have δ^{18}O $\simeq +7^{0}/oo$ at 300 ° C and $+13^{0}/oo$ at 100 °C. Similar equilibration temperatures would be predicted for ophiolites with a mineralogy including quartz, plagioclase, chlorite and clay minerals. However, in a closed system, the isotopic composition of the water will also be influenced by the interaction, according to the water/rock ratio, expressed as the relative atomic oxygen ratio (w/r). This is reasonably close to the water/rock volumetric ratio during interaction, and may be calculated for instantaneous bulk equilibrium from the equation

$$w/r = \frac{(\delta_r - \delta_r^0)}{(\delta_w^0 - (\delta_r - \Delta_{r-w}))}$$

where pre-interaction parameters are indicated by the superscript zero and all others refer to equilibrium. Δ_{r-w} is the rock–water fractionation (as in Fig. 15.11).

However, although this equation is useful in assessing relative variations in water/rock ratios and interaction temperatures, it has serious limitations for practical application. If we consider an ophiolite with $\delta^{18}O = +10^0/_{00}$ as having resulted from a basalt with a primary $\delta^{18}O$ of $+5^0/_{00}$ exchanging with seawater ($0^0/_{00}$) at 100 °C, the bulk volumetric water/rock ratio would be calculated as 1.7. Since the porosity of basalts is only of the order of 1%, this confirms a fact which we might well have expected, namely that interaction does not occur as an instantaneous closed-system event but during the prolonged passage of water flowing through the rock (Spooner *et al*. 1977). This is a complex problem since both the isotopic composition of the water and the temperature of interaction are likely to change during the exchange. Moreover, in a flow system we have to define water/rock ratios more carefully, distinguishing between bulk integrated ratios for the whole system and specific ratios for the amount of water which has exchanged with a given volume of rock. For example, considering a column with a cross section of 1 cm^2 within a vertical laminar flow system 100 m high, a bulk water/rock ratio of 1 : 1 would correspond to 10^4 cc of water, *all* of which will pass through every 1 cc of rock in the column. In the model of Spooner *et al*. (1977) for the Jurassic ophiolites of East Liguria, O-isotope shifts indicate volumetric water/rock ratios of around 5×10^4 relative to 1 cc of rock. Values derived from the degree of hydration and Fe-oxidation observed in the same rocks are of the order of 10^3 and 10^7 respectively. The discrepancies between these various estimates is not surprising in view of the probability that none of the three types of interaction is likely to be an equilibrium process in a flow system and that each may well proceed to a different 'blocking temperature' during cooling. They must all be regarded as minimum estimates for the actual volume of water which passed through the rocks. Thus it is not surprising that ophiolite complexes contain ore bodies representing very high concentrations of metals present at only trace levels in brines (e.g. Cu).

^{18}O-depleted igneous rocks. These are perhaps the most surprising result of O- and H-isotope investigations. The rocks concerned are high-level intrusive complexes, notably stocks and bosses of granophyre and associated mafic intrusions and lavas, which show a range of ^{18}O-depletion from just below 'normal' values down to $-10^0/_{00}$. δD is correspondingly decreased to values down to $-160^0/_{00}$. Typical examples are the Tertiary igneous complexes of the Hebridean province of Scotland, comparable complexes in the western San Juan Mountains, Colorado, and even parts of the Skaergaard Intrusion, East Greenland. The economically important porphyry copper ore deposits are also associated with this phenomenon (Taylor 1974).

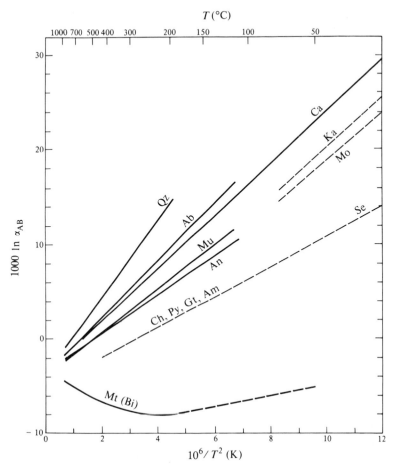

Figure 15.11 Equilibrium mineral–water fractionations for oxygen as a function of temperature (cf. Fig. 15.8a). Mineral abbreviations as before, except Ca = calcite, Ka = kaolinite, Mo = montmorillonite, Ch = chlorite, Gt = garnet, Am = amphibole, Se = serpentinite. Dotted lines are uncertain. Mostly after Taylor (1974). It is probable that most rock compositions approximate (for this diagram) to that of muscovite, so that equilibration with seawater below about 250 °C will result in whole-rock $+^{18}O$ values higher than $+5‰$. On the other hand, equilibration with ^{18}O-depleted meteoric water can result in negative $\delta^{18}O$ values at all temperatures above about 100 °C.

The only natural reservoir which contains light oxygen (relative to SMOW) is meteoric water. The combination of Rayleigh distillation as water vapour moves from the equator to the poles and the increasing equilibrium fractionation with decreasing temperature results in a steady latitude-dependent depletion of precipitated rain water from $\delta^{18}O = 0‰$ to a minimum of about $-55‰$ at the South Pole. δD

decreases correspondingly to less than $-400^{0}/oo$, according to the relationship $\delta D = 8\delta^{18}O + 5$ (Epstein *et al.* 1970). Tertiary meteoric waters at the latitudes of the above-mentioned igneous complexes would have had $\delta^{18}O \simeq -12^{0}/oo$ and $\delta D \simeq -85^{0}/oo$. This leads to the previously unexpected conclusion that the oxygen in many high-level intrusive and volcanic complexes has very largely exchanged with heated meteoric groundwater. As in the case of ophiolites discussed above, minimum bulk volumetric water/rock ratios estimated from the instantaneous equilibrium equation are generally greater than unity, indicating flow systems. This is consistent with the fact that the water discharged in present day geothermal areas (e.g. Iceland and New Zealand) is also heavy-isotope depleted, showing that it is mostly of meteoric origin, though with a slight increase in ^{18}O presumably caused by such interaction with volcanic rocks.

Analysis of separated minerals indicates that much of the isotope exchange occurs as a sub-solidus disequilibrium process. Thus quartz is characteristically only a little below normal igneous values for $\delta^{18}O$ (about $+8^{0}/oo$), whereas feldspars are very strongly affected, with $\Delta_{quartz-feldspar}$ up to $12^{0}/oo$, (cf. Fig. 15.8). This may be partly a kinetic effect associated with exchange at temperatures down to about 300 °C. In the most detailed studies the degree of ^{18}O (and D)-depletion has been shown to increase consistently towards the larger intrusive centres in a complex (e.g. Forester & Taylor 1976, 1977). Rearrangement of the exchange equation given above gives the final $\delta^{18}O$ value of hydrothermally altered rock:

$$\delta_r = \frac{\delta_r^0 + w/r(\delta_w^0 + \Delta_{r-w})}{1 + w/r}$$

from which it follows that more negative values of δ_r can result from either decreased Δ_{r-w} (corresponding to higher exchange temperatures) or increased water/rock ratios. (The reader should check this by substitution, taking $\delta_r^0 = +5.5^{0}/oo$, $\delta_w^0 = -12^{0}/oo$ and $\Delta_{r-w} = 2.5 \times 10^6/T^2 - 4$.) In practice, both effects are usually necessary to explain the observations, so that a recirculating flow system in which upflow is concentrated close to the core of the intrusions is indicated (Fig. 15.12). Zones of ^{18}O-depletion in the surrounding rocks (both volcanic and sedimentary) correspond roughly in form to metamorphic zones (e.g. in Mull; Forester & Taylor 1976), but extending out at least as far as the weakest petrographic signs of pneumatolysis (slight clouding of feldspars). This is especially true in jointed basalts which are efficient aquifers. Some rocks, notably the early granophyres of Skye (Forester & Taylor, 1977), seem to have been intruded as low-^{18}O magmas.

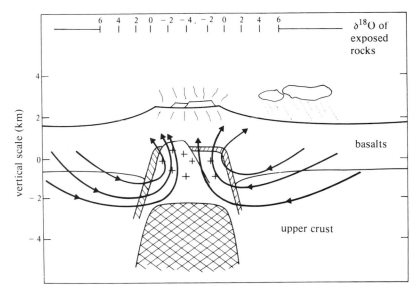

Figure 15.12 Schematic cross section through a high level igneous complex such as those of the Scottish Hebrides (modified from Taylor and Forester 1971). The complex is formed in response to the diapiric rise of a cylinder of mafic or ultramafic composition (cross-hatched). The first stage consists of basaltic volcanism to form the plateau basalts and early cone sheets (not shown). Subsequently acidic plutons and ring-dykes (shaded) are emplaced at local centres. Flow lines are indicated for the circulation of meteoric water in a hydro-thermal system driven by the heat of the cooling complex. Hypothetical contour points for whole-rock $\delta^{18}O$ values are marked at the top. The degree of ^{18}O-depletion increases towards the centre because interaction temperatures are highest there and also because water–rock ratios are highest in the uprising limb of the cell. The vertical scale (exaggerated by about two) is relative to zero for the present level of erosion in the Scottish Hebrides.

These could well arise through the fusion or assimilation of older ^{18}O-depleted rocks.

This is clearly a phenomenon of great geological significance. Hydrothermal interaction at low latitudes need not result in obvious ^{18}O- or D-depletion, but may nevertheless be pervasive. Taylor and Forester (1971) suggested that the trace element chemistry of rocks involved in hydrothermal systems might be considerably modified and that this might explain the high initial $^{87}Sr/^{86}Sr$ ratios of the Tertiary granites of N.W. Scotland compared to the basic rocks (Moorbath & Bell 1965). However, there is no direct evidence to support this, and the Sr-isotope compositions of the basic volcanics should have been affected to the same extent, since their ^{18}O-depletions are at least as great. It should also be remembered that the contents of trace components in meteoric – hydrothermal waters are many orders of magnitude lower than in seawater.

Applications to petrogenesis – carbon and sulphur. The application of these to petrogenetic problems has been extremely limited. Marine limestones have $\delta^{13}C$ values close to that of seawater (0‰ PDB). Carbonatites and diamond fall within a narrow range of $\delta^{13}C$ (ca. -3 to $-8‰$) which may be regarded as characteristic of a primary igneous origin. Secondary carbonates and reduced carbon in igneous rocks show a much greater range of depletion. Most published S-isotope data refer to ore genesis. Sulphur in basic igneous rocks differs little from that in meteorites (0‰ CD), but that in granites is much more variable, as is true for sedimentary and metamorphic rocks. Coleman (1977) has shown that for the New England Batholith of Australia there is a distinction between the 'I'-type (see previous section) granites ($\delta^{34}S = 3.6$ to $-5.0‰$) and 'S'-type granites (less than $-5‰$ or greater than $+5‰$) but the mechanisms responsible, and their implications, are not yet clear. Detailed analysis of trace carbon and sulphur in igneous rocks should be a useful area for future work.

Notes

1. In Table 15.1 each decay scheme is characterised by a half-life as well as a decay constant. The half-life is the time required for half of the radio-active atoms originally present to decay. Since in general:

$$t = \frac{1}{\lambda} \log_e \left[1 + \frac{N_D}{N} \right]$$

where t is the time in years, λ is the decay constant, N_D is the number of daughter atoms present, and N the number of parent atoms still present, the half-life (where $N_D = N$) is given by

$$t_{1/2} = \frac{1}{\lambda} \log_e 2 = \frac{0.693}{\lambda}$$

2. Since this chapter was written a considerable number of Sm-Nd studies of continental igneous rocks have appeared. Overall these indicate an essentially flat chondrite normalised REE pattern for the subcontinental mantle, in contrast to the light-depleted sub-oceanic mantle.

   For a general summary of this and other aspects see O'Nions, R. K., Carter, S. R., Evensen, N. M., and Hamilton, P. J., Geochemical and cosmochemical application of Nd-isotopes, *Ann. Rev. Earth Planet. Sci.* (1979, in press).

Further reading

Faure, G. 1977. *The principles of isotope geology*. New York: Wiley.
Faure, G., and J. L. Powell, 1972. *Strontium isotope geology*.
Doe, B. R. 1970. *Lead isotopes*.
Hoefs, J. 1973. *Stable isotope geochemistry*.

The last three books are volumes 5, 3 and 9 in the monograph series 'Minerals, rocks and inorganic materials' published by Springer-Verlag, Berlin.

Exercises

1. A single whole-rock sample of pegmatitic granite, intruded into 2500 Ma basement rocks which have an average $^{87}Sr/^{86}Sr$ ratio of 0.730, has itself a measured $^{87}Sr/^{86}Sr$ ratio of 0.780 and a $^{87}Rb/^{86}Sr$ ratio of 7.5. Calculate model ages for the granite according to whether it was derived from a mantle source region or by re-melting of the basement. (For simplicity you may assume that the mantle has always had a constant $^{87}Sr/^{86}Sr$ ratio of 0.700 and that the basement rocks were formed rapidly from their source.) In the crustal origin model what is the relative increase in Rb/Sr ratio during melting?

2. Calculate the range of Pb-isotope compositions which would result from a two-stage evolutionary model of the Earth's mantle in which μ_1 had a value of 7.5 from 4570 Ma ago until 2000 Ma ago, at which time a discrete fractionation event produced a range of μ_2 values from 10 to 20. Draw any necessary data from Table 15.1 and p. 380–1. Compare the results with the field for oceanic volcanic rocks shown in Figure 15.7 and explain any major discrepancies.

3. Mineral phenocrysts from a chemically-homogeneous syenite body give the following oxygen isotope compositions:

	Quartz	*Feldspar*	*Biotite*
Rock 1	+6.5	+5.7	+4.0
Rock 2	+8.1	+6.9	+4.1
Rock 3	+9.0	+7.6	+4.4

Assuming as a first approximation that the temperature-dependence of equilibrium fractionation between two minerals is of the form:

$$1000 \ln\alpha_{AB} = K_{AB} \left(\frac{10^6}{T^2} \right)$$

assess whether the data are concordant with isotopic equilibrium for all these rocks. If the quartz–biotite geothermometer corresponds to 900 °C for Rock 1, calculate the corresponding temperatures for Rocks 2 and 3.

4. A composite gabbro–granophyre pluton intruded into late Precambrian metasediments displays the isotopic characteristics given in Table 15.4.

Deduce as much as possible about the petrogenesis of the complex, especially with regard to (a) the age of intrusion, (b) the source regions of the gabbro and granophyre magmas, (c) sub-solidus interaction, including temperatures and water/rock ratios, (d) the origin and conditions of formation of the mineralisation.

Table 15.4

	$Rb(ppm)$	$Sr(ppm)$	$^{87}Sr/^{86}Sr$	$\delta^{18}O\%$	$\delta^{34}S^0/_{00}$	$^{206}Pb/^{204}Pb$
Gabbro Quartz	–	–	–	+5.5	–	–
Feldspar (An_{60})	15	300	0.7047	–1.0	–	18.6
Biotite	400	10	0.8530	–	–	–
Magnetite	–	–	–	+1.0	–	–
Pyrite	–	–	–	–	+8.5	–
Galena	–	–	–	–	+4.0	15.6
Whole-rock	50	200	0.7054	+1.0	–	–
Granophyre						
Whole-rock	100	75	0.7250	+2.0	–	16.0

You may assume the following:

(1) $^{87}Rb/^{86}Sr \simeq 3 \times Rb/Sr$.
(2) A palaeolatitude such that precipitation at the time of emplacement would have had $\delta^{18}O = -5^0/_{00}$.
(3) A primary magmatic $\delta^{18}O$ value for the gabbro of $+6.0^0/_{00}$.
(4) That mineral–H_2O oxygen fractionations are described by the equation $1000 \ln \alpha = A \ (10^6/T^2)-B$ where the constants A and B are:

	Quartz	Plagioclase (*whole-rock*)	Magnetite
A	3.8	2.5	–1.5
B	3.5	3.3	3.5

APPENDICES

Appendix 1

Nomenclature of igneous rocks

The system of nomenclature outlined here is due to Streckeisen (1976) and resulted from a considerable debate amongst petrographers seeking uniformity. For the plutonic rocks it seems at the time of writing to be gaining a certain degree of acceptance.

The system is mineralogical and based essentially on the following parameters:

Q = silica minerals
A = alkali feldspars including albite ($An_0 - An_5$)
P = plagioclase $An_5 - An_{100}$
F = feldspathoids (leucite, pseudoleucite, nepheline, analcime, soda-lite, nosean, hauyne, cancrinite etc.)
M = mafic minerals (micas, amphiboles, pyroxenes, olivines), plus ore minerals, garnets, melilites, primary carbonates and accessories (zircon, apatite, sphene etc.)

Rocks with $M = 90\text{--}100\%$ are classified separately as 'mafitites' (ultra-mafic rocks). For the remaining rocks the parameters Q, A, and P or A, P, and Fa are recalculated at 100% for projection into the double $Q - A - P - F$ triangle (Fig. A1.1). The names applied to rocks falling in the numbered fields are given in Table A1.1 with their volcanic equivalents. Obviously as a system for naming volcanic rocks, this, like any mineralogical system, is far from perfect.

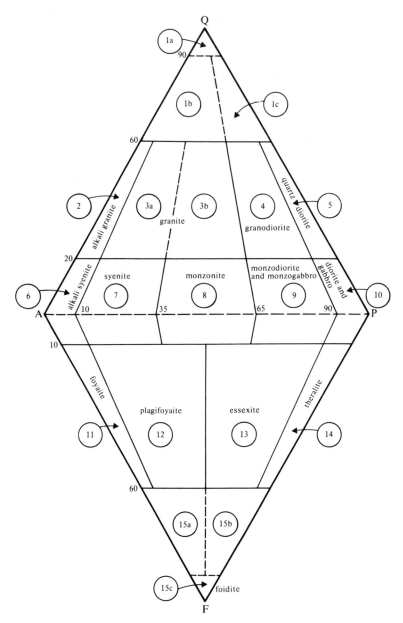

Figure A1.1   The double triangle QAPF showing fields of plutonic rock types (Streckeisen, 1976) as classified by modal mineralogy. For details see Table A1.1.

Table A1.1 Nomenclature of rocks in the Q–A–P and A–P–F triangles (after Streckeisen, 1976)

Field	Plutonic rocks	Volcanic rocks
1a	quartz rocks *sensu stricto*	–
1b	quartz–granite	–
1c	quartz–granodiorite	–
2	alkali granite	alkali rhyolite
3	granite	
3a	syenogranite	rhyolite
3b	monzogranite	rhyodacite
4	granodiorite	dacite
5	quartz–diorite	quartz–andesite
6	alkali syenite	alkali trachyte
7	syenite	trachyte
8	monzonite ($\equiv$ syenodiorite)	
9	monzodiorite and monzogabbro (= syenodiorite and syenogabbro)	trachyandesite and trachybasalt
10	diorite, gabbro, anorthosite[a]	andesite, basalt
11	foyaite[b]	phonolite[c]
12	plagifoyaite	tephritic phonolite
13	essexite	phonolitic tephrite
14	theralite	tephrite
15a	foyaitic foidite	phonolitic foidite
15b	theralitic foidite	tepritic foidite
15c	foidite[d] *sensu stricto*	extrusive foidite *sensu stricto*

[a]The distinction between diorite and gabbro is based on plagioclase composition (gabbro has plagioclase more calcic than An_{50}). Anorthosites have M < 10.
[b]Many plutonic rocks in fields 11 and 12 would normally be termed nepheline syenites.
[c]Most volcanic rocks in fields 11–15 have potassic equivalents (see Table 2.1).
[d]The commonest rocks in this group are the clinopyroxene–nepheline rocks, Urtite (M = 0–30), ijolite (M = 30–70) and Melteigite (M = 70–90). The approximate volcanic equivalent of ijolite is nephelinite (though most nephelinites contain appreciable normative feldspar).

The important ultramafic rocks (not shown in the QAPF diagram) are dunite (90–100% olivine), peridotite (30–90% olivine) and pyroxenite and hornblendite (0–30% olivine). The peridotites are divided into harzburgite (olivine + hypersthene), lherzolite (olivine + orthopyroxene + clinopyroxene) and wehrlite (olivine + clinopyroxene).

Average major element compositions and CIPW norms of common igneous rock types*

	Nepheline syenite	Phonolite	Syenite	Trachyte	Granite	Rhyolite	Adamellite
SiO_2	54.99	56.19	58.58	61.21	71.30	72.82	68.65
TiO_2	0.60	0.62	0.84	0.70	0.31	0.28	0.54
Al_2O_3	20.96	19.04	16.64	16.96	14.32	13.27	14.55
Fe_2O_3	2.25	2.79	3.04	2.99	1.21	1.48	1.23
FeO	2.05	2.03	3.13	2.29	1.64	1.11	2.70
MnO	0.15	0.17	0.13	0.15	0.05	0.06	0.08
MgO	0.77	1.07	1.87	0.93	0.71	0.39	1.14
CaO	2.31	2.72	3.53	2.34	1.84	1.14	2.68
Na_2O	8.23	7.79	5.24	5.47	3.68	3.55	3.47
K_2O	5.58	5.24	4.95	4.98	4.07	4.30	4.00
H_2O+	1.30	1.57	0.99	1.15	0.64	1.10	0.59
H_2O-	0.17	0.37	0.23	0.47	0.13	0.31	0.19
P_2O_5	0.13	0.18	0.29	0.21	0.12	0.07	0.19
CO_2	0.20	0.08	0.28	0.09	0.05	0.08	0.09
Total	99.69	99.86	99.74	99.94	100.07	99.96	100.05
Q	–	–	0.83	5.00	29.06	32.87	25.17
C	–	–	–	–	0.92	1.02	0.28
Or	32.98	30.96	29.29	29.41	24.50	25.44	23.66
Ab	29.45	35.48	44.34	46.26	31.13	30.07	29.36
An	3.78	1.50	7.24	7.05	8.04	4.76	11.55
Lc	–	–	–	–	–	–	–
Ne	21.77	16.50	–	–	–	–	–
Di	4.53	6.89	5.35	2.14	–	–	–
Wo	–	0.73	–	–	–	–	–
Hy	–	–	4.16	2.06	3.37	1.34	5.66
Ol	0.28	–	–	–	–	–	–
Mt	3.27	4.05	4.41	4.33	1.75	2.14	1.79
Il	1.13	1.18	1.60	1.34	0.58	0.54	1.03
Ap	0.30	0.41	0.70	0.49	0.28	0.17	0.44
Cc	0.45	0.17	0.64	0.20	0.12	0.17	0.20

	Granodiorite	Rhyodacite	Dacite	Tonalite	Diorite	Andesite
SiO_2	66.09	65.55	65.01	61.52	57.48	57.94
TiO_2	0.54	0.60	0.58	0.73	0.95	0.87
Al_2O_3	15.73	15.04	15.91	16.48	16.67	17.02
Fe_2O_3	1.38	2.13	2.43	1.83	2.50	3.27
FeO	2.73	2.03	2.30	3.82	4.92	4.04
MnO	0.08	0.09	0.09	0.08	0.12	0.14
MgO	1.74	2.09	1.78	2.80	3.71	3.33
CaO	3.83	3.62	4.32	5.42	6.58	6.79
Na_2O	3.75	3.67	3.79	3.63	3.54	3.48
K_2O	2.73	3.00	2.17	2.07	1.76	1.62
H_2O+	0.85	1.09	0.91	1.04	1.15	0.83
H_2O-	0.19	0.42	0.28	0.20	0.21	0.34
P_2O_5	0.18	0.25	0.15	0.25	0.29	0.21
CO_2	0.08	0.21	0.06	0.14	0.10	0.05
Total	99.90	99.79	99.78	100.01	99.98	99.93
Q	22.36	22.67	22.73	16.62	10.28	12.37
C	0.26	0.25	–	–	–	–
Or	16.11	17.72	12.82	12.24	10.42	9.60
Ab	31.73	31.05	32.07	30.67	29.96	29.44
An	17.34	15.04	20.01	22.58	24.40	26.02
Lc	–	–	–	–	–	–
Ne	–	–	–	–	–	–
Di	–	–	0.11	1.49	4.67	4.84
Wo	–	–	–	–	–	–
Hy	7.40	6.19	5.73	9.68	12.56	9.49
Ol	–	–	–	–	–	–
Mt	2.00	3.08	3.53	2.66	3.63	4.74
Il	1.03	1.14	1.09	1.40	1.80	1.65
Ap	0.42	0.59	0.34	0.58	0.68	0.50
Cc	0.19	0.47	0.14	0.33	0.23	0.11

	Monzonite	Latite	Trachy-andesite	Mugearite	Trachy-basalt	Hawaiite	Gabbro
SiO_2	62.60	61.25	58.15	50.52	49.21	47.48	50.14
TiO_2	0.78	0.81	1.08	2.09	2.40	3.23	1.12
Al_2O_3	15.65	16.01	16.70	16.71	16.63	15.74	15.48
Fe_2O_3	1.92	3.28	3.26	4.88	3.69	4.94	3.01
FeO	3.08	2.07	3.21	5.86	6.18	7.36	7.62
MnO	0.10	0.09	0.16	0.26	0.16	0.19	0.12
MgO	2.02	2.22	2.57	3.20	5.17	5.58	7.59
CaO	4.17	4.34	4.96	6.14	7.90	7.91	9.58
Na_2O	3.73	3.71	4.35	4.73	3.96	3.97	2.39
K_2O	4.06	3.87	3.21	2.46	2.55	1.53	0.93
H_2O+	0.90	1.09	1.25	1.27	0.98	0.79	0.75
H_2O-	0.19	0.57	0.58	0.87	0.49	0.55	0.11
P_2O_5	0.25	0.33	0.41	0.75	0.59	0.74	0.24
CO_2	0.08	0.19	0.08	0.15	0.10	0.04	0.07
Total	99.53	99.83	99.97	99.89	100.01	100.05	99.15
Q	14.02	14.26	7.80	–	–	–	0.71
C	–	–	–	–	–	–	–
Or	24.00	22.85	19.00	14.57	15.06	9.03	5.49
Ab	31.56	31.38	36.80	40.06	29.39	33.61	20.26
An	13.97	15.64	16.58	17.09	20.10	20.62	28.60
Lc	–	–	–	–	–	–	–
Ne	–	–	–	–	2.23	–	–
Di	3.78	2.05	3.95	6.19	11.85	10.94	13.70
Wo	–	–	–	–	–	–	–
Hy	6.01	4.57	6.06	2.19	–	0.10	22.13
Ol	–	–	–	4.14	8.28	9.32	–
Mt	2.78	3.91	4.73	7.07	5.36	7.16	4.36
Il	1.48	1.54	2.07	3.97	4.55	6.13	2.13
Ap	0.60	0.79	0.97	1.74	1.38	1.75	0.56
Cc	0.17	0.43	0.17	0.35	0.23	0.08	0.17

	Norite	Dolerite	Basalt	Basanite	Nephelinite	Tephrite
SiO_2	50.44	50.18	49.20	44.30	40.60	47.80
TiO_2	1.00	1.14	1.84	2.51	2.66	1.76
Al_2O_3	16.28	15.26	15.74	14.70	14.33	17.00
Fe_2O_3	2.21	2.86	3.79	3.94	5.48	4.12
FeO	7.39	8.05	7.13	7.50	6.17	5.22
MnO	0.14	0.19	0.20	0.16	0.26	0.15
MgO	8.73	6.78	6.73	8.54	6.39	4.70
CaO	9.41	9.24	9.47	10.19	11.89	9.18
Na_2O	2.26	2.56	2.91	3.55	4.79	3.69
K_2O	0.70	1.04	1.10	1.96	3.46	4.49
H_2O+	0.84	1.46	0.95	1.20	1.65	1.03
H_2O-	0.13	0.43	0.43	0.42	0.54	0.22
P_2O_5	0.15	0.27	0.35	0.74	1.07	0.63
CO_2	0.18	0.18	0.11	0.18	0.60	0.02
Total	99.86	99.64	99.95	99.89	99.89	100.01
Q	–	1.40	–	–	–	–
C	–	–	–	–	–	–
Or	4.15	6.12	6.53	11.61	3.16	26.56
Ab	19.14	21.63	24.66	12.42	–	8.81
An	31.49	27.12	26.62	18.38	7.39	16.55
Lc	–	–	–	–	13.57	–
Ne	–	–	–	9.55	21.95	12.14
Di	10.58	12.89	14.02	21.03	32.36	19.89
Wo	–	–	–	–	–	–
Hy	26.38	20.93	15.20	–	–	–
Ol	0.34	–	1.50	12.38	2.32	3.69
Mt	3.21	4.15	5.49	5.72	7.95	5.98
Il	1.90	2.67	3.49	4.77	5.05	3.35
Ap	0.36	0.63	0.82	1.74	2.51	1.46
Cc	0.41	0.42	0.26	0.40	1.37	0.05

	Anorthosite	Pyroxenite	Peridotite	Dunite
SiO_2	50.28	46.27	42.26	38.29
TiO_2	0.64	1.47	0.63	0.09
Al_2O_3	25.86	7.16	4.23	1.82
Fe_2O_3	0.96	4.27	3.61	3.59
FeO	2.07	7.18	6.58	9.38
MnO	0.05	0.16	0.41	0.71
MgO	2.12	16.04	31.24	37.94
CaO	12.48	14.08	5.05	1.01
Na_2O	3.15	0.92	0.49	0.20
K_2O	0.65	0.64	0.34	0.08
H_2O+	1.17	0.99	3.91	4.59
H_2O-	0.14	0.14	0.31	0.25
P_2O_5	0.09	0.38	0.10	0.20
CO_2	0.14	0.13	0.30	0.43
Total	99.80	99.83	99.46	98.58
Q	−	−	−	−
C	−	−	−	0.80
Or	3.86	3.75	2.02	0.47
Ab	23.16	7.76	4.15	1.69
An	49.71	13.54	8.32	1.17
Lc	−	−	−	−
Ne	1.89	−	−	−
Di	8.61	42.09	11.22	−
Wo	−	−	−	−
Hy	−	6.26	15.79	14.48
Ol	2.01	15.12	46.39	67.38
Mt	1.40	6.20	5.23	5.20
Il	1.22	2.79	1.19	0.18
Ap	0.21	0.90	0.23	0.47
Cc	0.31	0.28	0.67	1.00

*Analyses are selected from the compilation by Le Maitre (1976) based on 26 373 published analyses. Rock names are those used in the original publications, so irrespective of the highly varied classificatory criteria in the individual publications the analyses given here constitute a good guide to average usage of the terms. For explanation of CIPW norms see Appendix 3.

Norm calculations

The idea of the norm calculation was introduced into petrology by Cross, Iddings, Pirsson and Washington who devised a calculation by which rock analyses could be calculated into a hypothetical assemblage of standard minerals known as the **norm** to distinguish it from the mineral assemblage actually present which is termed the **mode** (Cross *et al.* 1903). The original purpose of the calculation was essentially taxonomic and an elaborate classificatory scheme based on the normative mineral percentages was also presented. The classification is thus entirely chemically based and groups together rocks of similar bulk chemical composition irrespective of their mineralogy. The original classification has fallen into disuse, probably because of its ponderous and artificial nomenclature involving a hierarchy of classes, sub-classes, orders, rangs, and sub-rangs, but the calculation itself remains in common use.

The essence of the calculation is elegant and is based on a number of simplifications, the most important of which are as follows

(a) The 'magma' is assumed to crystallise in an anhydrous condition so that no hydrous phases (e.g. hornblende, biotite) are formed.
(b) The ferromagnesian minerals are assumed to be free of Al_2O_3. Thus the Al_2O_3 content of the rock can be used to fix the amount of feldspar or feldspathoid in the norm, assuming that enough Na_2O, K_2O and CaO is available to satisfy the Al_2O_3.
(c) The magnesium/iron ratio of all the ferromagnesian minerals is assumed to be the same.
(d) Several mineral pairs are assumed to be incompatible, thus nepheline and quartz for example never appear in the same norm.

Using these basic principles the various oxides can be allotted to minerals in a particular sequence to calculate the normative mineral assemblage. The resultant norm often shows close correspondence with the modal mineralogy, particularly for dry basic rocks such as gabbros which have cooled slowly. Discrepancies may be considerable, however, for rocks rich in, say, hornblende or micas. It is, however, important to realise that the original purpose of the norm was not to achieve correspondence with the mode, but to try to indicate affinities which were otherwise masked by differences in grain size and mineralogy caused by differing water contents and cooling histories. To be able to compare gabbros with partly glassy basalts is obviously valuable.

Norm calculations are used for a variety of purposes today and there have been several modifications of the original Cross *et al.* system (known as the

CIPW system for short). The most useful function, however, is still the classificatory distinction between the silica-oversaturated and silica-undersaturated rocks. Another function has been demonstrated in Chapter 9 where the norm calculation is used to project the compositions of natural basalts into a phase diagram.

Calculation of the CIPW norm

Practically all norms calculated today are produced by computer, but it is essential to understand the nature of the calculation by working through some examples (see the exercises attached to Ch. 9). A form of the type illustrated in Figure A3.1 is useful for hand calculations. The following procedure is abbreviated for the purpose of calculating norms of most normal rocks. Full procedures are given in many older texts (e.g. Cross *et al.* 1903; Holmes 1930; Johannsen 1931). Note that some of the rules refer to normative minerals which are encountered rarely and are not shown on the form.

Rules for calculation according to the CIPW sequence. In the following the word 'amount' should be taken to mean the molecular proportion obtained by dividing the weight per cent of an oxide by its molecular weight.

1. The molecular proportion of each constituent is determined by dividing by the appropriate molecular weight. These values are inserted in the horizontal column marked mol. prop.
2. The amount of MnO is added to that of FeO.
3a. An amount of CaO equal to 3.33 that of P_2O_5 (or 3.00 P_2O_5 and 0.33 F, if the latter is present) is allotted for apatite. To do this the P_2O_5 amount is written in the bottom right hand box of the form (intersection of the apatite and P_2O_5 columns) and 3.33 times this amount is entered in the apatite horizontal column under CaO. (At the end of the calculation all the allotments in a particular vertical column must sum to the original molecular proportion for that oxide.)
3b. An amount of FeO equal to that of the TiO_2 is allotted for ilmenite. If there is an excess of TiO_2, an equal amount of CaO is to be allotted to it for provisional titanite, but only after the allotment of CaO to Al_2O_3 for anorthite (Rule 4c). If there is still an excess of TiO_2 it is calculated as rutile.
4a. An amount of Al_2O_3 equal to that of the K_2O is allotted for provisional orthoclase.
4b. An excess of Al_2O_3 over the K_2O is allotted to an equal amount of remaining Na_2O for provisional albite. If there is insufficient Al_2O_3, see Rule 4f.
4c. If there is an excess of Al_2O_3 over the $K_2O + Na_2O$ used in 4a and 4b, it is allotted to an equal amount of remaining CaO for anorthite.
4d. If there is an excess of Al_2O_3 over this CaO, it is calculated as corundum.
4e. If there is an excess of CaO over the Al_2O_3 of 4d, it is reserved for diopside and wollastonite (see Rules 7a and 7b).

Oxides		SiO$_2$	TiO$_2$	Al$_2$O$_3$	Fe$_2$O$_3$	FeO	MnO	MgO	CaO	Na$_2$O	K$_2$O	P$_2$O$_5$	Molecular wt (multiply by)	Norm
	wt per cent													
	mol. wt (divide by)	60	80	102	160	72	71	40	56	62	94	142		
	mol. prop.													
	Q (S)												60	
	or (KAS$_6$)												556	
Pl	ab (NAS$_6$)												524	
Pl	an (CAS$_2$)												278	
	lc (KAS$_4$)												436	
	ne (NAS$_2$)												284	
	C (A)												102	
	ac (NF^{+++}S$_4$)												462	
Di	wo (CS)												116	
Di	en (MS)												100	
Di	fs (F^{++}S)												132	
	wo (CS)												116	
Hy	en (MS)												100	
Hy	fs (F^{++}S)												132	
Ol	fo (M$_2$S)												140	
Ol	fa (F$_2^{++}$S)												204	
	mt (F^{++}F^{+++})												232	
	he (F^{+++})												160	
	il (F^{++}T)												152	
	ap (C$_{3.33}$P)												310	
	TOTALS													

Figure A3.1 Form used for calculation of CIPW norm.

4f. If in 4b there is an excess of Na_2O over Al_2O_3, it is to be reserved for acmite and possibly for sodium metasilicate. There is then no anorthite in the norm.

5a. To an amount of Fe_2O_3 equal to that of the excess of Na_2O over Al_2O_3 (see Rule 4f) is allotted an equal amount of Na_2O for acmite.

5c. If, as usually happens, there is an excess of Fe_2O_3 over remaining Na_2O, it is assigned to magnetite, an equal amount of FeO being allotted to it out of what remains from the formation of ilmenite (see Rule 3b).

5d. If there is still an excess of Fe_2O_3, it is calculated as haematite.

6. All the MgO and the FeO remaining from the previous allotments are added together and their relative proportions are ascertained.

7a. To the amount of CaO remaining after allotment in Rule 4e is allotted provisionally an equal amount of MgO + FeO to form diopside, the relative proportions of these two, as they occur in the remainder, being preserved.

7b. If there is an excess of CaO, it is reserved for provisional wollastonite.

7c. If there is an excess of MgO + FeO over that needed for diopside (7a), it is reserved for provisional hypersthene.

8a. Allot the necessary amount of silica to CaO to form titanite (Rule 3b, 1 : 1), to excess Na_2O to form acmite (Rule 5a, 4 : 1), to K_2O for provisional orthoclase (Rule 4a, 6 : 1), to Na_2O for provisional albite (Rule 4b, 6 : 1), to CaO for provisional anorthite (Rule 4c, 2 : 1), to CaO + (Mg, Fe)O for diopside (Rule 7a, 1 : 1), to excess CaO for wollastonite (Rule 7b, 1 : 1), and to (Mg, Fe)O for hypersthene (Rule 7c, 1 : 1).

The amounts of silica so assigned are subtracted from the total silica.

8b. If there is an excess of SiO_2, as there commonly is, it is calculated as quartz.

8c. If there is a deficiency of silica in 8a, the SiO_2 allotted to hypersthene (Rule 7c) is subtracted from the general sum of 8a and the remainder subtracted from the total SiO_2. If there is here an excess of SiO_2 more than enough to equal half the amount of (Mg, Fe)O of Rule 7c, it is alloted to the (Mg, Fe)O of 7c to form hypersthene and olivine, and is distributed according to equations (1) and (2).

Let x = the number of hypersthene molecules, y = the number of olivine molecules, M = the amount of available (Mg, Fe)O, S = the amount of available SiO_2; then

$$x = 2S - M \tag{1}$$

$$y = M - x \tag{2}$$

In this operation the relative proportions of MgO and FeO determined in Rule 6 and used in forming diopside (Rule 7a) are to be preserved. The fixed and provisional molecules of Rule 8a are calculated into their percentage weights (Rule 9).

If there is not enough silica to equal half the amount of (Mg, Fe)O of Rule 7c, all the (Mg, Fe)O of Rule 7c is calculated as olivine, SiO_2 equal to half its amount being assigned to it.

8d. If there is a deficiency of SiO_2 in 8c, the SiO_2 allotted to titanite (3b) is subtracted from the general sum of 8a, and the CaO and TiO_2 are calculated as perovskite ($CaO.TiO_2$).

8e. The sum of the SiO_2 needed to form the molecules of 8a is deducted from the total SiO_2, except that olivine is substituted for hypersthene and perovskite for titanite (sphene), and that albite is not included. If there is an excess of more than twice (and, of course, less than six times) that of the Na_2O for the provisional albite of 8a, this is distributed between albite and nepheline according to equations (3) and (4). If the excess is less than twice the Na_2O, it is taken care of in Rule 8f.

Let x = the number of albite molecules, y = the number of nepheline molecules, N = the amount of available Na_2O, S = the amount of available SiO_2; then

$$x = \frac{S - 2N}{4} \tag{3}$$

$$y = N - x \tag{4}$$

8f. If there is still a deficiency of SiO_2 – that is, in 8e, not enough to equal twice the amount of available Na_2O – all this Na_2O is allotted to nepheline and the K_2O is distributed between orthoclase and leucite as follows:

The sum of the SiO_2 needed for the molecules in 8a is subtracted from the total SiO_2, olivine being substituted for hypersthene, perovskite for titanite, and nepheline for albite, orthoclase being disregarded. If there is an excess of more than four times (and, of course, less than six times) that of K_2O, it is distributed between orthoclase and leucite according to equations (5) and (6). If the excess is less than four times the K_2O, it is taken care of in Rule 8g.

Let x = the number of orthoclase molecules, y = the number of leucite molecules, K = the amount of available K_2O, S = the amount of available SiO_2; then

$$x = \frac{S - 4K}{2} \tag{5}$$

$$y = K - x \tag{6}$$

8g. If there is still a deficiency of SiO_2 — that is, in 8f, not enough to equal four times the amount of K_2O – we have to distribute the CaO of wollastonite and diopside between these two and calcium orthosilicate, and the (Mg, Fe)O of diopside between diopside and olivine, according to available SiO_2. There are two possible cases.

The most common case is that in which there is no wollastonite or its amount is insufficient to satisfy the deficiency in SiO_2. Here, after allotting SiO_2 to form leucite (from all the K_2O), nepheline, anorthite, acmite, and olivine of Rule 8c, and possibly zircon and sodium metasilicate, the amount thus used is deducted from the total SiO_2, the residue being the available silica.

Let x = the number of new diopside molecules, y = the number of new olivine molecules, z = the number of calcium orthosilicate molecules, S = the amount of available SiO_2, M = the amount of (Mg, Fe)O (of provisional diopside), C = the amount of CaO of provisional diopside and wollastonite; then

$$x = \frac{2S - M - C}{2} \qquad (7)$$

$$y = \frac{M - x}{2} \qquad (8)$$

$$z = \frac{C - x}{2} \qquad (9)$$

In these three equations x = half the number of SiO_2 molecules in diopside, y = the number of SiO_2 molecules in olivine, and z = the number of SiO_2 molecules in calcium orthosilicate.

In the second case, where there is sufficient tentative wollastonite to meet the deficiency of SiO_2, the total amount of SiO_2 in the rock is subtracted from the sum of the SiO_2 which has been allotted to leucite, nepheline, anorthite, acmite, diopside, olivine, and to tentative wollastonite. The deficit is the number of molecules of necessary calcium orthosilicate, and also the amount of SiO_2 to be assigned to it. This requires twice as much CaO. The rest of the CaO remains in wollastonite and takes an equal amount of SiO_2, while the diopside remains unchanged.

9. All the allotments have now been made and the weight per cents of normative minerals are calculated by multiplying oxide amounts by the molecular weight of the minerals. To do this take the amount of any constituent which is present as a single molecule in the mineral formula and multiply this by the molecular weight to derive the weight per cent of that mineral; for example, if the figures in the orthoclase column are 0.320 SiO_2, 0.054 Al_2O_3 and 0.054 K_2O, the weight per cent of orthoclase is $0.054 \times 556 = 30.02$ (the number of orthoclase molecules being the same as the number of Al_2O_3 or K_2O molecules). The list of weight per cents of minerals should give the same total as the original analysis making allowance for the exclusion of H_2O, any other minor constituents that have been ignored, and rounding-off errors.

Effect of oxidation state on the norm. The oxidation state of Fe in an analysed rock has an important effect on the norm and does influence the silica saturation significantly. Hence when highly oxidised rocks are studied, assuming the oxidation is a secondary effect, a misleading result can be obtained.

Consider two rocks which are identical except that one has suffered secondary alteration involving an increase in the ratio of Fe^{3+}/Fe^{2+}. The oxidised rock will contain more normative magnetite and there will thus be less MgO + FeO available for diopside, hypersthene, and olivine. Since these are silicates, less SiO_2 will be required and there will thus be a relative excess of SiO_2 in the norm of the oxidised sample. Oxidation can change a rock from Ne-normative to Hy-normative or Q-normative. To overcome this problem it is common practice to adopt a standard oxidation state for analysed samples before calculating norms. This is in any case necessary if a separate analysis for Fe^{3+} and Fe^{2+} was not made. The standard best adopted is debatable but as good a choice as any often consists of choosing the Fe^{3+} Fe^{2+} ratio of the least oxidised sample of a group and adjusting the others to this value. Alternatively, Brooks (1976) has for example suggested standardising the analyses of basaltic rocks at an Fe_2O_3/FeO retio of 0.15.

Silica saturation in rocks. The great majority of rocks fall into one of the three major normative groups shown below:

1. *Silica-oversaturated* (Q + Hy in the norm).
2. *Silica-saturated* (Ol + Hy in the norm).
3. *Silica-undersaturated* (Ol + Ne in the norm).

The degree of silica saturation is highly dependent on the availability of SiO_2 Na_2O and K_2O in the analysis. Hence the alkali–silica diagram can be used to make an approximate prediction of this as shown in Figure A3.2. Notice that the silica-saturated field narrows to a negligible width in the more salic compositions. This is readily understood by thinking of the olivine + hypersthene assemblage in the norm as a 'buffer' which can take up different amounts of silica by varying the olivine/hypersthene ratio. When the rocks are rich in MgO and FeO a large amount of provisional hypersthene will be formed during the norm calculation. Quite large variations of silica can thus be accommodated without forming either quartz or nepheline. In salic rocks, however, there is very

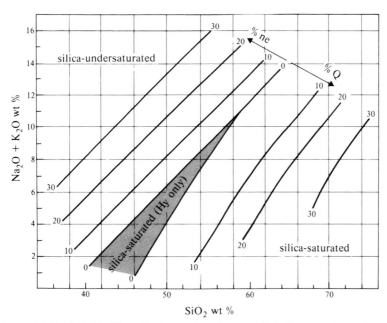

Figure A3.2 Alkali–silica diagram showing predicted normative character (CIPW weight per cent) of rocks.

little provisional hypersthene and hence only a very small probability that the silica content will be just sufficient to maintain this assemblage without the production of quartz or nepheline. Hence trachytic rocks are overwhelmingly either quartz- or nepheline-normative while many basalts are olivine – hypersthene normative.

Molecular norms

Several petrologists have argued that it is more useful to express normative mineralogy in molecular per cents rather than weight per cents, and it is certainly true that most petrological calculations are easier to perform in molecular rather than weight terms. Several systems have been devised, notable being those of Niggli (1954), Eskola (1954) and Barth (1955). The sequence of the calculation is flexible in the Niggli system and allows a greater choice of normative minerals which hence gives it applications in metamorphic and sedimentary petrology as well as in igneous petrology.

As an example, in the Eskola system the analysis is first recast into so called one-cation percentages. This is carried out by first dividing each oxide per cent by the *equilavent weight* of the oxide to produce a column of atomic proportions. The equivalent weight of the oxide is the same as the molecular weight when only one cation is present (e.g. SiO_2, FeO) and half the molecular weight when two cations are present (e.g. Al_2O_3, Na_2O). The atomic proportions are then totalled and recalculated as percentages (the one-cation percentages). Thus for example if the Si percentage is 49.5 this signifies that of every 100 cations in the rock 49.5 of them are Si.

The norm can then be calculated according to any preferred sequence but the proportions in which cations are allotted may differ from that in the CIPW weight norm. For example, in the weight norm orthoclase is made up of K_2O, Al_2O_3, and SiO_2 in the proportions $1 : 1 : 6$ whereas in the Eskola norm the proportions of K, Al, and Si are allotted as $1 : 1 : 3$ since the formula is $KAlSi_3O_8$. The difference between the 6 in the weight system and the 3 in the other is of course accounted for by the use of molecular weights as opposed to equivalent weights in the earlier stage of the calculation. Finally the normative minerals are derived in the Eskola norm by totalling the cations allotted to each mineral. For example if orthoclase is represented by allotments of 0.76 K, 0.76 Al, and 2.28 Si this represents 3.80% of orthoclase in the final norm, signifying that of every 100 molecules present 3.80 of them are orthoclase.

Appendix 4

Calculation of plotting parameters for O'Hara (1968) polybaric phase diagram

1. Oxide constituents of the analysis are divided by molecular weights (see Appendix 3) to derive molecular proportions.
2. Plotting parameters in the CMAS tetrahedron are calculated in the following form and then expressed as percentages:

$$C = (\text{mol.prop. } CaO - 3P_2O_5 + 2Na_2O + 2K_2O) \times 56.08$$
$$M = (\text{mol.prop. } FeO + MnO + NiO + MgO - TiO_2) \times 40.31$$
$$A = (\text{mol.prop. } Al_2O_3 + Cr_2O_3 + Fe_2O_3 + Na_2O + K_2O + TiO_2) \times 101.96$$
$$S = (\text{mol.prop. } SiO_2 - 2Na_2O - 2K_2O) \times 60.09$$

3. To calculate the projection parameters from olivine into CS–MS–A find the coefficient to balance the equation

$$\text{rock} + p \cdot \text{olivine} = x \cdot CS + y \cdot MS + z \cdot A$$

where p is the amount of olivine which must be added to or subtracted from the rock to bring it into the plane (p may be positive or negative) and x, y, and z expressed as percentages are the plotting parameters. The formulae for olivine, CS etc, must be expressed in the same form as the plotting parameters for the rock, hence for a rock with parameters calculated numerically as C = c, M = m, A = a, and S = s the equation has the format:

$$C_cM_mA_aS_s + p \cdot M_{57.3}S_{42.7} = x \cdot C_{48.3}S_{51.7} + y \cdot M_{40.1}S_{59.9} + z \cdot A_{100}$$

Then for the balance of C:

$$c = 48.3x$$

For M:

$$m + 57.3p = 40.1y$$

For A:

$$a = 100z$$

For S:

$$s + 42.7p = 51.7x + 59.9y$$

from which x, y, and z can be calculated.

4. The equation for projection from orthopyroxene into M_2S–C_2S_3–A_2S_3 is:

$$C_cM_mA_aS_s + p \cdot M_{40.1}S_{59.9} = x \cdot M_{57.3}S_{42.7} + y \cdot C_{38.4}S_{61.6} + z \cdot A_{53.1}S_{46.9}$$

5. The equation for projection from diopside into C_3A–M–S is:

$$C_cM_mA_aS_s + p \cdot C_{25.9}M_{18.6}S_{55.5} = x \cdot C_{62.3}A_{37.7} + y \cdot M_{100} + z \cdot S_{100}$$

Appendix 5

Some representative mineral analyses

	Pyroxenes					Olivines	
	1	2	3	4	5	6	7
SiO_2	57.73	50.08	52.92	52.84	51.92	39.87	30.56
TiO_2	0.04	0.64	0.50	0.22	0.77	0.03	0.72
Al_2O_3	0.95	1.23	2.80	0.44	1.85	–	0.09
Fe_2O_3	0.42	2.34	0.85	1.06	31.44	0.86	0.10
FeO	3.57	27.85	5.57	16.89	0.75	13.20	60.81
MnO	0.08	0.85	0.15	0.56	–	0.22	3.43
MgO	36.13	15.78	16.40	23.51	–	45.38	3.47
CaO	0.23	1.44	19.97	4.06	–	0.25	1.13
Na_2O	–	0.05	0.35	0.19	12.86	0.04	–
K_2O	–	0.02	0.01	–	0.19	0.01	–
H_2O+	0.52	–	0.10	–	0.17	0.33	–
H_2O-	0.04	–	0.07	0.22	–	0.10	–
Total	100.52	100.28	100.67	99.99	99.95	100.29	100.31

	Amphiboles			Biotite	Alkali feldspars	
	8	9	10	11	12	13
SiO_2	44.99	39.68	52.41	37.17	65.58	67.27
TiO_2	1.46	7.12	0.45	3.14	–	–
Al_2O_3	11.21	12.81	0.61	14.60	19.58	18.35
Fe_2O_3	3.33	4.04	14.37	3.75	0.21	0.92
FeO	13.17	8.79	14.82	26.85	–	
MnO	0.31	0.16	1.46	0.06	–	–
MgO	10.41	11.22	5.07	4.23	0.12	–
CaO	12.11	11.06	1.33	0.17	0.49	0.15
Na_2O	0.97	3.37	4.94	0.15	5.90	6.45
K_2O	0.76	1.04	2.10	8.25	7.88	7.05
H_2O+	1.48	0.78	2.02	1.35	0.23	0.08
H_2O-	0.04	0.15	0.10	–	0.14	0.08
Total	100.41	100.55	99.98	100.57	100.13	100.35

	Plagioclase		Nepheline	Leucite
	14	15	16	17
SiO_2	64.10	49.06	41.88	54.62
TiO_2	–	–	0.03	–
Al_2O_3	22.66	32.14	32.99	22.93
Fe_2O_3	0.14	0.27	0.74	0.26
FeO	0.17	–	–	0.26
MnO	–	–	0.07	–
MgO	0.25	0.20	–	–
CaO	3.26	15.38	0.78	0.08
Na_2O	9.89	2.57	16.11	0.66
K_2O	0.05	0.17	6.82	21.02
H_2O+	0.17	0.13	0.71	0.12
H_2O-	0.06	0.03	0.03	–
Total	100.75	99.95	100.16	99.95

	Ilmenite	*Magnetite*	*Apatite*	*Zircon*	*Sphene*
	18	19	20	21	22
SiO_2	0.51	0.27	–	32.51	30.44
TiO_2	50.02	tr	–	–	39.66
Al_2O_3	–	0.21	–	0.21	–
Fe_2O_3	4.19	68.85	–	0.08	–
FeO	42.18	30.78	0.21	–	0.14
MnO	1.44	–	1.52	–	0.05
MgO	0.46	tr	0.54	0.01	–
CaO	0.71	tr	52.40	0.22	27.20
Na_2O	–	–	–	–	0.37
K_2O	–	–	–	–	–
H_2O+	} 0.13	–		0.03	0.56
H_2O-		–	–	–	0.08
Total	99.64	100.11	100.60	100.12	100.26

Key

All analyses are taken from Deer *et al*. (1966). Numbers in parentheses in the list below refer to their table and analysis numbers.

1. Enstatite (Table 13, 1). Includes 0.46% Cr_2O_3, 0.35% NiO.
2. Ferrohypersthene (Table 13, 2).
3. Chromian augite (Table 13, 6). Includes 0.88% Cr_2O_3, 0.10% NiO.
4. Magnesian pigeonite (Table 13, 10).
5. Aegirine (Table 13, 12).
6. Olivine (chrysolite) (Table 1, 2).
7. Olivine (fayalite) (Table 1, 4).
8. Hornblende (Table 15, 7). Includes 0.17% P_2O_5.
9. Kaersutite (Table 15, 11). Includes 0.33% F.
10. Riebeckite (Table 15, 14). Includes 0.30% F.
11. Biotite (Table 18, 6). Includes 0.85% F.
12. Microcline perthite (Table 30, 5).
13. Sanidine (Table 30, 9).
14. Oligoclase (Table 31, 3).
15. Bytownite (Table 31, 6).
16. Nepheline (Table 35, 2).
17. Leucite (Table 36, 1).
18. Ilmenite (Table 44, 1).
19. Magnetite (Table 46, 4).
20. Apatite (Table 50, 1). Includes 40.98% P_2O_5, 1.15% F, 3.74% Cl.
21. Zircon (Table 2, 1). Includes 67.02% ZrO_2 + HfO_2, 0.04% REE oxides.
22. Sphene (Table 3, 1). Includes small amounts of Zr, Nb, Ta, V and REE.

Answers to exercises

Chapter 2

2. Analyses given in Table 2.4 represent the following rocks (names are those given by the authors quoted):

1. Phonolite, Nyamaji volcano, Kenya (Le Bas 1970).
2. Rhyolite, Andes, Chile (Zeil & Pichler 1967).
3. Average leucitite, Roccamonfina, Italy (Appleton 1972).
4. Trachyte, Nairobi, Kenya (Saggerson 1970).
5. Hawaiite/mugearite, Ross volcano, Auckland Is., New Zealand (Wright 1970).
6. Oceanite, Kartala volcano, Grande Comore (Strong 1972).
7. Trachybasalt, Tristan da Cunha (Baker *et al.* 1964).
8. Andesite, Andes, Chile (Zeil & Pichler 1967).

4. (a) 64%.
   (b) Any of the other oxides except SiO_2 could be used as the basis for a similar variation diagram though the sequence of the three most basic rocks would change slightly.
   (c) K_2O. This oxide is enriched through the series by a factor of 2.6, only slightly lower than the value of 2.76 for P_2O_5. Na_2O by way of contrast shows enrichment only by a factor of 1.33. The sequence thus shows a typical increase of K/Na ratio with fractionation.

Chapter 3

1. The system has four components CaO, MgO, Al_2O_3 and SiO_2. At fixed pressure three phases + liquid still has one degree of freedom so melting is not isothermal.
2. The diagram determined experimentally is given by Bowen (1914). The temperature of the eutectic is about 1362 °C and the liquid has composition about Di_{82}.
3. (a) An_{81}.
   (b) An_{13}.
   (c) liquid is An_{20}, solid is An_{57}.
   (d) An_0 at 1120 °C.
4. (a) At low temperatures composition Ne_{90} consists of one phase only, a nepheline containing 10% of An in solution. On heating the first change is encountered at 1293 °C when a carnegieite solid solution (composition approximately Ne_{98}) begins to exsolve. Between this temperature and 1353 °C both carnegieite and nepheline become more calcic. When

1353 °C is *first* reached the system consists of 67% carnegieite (comp. $Ne_{95.5}$) and 33% of nepheline (comp. Ne_{82}).

If more heat is now supplied the system begins to melt and *the temperature ceases to rise* (the system is now univariant and consists of three phases) until all the nepheline has disappeared. The system now consists of 78% carnegieite (same composition as above) and 22% liquid (comp. Ne_{71}).

Continued heating results in the progressive melting of the remaining carnegieite. Both carnegieite and liquid become more sodic until at 1472 °C the last remnant of carnegieite (comp. about $Ne_{98.5}$) disappears.

(b) We write down the state of the system, with phase proportions, for temperature incrementally above and incrementally below those of the univariant reaction (see above), i.e.

$$78 \text{ Car} + 22 \text{ L} = 60 \text{ Car} + 40 \text{ Ne}$$
$$\text{(above)} \qquad\qquad \text{(below)}$$

which reduces to:

$$18 \text{ Car} + 22 \text{ L} = 40 \text{ Ne}$$

The general expression for what has happened is

$$\text{carnegieite} + \text{liquid} = \text{nepheline}$$

This reaction demonstrates how a phase may react with the liquid and diminish in amount on cooling, or alternatively may increase in amount during heating. Such reactions are sometimes termed **peritectics**.

(c) L = nepheline + anorthite at 1300 °C. This is an example of a eutectic reaction (both solid phases increase in amount down-temperature).

(d) $Ne_{65}An_{35}$.

(e) $Ne_{54}An_{46}$.

5. Conversion of the analysis to mole per cent gives MgO = 13.5, FeO = 4.1 and Fe_2O_3 = 2.9. Using the first two figures Figure 3.18 gives a liquidus temperature of about 1205 °C and an olivine composition of about Fo_{92}. Converting all the Fe_2O_3 to additional FeO gives a new FeO figure of 9.9. Using this the liquidus is found as about 1235 °C and the olivine composition is Fo_{82}. The liquidus temperature experimentally determined for this rock under Ni/NiO buffer (Krishnamurthy & Cox 1977, specimen 312) is 1220 ± 7 °C, a satisfactory correspondence. Since the rock is almost certainly affected by some post-crystallisation oxidation of iron the olivine composition Fo_{85} actually present in small quantities seems likely to be the equilibrium olivine rather than an obviously xenocrystic type.

Chapter 4

1. Figure E.1 shows: Alkemade lines dashed, down-temperature arrows on divariant equilibria with double bar indicating thermal maxima.

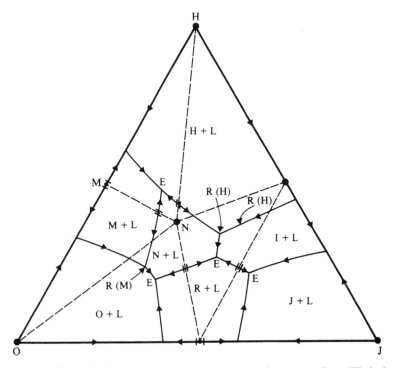

Figure E.1 Alkemade lines, temperature arrows, and nature of equilibria in system H–O–J.

Resorptional equilibria are labelled R with the resorbed phase in parentheses. Ternary eutectics are labelled E.

2. (a) Figure E.2 gives the isothermal section for 1050 °C.
   (b) The range is bounded by three lines, i.e. from AC to the 1040 °C eutectic, from AC to BC as far as the boundary curve A + AC + L, and the boundary curve A + AC + L. These compositions lie inside the sub-solidus triangle A–BC–AC so they must contain A when solid. However, the tie-line from AC to the liquid on the A + AC + L boundary curve sweeps across this area so all early A must be resorbed. Liquids then proceed to the eutectic at 1040 °C (where A begins to crystallise again) via the AC + L field and the BC + AC + L boundary.
   (c) Mineralogical composition is $A_{56}BC_{18}B_{26}$ before melting. Liquid composition is $A_{38}B_{41}C_{21}$, residual solid is $A_{71}B_{29}$, temperature is 1090 °C.

3. Figure E.3 shows the approximate form of the pseudobinary.

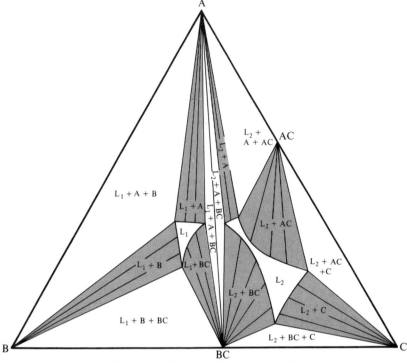

Figure E.2   Isothermal section for system A–B–C at 1050 °C.

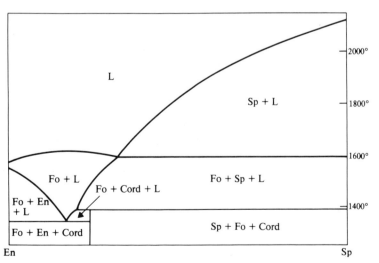

Figure E.3   Sketch pseudobinary for enstatite–spinel join in $MgO–Al_2O_3–SiO_2$.

Chapter 5

1. The isothermal section and liquidus projection are given in Figure E.4.
2. (a) Pure Y plus an XZ solid solution of composition approximately $X_{82}Z_{18}$.
   (b) Approximately $X_{47}Y_{33}Z_{20}$.
   (c) Approximately $X_{32}Y_{31}Z_{37}$.
   (d) Approximately 12%.
   (e) Minimum at about $X_{27}Y_{31}Z_{42}$.

Chapter 6

1. (a) SiO_2 47.69, TiO_2 0.15, Al_2O_3 15.89, Fe_2O_3 0.77, FeO 7.22, Mno 0.11, MgO 18.43, CaO 7.68, Na_2O 1.59, K_2O 0.15.
   (b) Orthopyroxene, 15%. Initial inspection should suggest this as a likely answer. The lower Al_2O_3 and CaO in the cumulus-enriched rock (B) suggests that plagioclase is not involved in significant amounts. The higher SiO_2 of the cumulate suggests orthopyroxene rather than olivine. In this case the suggested orthopyroxene solution could have been tested numerically, but in general a graphical treatment is best for exploring possibilities, e.g. that orthopyroxene, although dominant, is not the only phase concerned.
2. (b) The variation diagrams show on first inspection a marked inflection (coinciding roughly with analysis C) in MgO, Al_2O_3, CaO and FeO.

   The trend A–C has been formed by the removal of what we shall term the *first* extract. This so obviously coincides with olivine only that it requires little further discussion. Note the characteristic 'fanning' behaviour of the plots for elements not included in olivine (Ca, Al, Na, K, P, Ti). Olivine control should be identified readily in all sorts of variation diagrams involving two oxides at a time. It is characterised in other plots by constant Ca/Al ratios, constant Na/K ratios, etc. (straight lines back projecting through the origin). The second extract (analyses C–E) clearly involves a phase or phases rich in Ca and Al but still containing some Mg and Fe. Obvious possible combinations are Ol + Pl, Pl + Cpx, and Ol + Pl + Cpx. It is not very likely on general grounds that olivine has disappeared at the same time as the new phase or phases appeared so the first and third of these seem the most likely possibilities.

   Figure E.5 shows extract triangles for Ol–Cpx–Pl mixtures for the elements Al, Ca and Mg. These are superimposed on an equilaterial triangle to show that Ol + Pl mixtures are consistent with the data so far. However, Fe is not consistent with this (the Ol + Pl mixture would make it rise slightly but it stays steady) but since Ti also falls at this stage it seems likely that a titaniferous ore mineral has entered the crystallisation sequence. This would impoverish iron more than expected by the Ol + Pl hypothesis. Finally, since P falls off in the last analysis the possibility of apatite entering as a fractionating phase must be considered.

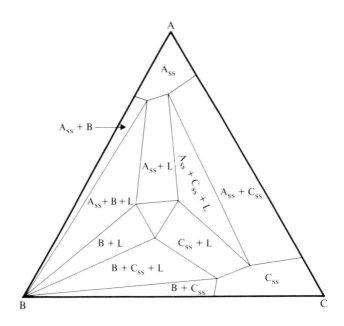

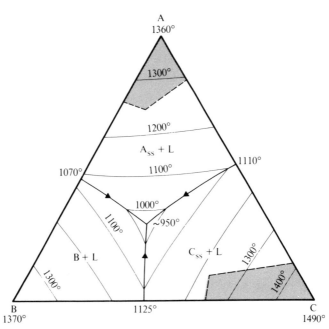

Figure E.4   Isothermal section at 1000 °C and liquidus projection for system A–B–C. Solid solution fields are shaded.

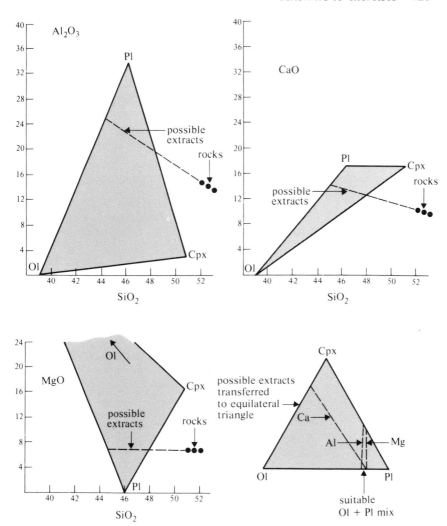

Figure E.5 Extract triangles.

Summarising, the sequence of fractionating phases would appear to be (1) olivine, (2) olivine + plagioclase + ilmenite, (3) olivine + plagioclase + ilmenite + apatite.

(c) Phenocrysts of olivine might well be present in all the rocks; Cpx phenocrysts should be absent in all; plagioclase phenocrysts should be present in C–E (or possibly only D and E) together with ilmenite; apatite should be present in D and E or perhaps just E.

Chapter 8

1. Since P (see Fig. E.6) lies in the sub-solidus tetrahedron A–B–AC–D the final liquid of crystallisation must be Q (i.e. the liquid equilibrating with A + B + AC + D). Crystallisation starts with precipitation of D followed by C + D. While C and D crystallise the liquid moves in projection across the surface C + D + L to meet the curve AC + C + D + L at R. The liquid moves down the boundary curve to S where resorption of C is complete since at this stage the bulk composition has come to lie on the bounding face AC + D + L of the tetrahedron AC + C + D + L. The liquid now leaves the curve and tracks across the AC + D + L surface directly to Q where it solidifies with the co-precipitation of A + B + AC + D.

2. The equilibrium is likely to be resorptional for B. The back-tangent cuts the plane A–B–C on the B-poor side of BC–ABC, co-ordinates of this point being approximately $B_{-11}BC_{64}ABC_{47}$.

3. The pyroxenes must be SiO_2-undersaturated relative to the plane En–An–Di. Hytönen & Schairer conclude that solid solution towards Ca-Tschermak's and Mg-Tschermak's molecule (CAS and MAS) is involved, in addition to the well-known mutual solid solution of Di and En. This involves substitution of Al in both octahedral and tetrahedral positions, e.g. $Ca\,Mg\,Si_2O_6$ (Di) shows solution towards $Ca\,Al\,AlSiO_6$ (Ca-Tschermak's).

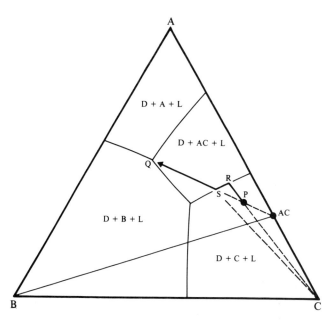

Figure E.6 The liquid path is shown by the heavy arrowed lines P–R–S–Q. Dotted lines are projections of four-phase tetrahedra involving D, the phase from which projection is made.

Reference to Figure 8.2 shows that CAS and MAS lie in the plane CS–MS–A (on the Al-rich side of the pyrope–grossular join) which is SiO_2-poor relative to En–An–Di.

Chapter 9

1. Calculated norms should correspond closely with those given below. Rounding-off errors and choice of molecular weights used in the calculations will cause small variations.

   Rock A. Weight norm: Q 4.80, Or 3.89, Ab 16.77, An 29.19, Di 20.07 (consisting of Wo 10.21, En 5.50, Fs 4.36), Hy 19.94 (consisting of En 11.10, Fs 8.84), Mt 2.09, Il 2.13, Ap 0.29.
   Molecular norm: Q 5.12, Or 3.80, Ab 18.30, An 29.93, Di 19.80 (Wo 9.90, En 6.38, Fs 3.52), Hy 19.10 (En 12.28, Fs 6.82), Mt 2.14, Il 1.56, Ap 0.25.

   Rock B. Weight norm: Or 6.12, Ab 9.43, An 27.52, Ne 4.83, Di 37.08 (Wo 19.02, En 11.20, Fs 6.86), Ol 2.56 (Fo 1.54, Fa 1.02), Mt 5.34, Il 4.10, Ap 1.12.
   Molecular norm: Or 6.35, Ab 11.70, An 28.55, Ne 5.13, Di 38.24 (Wo 19.12, En 14.80, Fs 4.32), Ol 0.64 (Fo 0.50, Fa 0.14), Mt 5.30, Il 3.12, Ap 1.00.

   Predictions from Figure A3.2 are that Rock A should contain about 6% Q and that B should be approximately critically undersaturated, i.e. no Ne, no Hy (such rocks are extremely rare since they contain just sufficient SiO_2 to convert all the provisional Hy to Ol). The first prediction is good, the second rather poor. Rock B is rather unusual in containing high CaO (see Fig. 2.3b) which results in high normative Di taking up substantial SiO_2 so that more Ne appears in the norm than expected. Most rocks give better correspondence than this. The reader is recommended to make further tests with any norms available.

2. Calculated projection parameters are Ol – 39.3, Di – 31.8, Pl – 28.8, indicating the crystallisation sequence – Ol followed by Cpx, followed by Pl, with reasonably large temperature intervals between the appearance of each phase and the next. The experimentally determined results for this rock (MK of Tilley *et al.* 1964) are Ol *in* 1420 °C, Cpx *in* 1210 °C and Pl *in* 1165 °C.

3. The co-ordinates in CMAS are C – 15.9, M – 11.3, A – 23.8, S – 49.0 and in CS–MS–A (olivine projection) are CS – 27.0, MS – 53.7, A – 19.4. Plotted Figure 9.9 this rock projects onto the Ol + Cpx + Pl + L boundary indicating that *if* olivine is the first phase to crystallise then Cpx and Pl should enter the crystallisation sequence together. In fact this rock (ML1935 of Tilley *et al.* 1964) shows Ol + Cpx *in* 1180 °C, Pl *in* 1160 °C so that it does in fact lie almost *on* the boundary concerned. The second part of the question is solved by seeing whether the projected point coincides with the boundary Ol + Opx + Ga + L at any pressure. The lower diagram in Figure 9.9

suggests that it might coincide at a pressure of somewhat above 30 kbar. A liquid originating here at say 35 kbar if suddenly decompressed (transported upwards) would find itself overtaken by the olivine field immediately and could thus evolve into the composition given by olivine-loss, having become by coincidence near-cotectic with regard to the three main low pressure phases.

Chapter 11

1. (a) (i) MgO = 12.2, phenocrysts nil.
   (ii) MgO contents (whole rocks) are 14.1%, 16%, and 16% respectively. Phenocryst contents are 5%, 10% and 10%.
   (b) 80%.

For elaborated versions of this type of calculation see Krishnamurthy and Cox (1977).

Chapter 14

1. To assess trace element equilibrium between the nodule minerals and host basalt we calculate apparent distribution coefficients (conc. in mineral/conc. in basalt):

	Ce	Yb	K	Rb	Sr
Cpx	0.090	0.325	0.004	0.003	0.069
Olivine	0.005	0.008	0.006	0.015	0.003

These are very reasonable values (cf. Table 14.1) and only the high K and Rb contents of the olivine relative to the Cpx suggest some alteration. The establishment of equilibrium permits the hypothesis but does not prove it. We then calculate bulk distribution coefficients for the assemblage on the basis of it representing the partial melt residue ($D = 0.8 \times K_D Ol + 0.2 \times K_D Cpx$):

	Ce	Yb	K	Rb	Sr
D	0.022	0.072	0.006*	0.013*	0.016

(*estimates of 0.002 preferred for these as explained above).

Rearranging the equilibrium melting equation to give F

$$F = \frac{(C_0/C_L - D)}{(1 - D)}$$

for Ce concentrations of 1 and 2 in the source this yields F = 0.025 and 0.075 respectively, representing a mean value of 5% melting. Corresponding figures for Yb are 1.3% and 10.2%. Substituting these data back into the equation to give C_L/C_0 and hence C_0 we obtain approximately $K = 240 \pm 175$ ppm, Rb = 0.6 ± 0.3 ppm and Sr = 20 ± 15 ppm (errors due to uncertainty in F only).

2. First calculate D for each stage.

	Ce	Yb	K	Rb	Sr	Ni	Eu	Eu*
I	0.060	2.14	0.0015	0.001	0.036	1.2	0.26	0.33
II	0.051	0.066	0.061	0.022	0.675	5.4	0.13	0.079

The effect of the second stage on Ce/Yb ratios will be negligible, so for the first stage we must effect an increase of $18/12 \div 2.5/7.5 = 4.5$. The increase, assuming Rayleigh fractionation is given by $F^{(D_{Ce} - D_{Yb})}$ and this yields (by taking logs) $F = 0.48$. Thus the first stage must involve crystallisation of 48% of the original liquid, yielding a starting composition for the second stage of:

Ce_N	Yb_N	K	Rb	Sr	Ni	Eu/Eu^*
5.0	3.3	625	1.0	100	86	1.1

The amount of crystallisation necessary in the second stage is then calculated from: Ce enrichment $= 18/5.0 = F^{(0.051 - 1)}$ yielding $F = 0.26$, i.e. 74% solidification. This yields a final composition of

Ce_N	Yb_N	K	Rb	Sr	Ni	Eu/Eu^*
18	11	2210	3.7	155	0.2	1.0

The results show a major discrepancy for Ni, which is far too low, and K and Rb which are over two times too high. The Rb/Sr ratio (0.024 instead of 0.010) is particularly difficult to reproduce as it is not significantly different in the two basalts. On the other hand; reducing the amount of plagioclase fractionation would leave a slightly positive Eu-anomaly from the first stage. Ni could be adjusted by mixing with more of the primitive magma (as in open-system crystal fractionation) but the Rb/Sr would be controlled by the second stage product. Thus the only circumstances under which the crystal fractionation hypothesis could be defended would be to postulate an additional process of metasomatic enrichment in K and Rb.

3. The data will be found to fit mixing curves (rather better than can be expected in practice!). Rb/Sr shows the greatest variation and using the Rb/Sr versus Zr/Y plot, limiting values of Zr/Y can be extrapolated at Rb/Sr $= 0$ and infinity to give 10.6 and 4.7 respectively (using the two most extreme samples to calculate A $= 52\,000$; B $= -11\,000$; C $= -1550$; and D $= 16\,300$ as in the text). Limiting compositions can be obtained by plotting Rb/Sr versus each of the elements in turn:

	Rb	Sr	Y	Zr
Basic end-member	(0)	534	14	148
Acid end-member	235	(0)	44	208

Alternatively these values are obtained, rather more easily in terms of mathematics, by plotting each element against Rb extrapolated to 0 for the basic end-member and against Sr extrapolated to 0 for the acid end-member. The data are consistent with assimilation of older differentiated material by an alkali basalt magma but do not alone rule out a simple crystal fractionation relationship between the end-members, although the slow increase of Zr relative to Rb suggest that this is unlikely.

Chapter 15

1. Solving the isochron equation for the case of mantle derivation we have $0.780 - 0.700 = 7.5 \ (e^{\lambda t_1} - 1)$ where $\lambda = 1.42 \times 10^{-11}$ years^{-1}, whence $t_1 = 747$ Ma. For the crustal derivation case we first have to define a Sr-isotope growth line for the crustal segment concerned. $0.730 - 0.700 = R \ (e^{\lambda t_2} - 1)$ where $t_2 = 2.5 \times 10^9$ years, whence R, the crustal ^{87}Rb/^{86}Sr ratio, $= 0.830$. We can then determine the intersection of this with the growth line for the granite at t_3 with an 'initial' ^{87}Sr/^{86}Sr ratio of x:

$$0.780 - x = 7.5 \ (e^{\lambda t_1} - 1)$$
$$0.730 - x = 0.83 \ (e^{\lambda t_3} - 1)$$

whence $t_3 = 526$ Ma, a significantly younger age than t_2. In the latter case the Rb/Sr ratio increases to a very close approximation, by the same factor as the ^{87}Rb/^{86}Sr ratio, that is by a factor of 9. If Rb is regarded as an incompatible element and Sr is buffered by plagioclase in the source during melting this would imply around 10% melting to produce the granite, not allowing for crystal fractionation (see Ch. 14).

2. The multistage equations for Pb-isotope evolution will be:

$$(^{206}\text{Pb}/^{204}\text{Pb})_T = 9.307 + 7.5(e^{\lambda t_0} - e^{\lambda t_1}) + \mu_2(e^{\lambda t_1} - 1)$$

and

$$(^{207}\text{Pb}/^{204}\text{Pb})_T = 10.294 + \frac{7.5}{137.88}(e^{\lambda' t_0} - e^{\lambda' t_1}) + \frac{\mu_2}{137.88}(e^{\lambda' t_1} - 1)$$

where $\lambda = 1.55125 \times 10^{-11}$ years^{-1}, $\lambda' = 9.8485 \times 10^{-11}$ years^{-1}, $t_0 = 4.57 \times 10^9$ years, $t_1 = 2.0 \times 10^9$ years and $t_2 = 0$. Substitution of $\mu_2 = 10$ and then 20 yields a range of present day ^{206}Pb/^{204}Pb values of 17.62 to 21.26 and ^{207}Pb/^{204}Pb ratios of 14.95 to 15.40. Comparison with Figure 15.7 shows that whereas the former is similar to the range shown by mantle-derived magmas the predicted ^{207}Pb/^{204}Pb ratios are too low. This demonstrates that the choice of μ_1 was too low, since almost all the radiogenic ^{207}Pb accumulated during the early stages of the Earth's history when ^{235}U was relatively abundant.

3. Since for equilibrium $\delta_A - \delta_B \approx K_{AB}(10^6/T^2)$ and also $\delta_B - \delta_C \approx K_{BC}(10^6/T^2)$ there should be a straight line passing through the origin for $\delta_A - \delta_B$ versus $\delta_A - \delta_C$. This is demonstrated in the present case, for example, by plotting Δ_{Q-B} versus Δ_{F-B}.

   For Rock 1 we have $K_{QB} = 2.5 \ (1173^2/10^6) = 3.44$, whence we derive temperatures of 654 °C and 592 °C for Rocks 2 and 3 respectively. These are clearly not crystallisation temperatures and suggest metamorphic re-equilibration.

4. $Rb - Sr$. The minerals and whole-rock data for the gabbro define an isochron age of about 85 Ma with an initial ^{87}Sr/^{86}Sr ratio of 0.7045. The initial ^{87}Sr/^{86}Sr ratio of the granophyre at this time would have been 0.7202. This suggests a mantle source for the gabbro and a crustal one (or at least substantial contamination) for the granophyre. The granophyre might have been

derived directly from 1000 Ma-old basement if Rb/Sr ratios increased by a factor of about four resulting from partial melting and fractional crystallisation.

Oxygen. The mineral–mineral fractionations may be estimated as the difference in their individual mineral–H_2O fractionations. Thus ($\Delta_{Qz-Mt} = (3.8 - (-1.5))$ $(10^6/T^2) - (3.5 - 3.5)$, whence the observed fractionations yield a Qz–Mt temperature of 812 °C and a Qz–Plag temperature of 174 °C. This, and the low whole-rock $\delta^{18}O$ value, demonstrates meteoric water interaction. Water/rock atomic oxygen ratios are given by substitution in the equation given in the text as at least 1.0 at 500 °C, or 31.3 at 250 °C. The minimum interaction temperature is given when $w/r = \infty$, at 245 °C. The granophyre is also probably involved in the hydrothermal system, but its primary $\delta^{18}O$ value is not known.

Sulphur. The pyrite–galena fractionation of $+4.5°/oo$ corresponds to an equilibration temperature of about 200–250 °C (from Fig. 15.8b).

Lead. The isotopic composition of Pb in both galena and the granophyre suggests an origin in the Precambrian crust, whereas that in the gabbro feldspar is as expected for a young mantle-derived igneous rock. This suggests leaching of Pb (and perhaps S) from the crust by recirculating hydrothermal water during emplacement of the igneous complex as the mode of formation of the mineralisation.

References

Allegre, C. J., Treuill, M., Minster, J.-F., Minster, B., and F. Albarede 1977. Systematic use of trace elements in igneous processes. Part 1: Fractional crystallization processes in volcanic suites. *Contr. Mineral. Petrol.*, **60**, 57–76.

Andersen, O. 1915. The system anorthite–forsterite–silica. *Amer. J. Sci.*, 4th series, **39**, 407–54.

Anderson, A. T., Clayton, R. N., and T. K. Mayeda 1971. Oxygen isotope geothermometry of mafic igneous rocks. *J. Geol.*, **79**, 715–29.

Appleton, J. D. 1972. Petrogenesis of potassium-rich lavas from the Roccamonfina volcano, Roman region, Italy. *J. Petrol.*, **13**, 425–56.

Armstrong, R. L. 1968. A model for the evolution of strontium and lead isotopes in a dynamic Earth. *Rev. Geophys.*, **6**, 175–99.

Armstrong, R. L., Taubeneck, W. H., and P. O. Hales 1977. Rb–Sr and K–Ar geochronometry of Mesozoic granitic rocks and their Sr isotopic composition, Oregon, Washington and Idaho. *Bull. Geol. Soc. Amer.*, **88**, 397–411.

Atkins, F. B. 1965. *The pyroxenes of the Bushveld Igneous Complex*. Unpub. D.Phil. thesis, Oxford University.

Baker, P. E. 1968. Petrology of Mt. Misery volcano, St. Kitts, West Indies. *Lithos*, **1**, 124–50.

Baker, P. E., Gass, I. G., Harris, P. G., and R. W. Le Maitre 1964. The vulcanological report of the Royal Society expedition to Tristan da Cunha, 1962. *Phil. Trans. R. Soc. Lond.*, A., **256**, 439–578.

Barberi, F., Ferrara, G., Santacroce, R., Treuil, M., and J. Varet 1975. A transitional basalt–pantellerite sequence of fractional crystallisation, the Boina centre (Afar rift, Ethiopia). *J. Petrol.*, **16**, 22–56.

Barth, T. F. W. 1955. Presentation of rock analyses. *J. Geol.*, **63**, 340–63.

Bartlett, R. W. 1969. Magma convection, temperature distribution, and differentiation. *Amer. J. Sci.*, **267**, 1067–82.

Beckinsale, R. D., Pankhurst, R. J., Skelhorn, R. R., and J. N. Walsh 1978. Geochemistry and petrogenesis of the Early Tertiary lava pile of the Isle of Mull, Scotland. *Contr. Mineral. Petrol.*, **66**, 415–27.

Bell, J. D. 1966. Granites and associated rocks of the eastern part of the Western Red Hills Complex, Isle of Skye. *Trans. R. Soc. Edinb.*, **66**, 307–43.

Bhattacharji, S., and C. H. Smith 1964. Flowage differentiation. *Science*, **145**, 150–3.

Bickle, M. J., Hawksworth, C. J., Martin, A., Nisbet, E. G., and R. K. O'Nions 1976. Mantle composition derived from the chemistry of ultramafic lavas. *Nature*, **263**, 577–80.

Biggar, G. M., and D. B. Clarke 1971. Aspects of phase equilibria in the inversion and exsolution of pyroxenes. *Indian Miner.*, **12**, 1–13.

Blake, D. H., Elwell, R. W. D., Gibson, I. L., Skelhorn, R. R., and G. P. L. Walker 1965. Some relationships resulting from the intimate association of acid and basic magmas. *Quart. J. Geol. Soc. Lond.*, **121**, 31–49.

Blank, H. R., and M. E. Gettings 1973. Subsurface form and extent of the Skaergaard Intrusion, East Greenland. *Trans. Amer. Geophys. Un.*, **54**, 507.

Blaxland, A. B., Breemen, van O., Emeleus, C. H., and J. G. Anderson 1978.

Age and origin of the major syenite centers in the Gardar province of south Greenland: Rb–Sr studies. *Bull. Geol. Soc. Amer.*, **89**, 231–44.

Bott, M. H. P., and J. Tuson 1973. Deep structure beneath the Tertiary volcanic regions of Skye, Mull and Ardnamurchan, north-west Scotland. *Nature Phys. Sci.*, **242**, 114–16.

Bottinga, Y., and D. F. Weill 1970. Densities of liquid silicate systems calculated from partial molar volumes of oxide components. *Amer. J. Sci.*, **269**, 169–82.

Bottinga, Y., and D. F. Weill 1972. The viscosity of magmatic silicate liquids: a model for calculation. *Amer. J. Sci.*, **272**, 438–75.

Bowen, N. L. 1912. The system nephelite–anorthite. *Amer. J. Sci.*, 4th series, **33**, 551–73.

Bowen, N. L. 1913. The melting phenomena of the plagioclase feldspars. *Amer. J. Sci.*, 4th series, **34**, 577–99.

Bowen, N. L. 1914. The ternary system diopside–forsterite–silica. *Amer. J. Sci.*, 4th series, **38**, 207–64.

Bowen, N. L. 1915. The crystallisation of haplobasaltic, haplodioritic, and related magma. *Amer. J. Sci.*, 4th series, **40**, 161–85.

Bowen, N. L. 1928. *The evolution of the igneous rocks.* Princeton University Press.

Bowen, N. L., and O. Anderson 1914. The system $MgO-SiO_2$. *Amer. J. Sci.*, 4th series, **37**, 487–500.

Bowen, N. L., and J. F. Schairer 1935. The system $MgO-FeO-SiO_2$. *Amer. J. Sci.*, **29**, 151–217.

Boyd, F. R., and G. M. Brown 1968. Electron-probe study of exsolution in pyroxenes. *Yb. Carnegie Instn. Wash.*, **66**, 353–9.

Boyd, F. R., and J. L. England 1960. The quartz–coesite transition. *J. Geophys. Res.*, **65**, 749–56.

Bravo, M. S., and M. J. O'Hara 1975. Partial melting of phlogopite-bearing synthetic spinel- and garnet-lherzolites. *Phys. Chem. Earth*, **9**, 845–54.

Brooks, C. K. 1976. The Fe_2O_3/FeO ratio of basalt analyses: an appeal for a standardised procedure. *Bull. Geol. Soc. Denmark*, **25**, 117–20.

Brooks, C., and W. Compston 1965. The age and initial Sr^{87}/Sr^{86} of the Heemskirk granite, western Tasmania. *J. Geophys. Res.*, **70**, 6249–6262.

Brooks, C., Hart, S. R., Hofmann, A., and D. E. James 1976b. Rb–Sr mantle isochrons from oceanic regions. *Earth Plan. Sci. Lett.*, **32**, 51–61.

Brooks, C. James, D. E., and S. R. Hart 1976a. Ancient lithosphere: its role in young continental volcanism. *Science*, **193**, 1086–1094.

Brown, G. M. 1956. The layered ultrabasic rocks of Rhum, Inner Hebrides. *Phil. Trans. R. Soc. Lond.*, B, **240**, 1–53.

Brown, G. M., Holland, J. G., Sigurdsson, H., Tomblin, J. F., and R. J. Arculus 1977. Geochemistry of the Lesser Antilles volcanic island arc. *Geochim. Cosmochim. Acta*, **41**, 785–802.

Brown, G. M., and J. F. Schairer 1968. Melting relations of some calc-alkaline volcanic rocks. *Yb. Carnegie Instn. Wash.*, **66**, 460–7.

Bryan, W. B., Finger, L. W., and F. Chayes 1969. Estimating proportions in petrographic mixing equations by least squares approximation. *Science*, **163**, 926–7.

Byers, F. M., Orkild, P. P., Carr, W. J., and W. D. Quinlivan 1968. Timber Mountain tuff, southern Nevada, and its relation to cauldron subsidence. In *Nevada test site*, ed. E. B. Eckel, *Mem. Geol. Soc. Amer.*, **110**, 87–99.

Cameron, E. G. 1969. Postcumulus changes in the Eastern Bushveld Complex. *Amer. Min.*, **54**, 754–79.

Carmichael, I. S. E., Turner, F. J., and J. Verhoogen 1974. *Igneous petrology.* New York: McGraw-Hill.

Carter, S. R., and M. J. Norry 1976. Genetic implications of Sr isotopic data from the Aden volcano, South Arabia. *Earth Plan. Sci. Lett.*, **31**, 161–6.

Cawthorn, R. G., Curran, E. B., and R. J. Arculus 1973. A petrogenetic model for the origin of the calc-alkaline suite of Grenada, Lesser Antilles. *J. Petrol.*, **14**, 327–38.

Chappell, B. W., and A. J. R. White 1974. Two contrasting granite types. *Pacific Geol.*, **8**, 173–4.

Chayes, F. 1964. Variance–covariance relations in Harker diagrams of volcanic suites. *J. Petrol.*, **5**, 219–37.

Church, S. E., and M. Tatsumoto 1975. Lead isotope relations in oceanic ridge basalts from the Juan de Fuca–Gorda Ridge area, N.E. Pacific Ocean. *Contr. Mineral. Petrol.*, **53**, 253–79.

Church, S. E., and G. R. Tilton 1973. Lead and strontium isotopic studies in the Cascade Mountains: bearing on andesite genesis. *Bull. Geol. Soc. Amer.*, **84**, 431–54.

Clark, S. P., and A. E. Ringwood 1964. Density distribution and constitution of the mantle. *Rev. Geophys.*, **2**, 35–88.

Coleman, M. 1977. Sulphur isotopes in petrology *J. Geol. Soc. Lond.*, **133**, 593–608.

Coombs, D. S., and J. F. G. Wilkinson 1969. Lineages and fractionation trends in undersaturated volcanic rocks from the East Otago province (New Zealand) and related rocks. *J. Petrol.*, **10**, 440–501.

Cox, K. G. 1978. Komatiites and other high-magnesia lavas: some problems. *Phil. Trans. R. Soc. Lond.*, A, **288**, 599–609.

Cox, K. G., and J. D. Bell 1972. A crystal fractionation model for basaltic rocks of the New Georgia Group, British Solomon Islands. *Contr. Mineral. Petrol.*, **37**, 1–13.

Cox, K. G., Gass, I. G., and D. I. J. Mallick 1969. The evolution of the volcanoes of Aden and Little Aden, South Arabia. *Quart. J. Geol. Soc. Lond.*, **124**, 283–308.

Cox, K. G., Gass, I. G., and D. I. J. Mallick 1970. The peralkaline volcanic suite of Aden and Little Aden, South Arabia. *J. Petrol.*, **11**, 433–61.

Cox, K. G., Gass, I. G., and D. I. J. Mallick 1977. The western part of the Shuqra volcanic field, South Yemen. *Lithos*, **10**, 185–91.

Cox, K. G., and B. G. Jamieson 1974. The olivine-rich lavas of Nuanetsi: a study of polybaric magmatic evolution. *J. Petrol.*, **15**, 269–301.

Cross, W., Iddings, J. P., Pirsson, L. V., and H. S. Washington 1903. *Quantitative classification of igneous rocks.* University of Chicago Press.

Dawson, J. B. 1971. Advances in kimberlite geology. *Earth Sci. Rev.*, **7**, 187–214.

De Paulo, D. J., and G. J. Wasserburg 1976. Nd isotopic variations and petrogenetic models. *Geophys. Res. Lett.*, **3**, 249–52.

Deer, W. A., Howie, R. A., and J. Zussman 1966. *An introduction to the rock forming minerals.* London: Longman.

Dickinson, D. R., Dodson, M. H., and I. G. Gass 1969. Correlation of initial $^{87}Sr/^{86}Sr$ with Rb/Sr in some late Tertiary volcanic rocks of South Arabia. *Earth Plan. Sci. Lett.*, **6**, 84.

Doe, B. R. 1970. *Lead isotopes*. Berlin: Springer-Verlag, 137 pp.

Donaldson, C. H. 1974. Olivine crystal types in harrisitic rocks of the Rhum pluton and in Archaean spinifex rocks. *Bull. Geol. Soc. Amer.*, **85**, 1721–6.

Drake, M. J. 1975. The oxidation state of europium as an indicator of oxygen fugacity. *Geochim. Cosmochim. Acta*, **39**, 55–64.

Drake, M. J., and D. F. Weill 1975. Partition of Sr, Ba, Ca, Y, Eu^{2+}, Eu^{3+}, and other REE between plagioclase feldspar and magmatic liquid: an experimental study. *Geochim. Cosmochim. Acta*, **39**, 689–712.

Duncan, R. A., and W. Compston 1976. Sr-isotopic evidence for an old mantle source region for French Polynesian volcanism. *Geology*, **4**, 728–32.

Dunham, A. C. 1970. The emplacement of the Tertiary igneous complex of Rhum. In *Mechanism of igneous intrusion*, eds. G. Newall and N. Rast. *Geological Journal* Special Issue 2.

Edgar, A. D. 1973. *Experimental petrology*. Oxford: Clarendon Press.

Ehlers, E. G. 1972. *The interpretation of geological phase diagrams*. San Francisco: W. H. Freeman and Company.

Eitel, W. 1965. *Silicate science*. Vol. III. *Dry silicate systems*. New York: Academic Press.

Elder, J. 1976. *The bowels of the Earth*. Oxford University Press.

Elsdon, R. 1971. Crystallisation history of the Upper Layered Series, Kap Edvard Holm, East Greenland. *J. Petrol.*, **12**, 499–521.

Ernst, W. G. 1976. *Petrologic phase equilibria*. San Francisco: W. H. Freeman and Company.

Epstein, S., Sharp, R. P., and A. J. Gow 1970. Antarctic ice sheet: stable isotope analysis of Byrd station cores and interhemispheric climactic implications. *Science*, **168**, 1570–2.

Eskola, P. 1954. A proposal for the presentation of rock analyses in ionic percentages. *Annals of the Finnish Academy of Science*, Series AIII, **38,** 3–15.

Ewart, A. 1965. Mineralogy and petrogenesis of the Whakamaru ignimbrite in the Maraetai area of the Taupo volcanic zone, New Zealand. *N.Z. J. Geol. Geophys.*, **8**, 611–77.

Faure, G. 1977. *The principles of isotope geology*. New York: John Wiley.

Faure, G., Bowman, J. R., Elliott, D. H., and L. M. Jones 1974. Strontium isotope composition and petrogenesis of the Kirkpatrick basalt, Queen Alexandra Range, Antarctica. *Contr. Mineral. Petrol.*, **48**, 153–69.

Faure, G., and J. L. Powell 1972. *Strontium isotope geology*. Berlin: Springer-Verlag, 188 pp.

Ferguson, J. 1964. Geology of the Ilímaussaq alkaline intrusion, South Greenland. *Medd. om Grønland*, **172**, 5–82.

Forbes, R. B., and H. Kuno 1965. *The regional petrology of peridotite inclusions and basaltic host rocks*. Int. Union Geol. Sci. Upper Mantle Symposium, New Delhi 1964, 161–79.

Forester, R. W., and H. P. Taylor Jr. 1976. ^{18}O-depleted igneous rocks from the Tertiary complex of the Isle of Mull, Scotland. *Earth Plan. Sci. Lett.*, **32**, 11–17.

Forester, R. W., and H. P. Taylor Jr. 1977. [18]O/^{16}O, D/H and ^{13}C/^{12}C studies of the Tertiary igneous complex of Skye, Scotland. *Amer. J. Sci.*, **277**, 136–77.

Fraser, D. G. 1977. Thermodynamic properties of silicate melts. In *Thermodynamics in geology*, ed. D. G. Fraser, pp. 301–25. Dordrecht: D. Reidel Publishing Co.

Friedman, I., and J. R. O'Neil 1977. Compilation of stable isotope fractionation factors of geochemical interest. In *Data of geochemistry*, ed. M. Fleischer. U.S. Geol. Surv. Prof. Pap., 440-KK.

Fudali, R. F. 1963. Experimental studies bearing on the origin of pseudoleucite and associated problems of alkalic rock systems. *Bull. Geol. Soc. Amer.*, **74**, 1101–26.

Gandy, M. K. 1975. The petrology of the Lower Old Red Sandstone lavas of the eastern Sidlaw Hills, Perthshire, Scotland. *J. Petrol.*, **16**, 189–211.

Gass, I. G., Mallick, D. I. J., and K. G. Cox 1973. Volcanic islands of the Red Sea. *J. Geol. Soc. Lond.*, **129**, 275–310.

Gast, P. W. 1968. Trace element fractionation and the origin of tholeiitic and alkaline magma types. *Geochim. Cosmochim. Acta*, **32**, 1057–86.

Gibson, I. L. 1970. A pantelleritic welded ash-flow tuff from the Ethiopian rift-system. *Contr. Mineral. Petrol.*, **28**, 89–111.

Giles, D. L., and E. F. Cruft 1968. Major and minor element variation in zoned ash-flows and their biotites. *Geol. Soc. Amer. Spec. Pap.*, **101**, 86.

Goode, A. D. T. 1976. Small scale primary cumulus igneous layering in the Kalka Layered Intrusion, Giles Complex, Central Australia. *J. Petrol.*, **17**, 379–97.

Goode, A. D. T. 1977. Flotation and remelting of plagioclase in the Kalka Intrusion, Central Australia: petrological implications for anorthosite genesis. *Earth Plan. Sci. Lett.*, **34**, 375–80.

Green, D. H. 1973. Experimental melting studies on a model upper mantle composition at high pressure under water-saturated and water-undersaturated conditions. *Earth Plan. Sci. Lett.*, **19**, 37–53.

Green, D. H., and A. E. Ringwood 1967. The genesis of basaltic magmas. *Contr. Mineral. Petrol.*, **15**, 103–90.

Greig, J. W. 1927. Immiscibility in silicate melts. *Amer. J. Sci.*, 5th series, **13**, 1–44.

Grutzeck, M., Kridelbaugh, S., and D. F. Weill 1974. The distribution of Sr and REE between diopside and silicate liquid. *Geophys. Res. Lett.*, **1**, 273–5.

Gulson, B. L., and T. E. Krogh 1973. Old lead components in the young Bergell massif, southeast Swiss Alps. *Contr. Mineral. Petrol.*, **40**, 239.

Gunn, B. M., and F. Mooser 1971. Geochemistry of the volcanics of Central Mexico, *Bull. Volc.*, **34**, 577–616.

Hamilton, D. L., and W. S. MacKenzie 1965. Phase equilibria studies in the system $NaAlSiO^4$(nepheline)–$KAlSiO^4$(kalsilite)–SiO^2–H^2O. *Miner. Mag.*, **34**, 214–31.

Hamilton, E. I. 1963. The isotopic composition of strontium in the Skaergaard intrusion, East Greenland. *J. Petrol.*, **4**, 383–91.

Hamilton, P. J. 1977. Isotope and trace element studies of the Great Dyke and Bushveld mafic phase and their relation to Early Proterozoic magmatism. *J. Petrol.*, **18**, 24–52.

Hanson, G. N. 1977. Geochemical evolution of the suboceanic mantle. *J. Geol. Soc. Lond.*, **134**, 235–53.

Harker, A. 1909. *The natural history of igneous rocks*. New York: Macmillan.

Harris, P. G., Hutchison, R., and D. K. Paul 1972. Plutonic xenoliths and their relation to the upper mantle. *Phil. Trans. R. Soc. Lond.*, A, **271**, 313–23.

Harry, W. T., and C. H. Emeleus 1960. *Mineral layering in some granite intrusions of S. W. Greenland*. Rep. 21st Int. Geol. Congr. (Norden), pt. 14, 172–81.

Hart, S. R. 1971. K, Rb, Cs, Sr and Ba contents and Sr isotope composition of ocean floor basalts. *Phil. Trans. R. Soc. Lond.*, A, **268**, 573–88.

Hart, S. R., and C. J. Brooks 1977. The geochemistry and evolution of early Precambrian mantle. *Contr. Mineral. Petrol.*, **61**, 109–28.

Hawkesworth, C. J., O'Nions, R. K., Pankhurst, R. J., Hamilton, P. J., and N. M. Evensen 1977. A geochemical study of island-arc and back-arc tholeiites from the Scotia Sea. *Earth Plan. Sci. Lett.*, **36**, 253–62.

Heier, K. S. 1973. Geochemistry of granulite facies rocks and problems of their origin. *Phil. Trans. R. Soc. Lond.*, A, **273**, 429–42.

Hellman, P. L., and P. Henderson 1977. Are rare earth elements mobile during spilitisation? *Nature*, **267**, 38–40.

Helz, R. T. 1976. Phase relations of basalts in their melting ranges at P_{H_2O} = 5 kb. Part II. Melt compositions. *J. Petrol.*, **17**, 139–93.

Henderson, P. 1970. The significance of the mesostasis of basic layered igneous rocks. *J. Petrol.*, **11**, 463–73.

Hess, H. H. 1960. The Stillwater Igneous Complex. *Mem. Geol. Soc. Amer.*, **80**.

Hill, P. G. 1974. *The petrology of the Aden volcano, People's Democratic Republic of Yemen*. Unpublished Ph.D. thesis, University of Edinburgh.

Hoefs, J. 1973. *Stable isotope geochemistry*. Berlin: Springer-Verlag.

Hofmann, A. W., and S. R. Hart 1975. An assessment of local and regional isotopic equilibrium in a partially molten mantle. *Yb. Carnegie Instn. Wash.*, **74**, 195–210.

Holmes, A. 1930. *Petrographic methods and calculations*. London: Thomas Murby.

Holmes, A., and H. F. Harwood 1937. The volcanic area of Bufumbira. Part II. The petrology of the volcanic field of Bufumbira, south-west Uganda and of other parts of the Birunga field. *Geol. Surv. Uganda Mem.*, **3**.

Hytönen, K., and J. F. Schairer 1961. The plane enstatite–anorthite–diopside and its relation to basalts. *Yb. Carnegie Instn. Wash.*, **60**, 125–41.

Ingersoll, L. R., Zobel, O. J., and A. C. Ingersoll 1954. *Heat conduction with engineering, geological and other applications*. Madison: University of Wisconsin Press.

Irvine, T. N. 1974. Petrology of the Duke Island ultramafic complex, southeastern Alaska. *Mem. Geol. Soc. Amer.*, **138**.

Ito, K., and G. C. Kennedy 1967. Melting and phase relations in a natural peridotite to 40 kilobars. *Amer. J. Sci.*, **265**, 519–38.

Jackson, E. D. 1961. *Primary textures and mineral associations in the ultramafic zone of the Stillwater Complex, Montana*. U.S. Geol. Surv. Prof. Pap., 358.

Jackson, E. D. 1967. Ultramafic cumulates in the Stillwater, Great Dyke, and Bushveld Intrusions. In *Ultramafic and related rocks*, ed. P. J. Wyllie. New York: John Wiley.

Jackson, E. D. 1971. The origin of ultramafic rocks by cumulus processes. *Fortschr. Miner.*, **48**, 128–74.

Jaeger, J. C. 1968. Cooling and solidification of igneous rocks. In *Basalts. The Poldervaart treatise on rocks of basaltic composition*, ed. H. H. Hess and A. Poldervaart, Vol. 2, pp. 503–36. New York: Interscience.

Johannsen, A. 1931. *A descriptive petrography of the igneous rocks*. Vol. 1. University of Chicago Press.

Katsui, Y. 1963. Evolution and magmatic history of some Krakatoan calderas in Hokkaido, Japan. *J. Fac. Sci. Hokkaido University* Series IV, **11**, 631–50.

Kay, J. M., and R. M. Nedderman, 1974. *An introduction to fluid mechanics and heat transfer*. London: Cambridge University Press.

Kay, R. W., and P. W. Gast 1973. The rare earth content and origin of alkali-rich basalts. *J. Geol.*, **81**, 653–82.

Kay, R. W., and R. G. Senechal 1976. The rare earth geochemistry of the Troodos ophiolite complex. *J. Geophys. Res.*, **81**, 964–70.

Kinoshita, W. T., Swanson, D. A., and D. B. Jackson 1974. The measurement of crustal deformation related to volcanic activity at Kilauea volcano, Hawaii. In *Physical volcanology*, eds. L. Civetta, P. Gasparini, G. Luongo and A. Rapolla. Amsterdam: Elsevier Scientific Publishing Company.

Kistler, R. W., and Z. E. Peterman 1973. Variations in Sr, Rb, K, Na and initial Sr^{87}/Sr^{86} in Mesozoic granitic rocks and intruded wall rocks in central California. *Bull. Geol. Soc. Amer.*, **84**, 3489–512.

Krishnamurthy, P., and K. G. Cox 1977. Picrite basalts and related lavas from the Deccan Traps of western India. *Contr. Mineral. Petrol.*, **62**, 53–75.

Kushiro, I. 1972a. Effect of water on the composition of magmas formed at high pressures. *J. Petrol.*, **13**, 311–34.

Kushiro, I. 1972b. Determination of liquidus relations in synthetic silicate systems with electron-probe analysis: the system forsterite–diopside–silica at 1 atmosphere. *Amer. Min.*, **57**, 1260–71.

Kushiro, I. 1973a. Regularities in the shift of liquidus boundaries in silicate systems and their significance in magma genesis. *Yb. Carnegie Instn. Wash.*, **72**, 497–502.

Kushiro, I. 1973b. The system diopside–anorthite–albite: determination of compositions of coexisting phases. *Yb. Carnegie Instn. Wash.*, **72**, 502–7.

Kushiro, I., Shimizu, N., Nakamura, Y., and S. Akimoto 1972. Composition of co-existing liquid and solid phases formed upon melting of natural garnet and spinel lherzolites at high pressures: a preliminary report. *Earth Plan. Sci. Lett.*, **14**, 19–25.

Lambert, I. B., and P. J. Wyllie 1968. Stability of hornblende and a model for the low velocity zone. *Nature*, **215**, 1240–1.

Lancelot, J. R., and C. J. Allegre 1974. Origin of carbonatitic magma in the light of the Pb–U–Th isotope system. *Earth Plan. Sci. Lett.*, **22**, 233–8.

Langmuir, C. H., Bender, J. F., Bence, A. E., and G. N. Hanson 1977a. Petrogenesis of basalts from the Famous area: Mid-Atlantic Ridge. *Earth Plan. Sci. Lett.*, **36**, 133–56.

Langmuir, C. H., Vocke, R. D., Hanson, G. N., and S. R. Hart 1977b. A general mixing equation: applied to the petrogenesis of basalts from Iceland and Reykjanes Ridge. *Earth Plan. Sci. Lett.*, **37**, 380–92.

Larsen, E. S., Jr. 1938. Some new variation diagrams for groups of igneous rocks. *J. Geol.*, **46**, 505–20.

Larsen, E. S., Jr. 1945. Time required for the crystallisation of the great batholith of southern and lower California. *Amer. J. Sci.*, **243-A** (Daly Volume), 399–416.

Le Bas, M. J. 1970. A combined central- and fissure-type phonolitic volcano in western Kenya. *Bull. Volc.*, **34**, 518–36.

Le Maitre, R. W. 1976. The chemical variability of some common igneous rocks, *J. Petrol.*, **17**, 589–637.

Leeman, W. P. 1976. Petrogenesis of McKinney (Snake River) olivine tholeiite in light of rare-earth element and Cr/Ni distributions. *Bull. Geol. Soc. Amer.*, **87**, 1582–6.

Lewis, W. K., Gilliland, E. R., and W. C. Bauer 1949. Characteristics of fluidised particles. *Ind. Eng. Chem.*, **41**, 1104–17.

Lindsley, D. H., Brown, G. M., and I. D. Muir 1969. *Conditions of the ferrowollastonite–ferrohedenbergite inversion in the Skaergaard Intrusion, East Greenland*. Miner. Soc. Amer. Spec. Pap., 2, 193–201.

Lipman, P. W. 1967. Mineral and chemical variations within an ash-flow sheet from Aso Caldera, southwestern Japan. *Contr. Mineral. Petrol.*, **16**, 300–27.

Lipman, P. W., Christiansen, R. L., and J. T. O'Connor 1966. *A compositionally zoned ash-flow sheet in southern Nevada*. U.S. Geol. Surv. Prof. Pap., 524-F.

Lofgren, G. 1974. An experimental study of plagioclase crystal morphology. *Amer. J. Sci.*, **274**, 243–73.

McBirney, A. R. 1975. Differentiation of the Skaergaard Intrusion. *Nature*, **253**, 691–4.

Macdonald, G. A., and T. Katsura 1964. Chemical composition of Hawaiian lavas. *J. Petrol.*, **5**, 82–133.

McIver, J. R. 1975. Aspects of some high magnesia eruptives in southern Africa. *Contr. Mineral. Petrol.*, **51**, 99–118.

Martin, R. C. 1965. Ignimbrite lithology at Whakamaru, New Zealand. *N. Z. J. Geol. Geophys.*, **8**, 680–701.

Mason, B. 1966. *Principles of geochemistry*, 3rd edition. New York: John Wiley.

Modreski, P. J., and A. L. Boettcher 1973. Phase relationships of phlogopite in the system $K_2O–MgO–CaO–Al_2O_3–SiO_2–H_2O$ to 35 kilobars: a better model for micas in the interior of the earth. *Amer. J. Sci.*, **273**, 385–414.

Moorbath, S. 1971. Measuring geological time. In *Understanding the Earth*, eds. I. G. Gass, P. J. Smith and R. C. L. Wilson, 41–51. Sussex: Artemis Press.

Moorbath, S. 1975. Evolution of Precambrian crust from strontium isotopic evidence. *Nature*, **254**, 395–8.

Moorbath, S., and J. D. Bell 1965. Strontium isotope abundance studies and rubidium–strontium age determinations on Tertiary igneous rocks from the Isle of Skye, northwest Scotland. *J. Petrol.*, **6**, 37–66.

Moorbath, S., and H. Welke 1969. Lead isotope studies on igneous rocks from the Isle of Skye, northwest Scotland. *Earth Plan. Sci. Lett.*, **5**, 217–30.

Morse, S. A. 1970. Alkali feldspars with water at 5 kb pressure. *J. Petrol.*, **11**, 221–53.

Murata, K. J. 1960. A new method of plotting chemical analyses of basaltic rocks. *Amer. J. Sci.*, **258-A** (Bradley Volume), 247–52.

Murata, K. J., and D. H. Richter 1966. *The 1959—60 eruptions of Kilauea volcano, Hawaii; chemistry of the lavas*. U.S. Geol. Surv. Prof. Pap., 537-A, pp. A1–26.

Mysen, B., and A. L. Boettcher 1975a. Melting of hydrous mantle: I. Phase relations of natural peridotite at high pressures and temperatures with controlled activities of water, carbon dioxide and hydrogen. *J. Petrol.*, **16**, 520–48.

Mysen, B., and A. L. Boettcher 1975b. Melting of hydrous mantle: II. Geochemistry of crystals and liquids formed by anatexis of mantle peridotite at high pressures and high temperatures as a function of controlled activities of water, hydrogen and carbon dioxide. *J. Petrol.* **16**, 549–93.

Mysen, B., and A. L. Boettcher 1976. Melting of hydrous mantle: III. Phase relations of a garnet websterite + H_2O at high pressures and temperatures. *J. Petrol.*, **17**, 1–14.

Naslund, H. R. 1976. Mineralogical variations in the upper part of the Skaergaard Intrusion, East Greenland. *Yb. Carnegie Instn. Wash.*, **75**, 640–4.

Nathan, H. D. and C. K. Van Kirk 1978. A model of magmatic crystallisation. *J. Petrol.*, **19**, 66–94.

Nesbitt, R. W. 1971. Skeletal crystal forms in the ultramafic rocks of the Yilgarn block, western Australia: evidence for an Archaean ultramafic liquid. *Geol. Soc. Aust. Spec. Pub.*, **3**, 331–50.

Nesbitt, R. W., and D. L. Hamilton 1970. Crystallisation of an alkali-olivine basalt under controlled P_{O_2}, P_{H_2O} conditions. *Phys. Earth Plan. Interiors*, **3**, 309–15.

Nicholls, I. A. 1974. Liquids in equilibrium with peridotitic mineral assemblages at high water pressures. *Contr. Mineral. Petrol.*, **45**, 289–316.

Nicholls, I. A., and A. E. Ringwood 1973. Effect of water on olivine stability in tholeiites and the production of silica-saturated magmas in the island-arc environment. *J. Geol.*, **81**, 285–300.

Nicholls, J., Carmichael, I. S. E., and J. C. Stormer 1971. Silica activity and P_{total} in igneous rocks. *Contr. Mineral. Petrol.*, **33**, 1–20.

Niggli, P. 1954. *Rocks and mineral deposits*. San Francisco: W. H. Freeman.

Nisbet, E. G., Bickle, M. J., and A. Martin 1977. The mafic and ultramafic lavas of the Belingwe Greenstone Belt, Rhodesia. *J. Petrol.*, **18**, 521–66.

O'Hara, M. J. 1965. Primary magmas and the origin of basalts. *Scott. J. Geol.*, **1**, 19–40.

O'Hara, M. J. 1968. The bearing of phase equilibria studies on the origin and evolution of basic and ultrabasic rocks. *Earth Sci. Rev.*, **4**, 69–133.

O'Hara, M. J. 1977. Geochemical evolution during fractional crystallisation of a periodically refilled magma chamber. *Nature*, **266**, 503–7.

O'Hara, M. J., and H. J. Yoder, Jr. 1967. Formation and fractionation of basic magmas at high pressures. *Scott. J. Geol.*, **3**, 67–117.

Ohmoto, H. 1972. Systematics of sulphur and carbon isotopes in hydrothermal ore deposits. *Econ. Geol.*, **67**, 551–78.

O'Neil, J. R., Shaw, S. E., and R. H. Flood 1977. Oxygen and hydrogen isotope compositions as indicators of granite genesis in the New England Batholith, Australia. *Contr. Mineral. Petrol.*, **62**, 313–28.

O'Nions, R. K., Evensen, N. M., Hamilton, P. J., and S. R. Carter 1978. Melting of the mantle past and present: isotope and trace element evidence. *Phil. Trans. R. Soc. Lond.*, A**288**, 547–59.

O'Nions, R. K., and K. Gronvold 1973. Petrogenetic relationships of acid and basic rocks in Iceland: Sr-isotopes and rare-earth elements in Late and Postglacial volcanics. *Earth Plan. Sci. Lett.*, **19**, 397–409.

O'Nions, R. K., Hamilton, P. J., and N. M. Evensen 1977. Variations in $^{143}Nd/^{144}Nd$ and $^{87}Sr/^{86}Sr$ ratios in oceanic basalts. *Earth Plan. Sci. Lett.*, **34**, 13–22.

O'Nions, R. K., and R. J. Pankhurst 1974a. Rare-earth element distribution in Archaean gneisses and anorthosites, Gothåb area, West Greenland. *Earth Plan. Sci. Lett.*, **22**, 328–38.

O'Nions, R. K., and R. J. Pankhurst 1974b. Petrogenetic significance of isotope and trace element variations in volcanic rocks from the Mid Atlantic. *J. Petrol.*, **15**, 603–34.

O'Nions, R. K., and R. J. Pankhurst 1976. Sr isotope and rare earth element geochemistry of DSDP Leg 37 basalts. *Earth Plan. Sci. Lett.*, **31**, 255–61.

O'Nions, R. K., and R. J. Pankhurst 1978. Early Archaean rocks and geochemical evolution of the earth's crust. *Earth Plan. Sci. Lett.*, **38**, 211–36.

O'Nions, R. K., Pankhurst, R. J. and K. Gronvold 1976. Nature and development of basalt magma sources beneath Iceland and the Reykjanes Ridge. *J. Petrol.*, **17**, 315–38.

O'Nions, R. K., and R. Powell 1977. The thermodynamics of trace element distribution. In *Thermodynamics in geology*, ed. D. G. Fraser, pp. 349–63. Nato Advanced Study Institute Series, Dordrecht: D. Reidel Publishing Co.

Osborn, E. F., and A. Muan 1960. The system $MgO–Al_2O_3–SiO_2$. Plate 3. In *Phase equilibrium diagrams of oxide systems*. American Ceramic Society and Edward Orton Jr. Ceramic Foundation, Columbus, Ohio.

Osborn, E. F., and J. F. Schairer 1941. The ternary system pseudowollastonite–akermanite–gehlenite. *Amer. J. Sci.*, **239**, 715–63.

Osborn, E. F., and D. B. Tait 1952. The system diopside–forsterite–anorthite. *Amer. J. Sci.* (Bowen Volume), 413–33.

Pankhurst, R. J. 1969. Strontium isotopic studies related to petrogenesis in the Caledonian basic igneous province of northeast Scotland. *J. Petrol.*, **10**, 115–43.

Pankhurst, R. J. 1974. Rb–Sr whole-rock chronology of Caledonian events in northeast Scotland. *Bull. Geol. Soc. Amer.*, **85**, 345–50.

Pankhurst, R. J. 1977a. Open system crystal fractionation and variation in incompatible elements in basalts. *Nature*, **268**, 36–38.

Pankhurst, R. J. 1977b. Strontium isotope evidence for mantle events in the continental lithosphere. *J. Geol. Soc.*, **134**, 255–68.

Pankhurst, R. J., and R. T. Pidgeon 1976. Inherited isotope systems and the source region pre-history of early Caledonian granites in the Dalradian Series of Scotland. *Earth Plan. Sci. Lett.*, **31**, 55−68.

Papanastassiou, D A., and G. J. Wasserburg 1969. The determination of small time differences in the formation of planetary objects. *Earth Plan. Sci. Lett.*, **5**, 361–76.

Pearce, J. A., and J. R. Cann 1973. Tectonic setting of basic volcanic rocks determined using trace element analysis. *Earth Plan. Sci. Lett.*, **19**, 290–300.

Peck, D. L., Wright, T. L., and J. G. Moor 1966. Crystallisation of tholeiitic basalt in Alae lava lake, Hawaii. *Bull. Volc.*, **29**, 629.

Peterman, Z. E., and C. E. Hedge 1971. Related strontium isotopic and chemical variations in oceanic basalts. *Bull. Geol. Soc. Amer.*, **82**, 493–9.

Phillips, E. R., 1974. Myrmekite – one hundred years later. *Lithos*, **7**, 181–94.

Powers, H. A., 1955. Composition and origin of basaltic magma of the Hawaiian islands. *Geochim. Cosmochim. Acta*, **7**, 77–107.

Pyke, D. R., Naldrett, A. J., and O. R. Eckstrand 1973. Archaean ultramafic flows in Munro Township, Ontario. *Bull. Geol. Soc. Amer.*, **84**, 955–78.

Quinlivan, W. D., and P. W. Lipman 1965. Compositional variation in some Cenozoic ash-flow tuffs, southern Nevada. *Geol. Soc. Amer. Spec. Pap.*, **82**, 342.

Ratté, J. C., and T. A. Steven 1964. *Magmatic differentiation in a volcanic sequence related to the Creede Caldera, Colorado*. U.S. Geol. Surv. Proof. Pap., 475–D, pp. 49–53.

Richard, P., Shimizu, N., and C. Allegre 1976. $^{143}Nd/^{146}Nd$, a natural tracer: an application to oceanic basalts. *Earth Plan. Sci. Lett.*, **31**, 269.

Robins, B. 1973. Crescumulate layering in a gabbroic body on Seiland, northern Norway. *Geol. Mag.*, **109**, 533–42.

Roddick, J. C., and W. Compston 1977. Strontium isotope equilibration: a solution to a paradox. *Earth Plan. Sci. Lett.*, **34**, 238–46.

Roeder, P. L., and R. F. Emslie 1970. Olivine–liquid equilibrium. *Contr. Mineral. Petrol.*, **29**, 275–89.

Roscoe, R. 1952. The viscosity of suspensions of rigid spheres. *Brit. J. Appl. Phys.*, **3**, 267–9.

Saggerson, E. P. 1970. The structural control and genesis of alkaline rocks in central Kenya. *Bull. Volc.*, **34**, 38–76.

Schairer, J. F. 1938. The system leucite–diopside–silica. *Amer. J. Sci.*, **35A**, 289.

Schairer, J. F., 1950. The alkali feldspar join in the system NaAlSiO₄–KAlSiO₄–SiO₂. *J. Geol.*, **58**, 512–17.

Schairer, J. F. 1957. Melting relations of the common-rock-forming silicates. *J. Amer. Ceram. Soc.*, **40**, 215–35.

Schairer, J. F., and N. L. Bowen 1947. The system anorthite–leucite–silica. *Comm. géol. Finlande Bull.*, **140**, 67–87.

Schilling, J.-G. 1973. Iceland mantle plume: geochemical study of Reykjanes Ridge. *Nature*, **242**, 565–71.

Schilling, J.-G. 1975. Rare-earth variations across 'normal segments' of the Reykjanes Ridge, 60°–53°N, Mid-Atlantic Ridge, 29°S, and East Pacific Rise, 2°–19°S, and evidence on the composition of the underlying low-velocity layer. *J. Geophys. Res.*, **80**, 1459–73.

Schmincke, H.-U. 1969. Ignimbrite sequence on Gran Canaria. *Bull. Volc.*, **33**, 1199–219.

Schmincke, H.-U., and M. Weibel 1972. Chemical study of rocks from Madeira, Porto Santo, and São Miguel, Terceira (Azores). *N. Jb. Miner. Abh.*, **117**, 253–81.

Shaw, D. M., Reilly, G. A., Muysson, J. R., Pattenden, G. E., and F. E. Campbell 1967. An estimate of the chemical composition of the Canadian Shield. *Can. J. Earth Sci.*, **4**, 829.

Shaw, H. R. 1965. Comments on viscosity, crystal settling and convection in granitic magmas. *Amer. J. Sci.*, **263**, 120–52.

Shaw, H. R., 1969. Rheology of basalt in the melting range. *J. Petrol.*, **10**, 510–35.

Shaw, H. R. 1972. Viscosities of magmatic silicate liquids: an empirical method of prediction. *Amer. J. Sci.*, **272**, 870–93.

Shimizu, N. 1975. Geochemistry of ultramafic inclusions from Salt Lake Crater, Hawaii, and from southern African kimberlites. *Phys. Chem. Earth*, **9**, 655–69.

Shimizu, N., and R. J. Arculus 1975. Rare earth element concentrations in a suite of basanitoids and alkali olivine basalts from Grenada, Lesser Antilles. *Contr. Mineral. Petrol.*, **50**, 231–40.

Simpson, E. S. W. 1954. On the graphical representation of differentiation trends in igneous rocks. *Geol. Mag.*, **91**, 238–44.

Smewing, J. D., and P. J. Potts 1976. Rare earth abundances in basalts and metabasalts from the Troodos massif, Cyprus. *Contr. Mineral. Petrol.*, **75**, 245–58.

Smith, R. L., and R. A. Bailey 1966. The Bandolier tuff: a study of ash-flow eruption cycles from zoned magma chambers. *Bull. Volc.*, **29**, 83–103.

Spooner, E. T. C. 1976. The strontium isotopic composition of seawater and seawater–oceanic crust interaction. *Earth Plan. Sci. Lett*, **31**, 167–74.

Spooner, E. T. C., Beckinsale, R. D., England, P. C., and A. Senior 1977. Hydration, ¹⁸O enrichment and oxidation during ocean floor hydrothermal metamorphism of ophiolitic metabasic rocks from E. Liguria, Italy. *Geochim. Cosmochim. Acta*, **41**, 857–71.

Spooner, E. T. C. and W. S. Fyfe 1973. Sub-sea-floor metamorphism, heat and mass transfer. *Contr. Mineral. Petrol.*, **42**, 287–304.

Stanton, R. L., and J. D. Bell 1969. Volcanic and associated rocks of the New

Georgia Group, British Solomon Islands. *Overseas Geol. Mineral Resources*, **10**, 113–45.

Streckeisen, A. 1976. To each plutonic rock its proper name. *Earth Sci. Rev.*, **12**, 1–33.

Strong, D. F. 1972. The petrology of the lavas of Grande Comore. *J. Petrol.*, **13**, 181–217.

Sun, S. S. and G. N. Hanson 1975. Evolution of the mantle: geochemical evidence from alkali basalts. *Geology*, **3**, 297–302.

Tatsumoto, M. 1978. Isotopic composition of lead in oceanic basalts and its implication to mantle evolution. *Earth Plan. Sci. Lett.*, **38**, 63–87.

Taylor, H. P. Jr. 1968. The oxygen isotope geochemistry of igneous rocks. *Contr. Mineral. Petrol.*, **19**, 1–71.

Taylor, H. P. Jr. 1974. The application of oxygen and hydrogen isotope studies to problems of hydrothermal alteration and ore deposition. *Econ. Geol.*, **69**, 843–83.

Taylor, H. P., Jr., and R. W. Forester 1971. Low-O^{18} igneous rocks from the intrusive complexes of Skye, Mull and Ardnamurchan, western Scotland. *J. Petrol.*, **12**, 465–97.

Taylor, S. R. 1964. The abundance of chemical elements in the continental crust – a new table. *Geochim. Cosmochim. Acta*, **28**, 1273–85.

Thompson, R. N. 1972a. The l-atmosphere melting patterns of some basaltic volcanic series. *Amer. J. Sci.*, **272**, 901–32.

Thompson, R. N. 1972b. Evidence for a chemical discontinuity near the basalt – 'andesite' transition in many an orogenic volcanic suite. *Nature*, **236**, 106–10.

Thompson, R. N., Esson, J., and A. C. Dunham 1972. Major element chemical variation in the Eocene lavas of the Isle of Skye, Scotland. *J. Petrol.*, **13**, 219–53.

Thompson, R. N., and M. J. Flower 1971. One-atmosphere melting and crystallisation relations of lavas from Anjouan, Comores archipelago, western Indian Ocean. *Earth Plan. Sci. Lett.*, **12**, 97–107.

Thompson, R. N., and C. E. Tilley 1969. Melting and crystallisation relations of Kilauean basalts of Hawaii. The lavas of the 1959–60 Kilauea eruption. *Earth Plan. Sci. Lett.*, **5**, 469–77.

Thorarinsson, S. 1950. The eruption of Mount Hekla 1947–1948. *Bull. Volc.*, **10**, 157–68.

Thorarinsson, S. 1970. The Lakagigar eruption of 1783. *Bull. Volc.*, ser. 2, **33**, 910–27.

Thorarinsson, S., and G. E. Sigvaldason 1972. The Hekla eruption of 1970. *Bull. Volc.*, **36**, 269–88.

Thornton, C. P., and O. F. Tuttle 1960. Chemistry of igneous rocks. I. Differentiation index. *Amer. J. Sci.*, **258**, 664–84.

Thorpe, R S., Potts, P. J. and M. B. Sarre 1977. Rare earth evidence concerning the origin of granites of the Isle of Skye, northwest Scotland. *Earth Plan. Sci. Lett.*, **36**, 111–20.

Tilley, C. E., and R. N. Thompson 1970. Melting and crystallisation relations of the Snake River basalts of southern Idaho, U.S.A. *Earth Plan. Sci. Lett.*, **8**, 79–92.

Tilley, C. E., Yoder, H. S., and J. F. Schairer 1963. Melting relations of basalts. *Yb. Carnegie Instn. Wash.*, **62**, 77–84.

Tilley, C. E., Yoder, H. S., and J. F. Schairer 1964. New relations on melting of basalts. *Yb. Carnegie Instn. Wash.*, **63**, 92–7.

Tilley, C. E., Yoder, H. S., and J. F. Schairer 1965. Melting relations of volcanic tholeiite and alkali rock series. *Yb. Carnegie Instn. Wash.*, **64**, 69–82.

Tilley, C. E., Yoder, H. S., and J. F. Schairer 1967. Melting relations of volcanic rock series. *Yb. Carnegie Instn. Wash.*, **65**, 260–9.

Tuttle, O. F., and N. L. Bowe 1958. Origin of granite in the light of experimental studies in the system $NaAlSi_3O_8$–$KAlSi_3O_8$–SiO–H_2O. *Geol. Soc. Amer. Mem.*, **74**.

Upton, B. G. J. 1974. The alkaline province of south-east Greenland. In *The alkaline rocks*, ed. H. Sorensen. London: John Wiley, 622 pp.

van Breemen, O., Hitchinson, J., and P. Bowden 1975. Age and origin of the Nigerian Mesozoic granites: a Rb–Sr isotopic study. *Contr. Mineral. Petrol.*, **50**, 157–72.

Viljoen, M. P., and R. P. Viljoen 1969a. *The geology and geochemistry of the lower ultramafic unit of the Onverwacht Group, and a proposed new class of igneous rocks*. Geol. Soc. S. Afr. Spec. Pub., 2 (The Upper Mantle Project), pp. 55–85.

Viljoen, M. P., and R. P. Viljoen 1969b. *Evidence of the existence of a mobile extrusive peridotite magma*. Geol. Soc. S. Afr. Spec. Pub., 2 (The Upper Mantle Project), pp. 87–112.

Vollmer, R. 1976. Rb–Sr and U–Th–Pb systematics of alkaline rocks from Italy. *Geochim. Cosmochim. Acta*, **40**, 283–96.

Wager, L. R., and G. M. Brown 1968. *Layered igneous rocks*. Edinburgh and London: Oliver & Boyd, 588 pp.

Wager, L. R., Brown, G. M., and W. J. Wadsworth 1960. Types of igneous cumulates. *J. Petrol.*, **1**, 73–85.

Wager, L. R., and W. A. Deer 1939. Geological investigations in East Greenland. Part III. The petrology of the Skaergaard intrusion, Kangerdlugssuaq, East Greenland. *Medd. om Grønland*, **105**, no. 4, 1–352.

Wager, L. R., Vincent, E. A., Brown, G. M. and J. D. Bell 1965. Marscoite and related rocks of the Western Red Hills Complex, Isle of Skye. *Phil. Trans. R. Soc. Lond.*, A, **257**, 273–307.

Walker, K. R. 1969. *The Palisades Sill, New Jersey: a reinvestigation*. Geol. Soc. Amer. Spec. Pap., 111.

Whittaker, E. J. W., and R. Muntus 1970. Ionic radii for use in geochemistry. *Geochim. Cosmochim. Acta*, **34**, 945–56.

Williams, H. 1942. *Geology of Crater Lake National Park, with a reconnaissance of the Cascade Range southward of Mount Shasta*. Carnegie Instn. Publ., 540.

Wilshire, H. G. 1967. The Prospect alkaline diabase–picrite intrusion, New South Wales, Australia. *J. Petrol.*, **8**, 97–163.

Winchester, J. A. and P. A. Floyd 1976. Geochemical magma type discrimination: application to altered and metamorphosed basic igneous rocks. *Earth Plan. Sci. Lett.*, **28**, 459–69.

Wood, B. J., and D. G. Fraser 1976. *Elementary thermodynamics for geologists*. London: Oxford University Press.

Wright, J. B. 1970. Contributions to the volcanic succession and petrology of the Auckland Islands, New Zealand. IV. Chemical analyses from the lower half of the Ross volcano. *Trans. R. Soc. N.Z., Earth Sci.*, **8**, 109–15.

Wright, T. L., and P. C. Doherty 1970. A linear programming and least squares computer method for solving petrologic mixing problems. *Bull. Geol. Soc. Amer.*, **81**, 1995–2008.

Wright, T. L., and R. Fiske 1971. Origin of the differentiated and hybrid lavas of Kilauea volcano, Hawaii. *J. Petrol.*, **12**, 1–65.

Wyllie, P. J. 1971. *The dynamic earth*. New York: John Wiley.

Yoder, H. S. 1971. Contemporaneous rhyolite and basalt. *Yb. Carnegie Instn. Wash.*, **69**, 141–5.

Yoder, H. S., Stewart, D. B., and J. R. Smith 1957. Ternary feldspars. *Yb. Carnegie Instn. Wash.*, **56**, 206–14.

Yoder, H. S., and C. E. Tilley 1962. Origin of basalt magmas: an experimental study of natural and synthetic rock systems. *J. Petrol*, **3**, 342–532.

Zartman, R. E. 1974. Lead isotopic provinces in the Cordillera of the western United States and their geologic significance. *Econ. Geol.*, **69**, 792–805.

Zeil, W., and H. Pichler 1967. Die känozoische Rhyolith-Formation mittleren Abschnitt der Anden. *Geol. Rundsch.*, **57**, 48–81.

Zielinski, R. A. 1975. Trace element evaluation of a suite of rocks from Réunion Island, Indian Ocean. *Geochim. Cosmochim. Act.*, **39**, 713–34.

Index